AF679839

Molecular Virology and Biology of Iridoviruses

Molecular Virology and Biology of Iridoviruses

Editor: Gus Romero

AMERICAN
MEDICAL PUBLISHERS
www.americanmedicalpublishers.com

AMERICAN
MEDICAL PUBLISHERS
www.americanmedicalpublishers.com

Cataloging-in-Publication Data

Molecular virology and biology of iridoviruses / edited by Gus Romero.
p. cm.
Includes bibliographical references and index.
ISBN 978-1-63927-732-2
1. Molecular virology. 2. Iridoviruses. 3. Virology. 4. Molecular microbiology.
5. DNA viruses. 6. Medical virology. I. Romero, Gus.
QR389 .M65 2023
579.2--dc23

© American Medical Publishers, 2023

American Medical Publishers,
41 Flatbush Avenue,
1st Floor, New York,
NY 11217, USA

ISBN 978-1-63927-732-2 (Hardback)

This book contains information obtained from authentic and highly regarded sources. Copyright for all individual chapters remain with the respective authors as indicated. All chapters are published with permission under the Creative Commons Attribution License or equivalent. A wide variety of references are listed. Permission and sources are indicated; for detailed attributions, please refer to the permissions page and list of contributors. Reasonable efforts have been made to publish reliable data and information, but the authors, editors and publisher cannot assume any responsibility for the validity of all materials or the consequences of their use.

Trademark Notice: Registered trademark of products or corporate names are used only for explanation and identification without intent to infringe.

Contents

Preface VII

Chapter 1 **Invertebrate Iridoviruses: A Glance over the Last Decade** 1
İkbal Agah İnce, Orhan Özcan, Ayca Zeynep Ilter-Akulke, Erin D. Scully and Arzu Özgen

Chapter 2 **Cloning of the Major Capsid Protein (MCP) of Grouper Iridovirus of Taiwan (TGIV) and Preliminary Evaluation of a Recombinant MCP Vaccine against TGIV** 26
Hsin-I Liu, Pinwen Peter Chiou, Hong-Yi Gong and Hsin-Yiu Chou

Chapter 3 **Geographic Distribution of *Epizootic haematopoietic necrosis virus* (EHNV) in Freshwater Fish in South Eastern Australia: Lost Opportunity for a Notifiable Pathogen to Expand its Geographic Range** 36
Joy A. Becker, Dean Gilligan, Martin Asmus, Alison Tweedie and Richard J. Whittington

Chapter 4 **Distribution and Phylogeny of Erythrocytic Necrosis Virus (ENV) in Salmon Suggests Marine Origin** 51
Veronica A. Pagowski, Gideon J. Mordecai, Kristina M. Miller, Angela D. Schulze, Karia H. Kaukinen, Tobi J. Ming, Shaorong Li, Amy K. Teffer, Amy Tabata and Curtis A. Suttle

Chapter 5 **Interaction between Two Iridovirus Core Proteins and their Effects on Ranavirus (RGV) Replication in Cells from Different Species** 67
Xiao-Tao Zeng and Qi-Ya Zhang

Chapter 6 ***Artemia* spp., a Susceptible Host and Vector for Lymphocystis Disease Virus** 80
Estefania J. Valverde, Alejandro M. Labella, Juan J. Borrego and Dolores Castro

Chapter 7 **Susceptibility of *Exopalaemon carinicauda* to the Infection with Shrimp Hemocyte Iridescent Virus (SHIV 20141215), a Strain of Decapod Iridescent Virus 1 (DIV1)** 90
Xing Chen, Liang Qiu, Hailiang Wang, Peizhuo Zou, Xuan Dong, Fuhua Li and Jie Huang

Chapter 8 **Localization of *Frog Virus 3* Conserved Viral Proteins 88R, 91R and 94L** 105
Emily Penny and Craig R. Brunetti

Chapter 9 **Ranaviruses Bind Cells from Different Species through Interaction with Heparan Sulfate** 113
Fei Ke, Zi-Hao Wang, Cheng-Yue Ming and Qi-Ya Zhang

Chapter 10 **IIV-6 Inhibits NF-κB Responses in *Drosophila*** 124
Cara West, Florentina Rus, Ying Chen, Anni Kleino, Monique Gangloff, Don B. Gammon and Neal Silverman

Chapter 11 **Description of a Natural Infection with Decapod Iridescent Virus 1 in Farmed Giant Freshwater Prawn, *Macrobrachium rosenbergii*......139**
Liang Qiu, Xing Chen, Ruo-Heng Zhao, Chen Li, Wen Gao, Qing-Li Zhang and Jie Huang

Chapter 12 **Pathogen Risk Analysis for Wild Amphibian Populations Following the First Report of a Ranavirus Outbreak in Farmed American Bullfrogs *(Lithobates catesbeianus)*......153**
Bernardo Saucedo, José M. Serrano, Mónica Jacinto-Maldonado, Rob S.E.W. Leuven, Abraham A. Rocha García, Adriana Méndez Bernal, Andrea Gröne, Steven J. van Beurden and César M. Escobedo-Bonilla

Chapter 13 **Characterization of a Novel Megalocytivirus Isolated from European Chub *(Squalius cephalus)*......168**
Maya A. Halaly, Kuttichantran Subramaniam, Samantha A. Koda, Vsevolod L. Popov, David Stone, Keith Way and Thomas B. Waltzek

Chapter 14 **Modelling Ranavirus Transmission in Populations of Common Frogs *(Rana temporaria)*......178**
Amanda L.J. Duffus, Trenton W.J. Garner, Richard A. Nichols, Joshua P. Standridge and Julia E. Earl

Chapter 15 **Evaluating the Within-Host Dynamics of *Ranavirus* Infection with Mechanistic Disease Models and Experimental Data......191**
Joseph R. Mihaljevic, Amy L. Greer and Jesse L. Brunner

Chapter 16 **Critical Role of an MHC Class I-Like/Innate-Like T Cell Immune Surveillance System in Host Defense against Ranavirus (Frog Virus 3) Infection......202**
Eva-Stina Isabella Edholm, Francisco De Jesús Andino, Jinyeong Yim, Katherine Woo and Jacques Robert

Chapter 17 **eDNA Increases the Detectability of Ranavirus Infection in an Alpine Amphibian Population......216**
Claude Miaud, Véronique Arnal, Marie Poulain, Alice Valentini and Tony Dejean

Permissions

List of Contributors

Index

Preface

This book has been a concerted effort by a group of academicians, researchers and scientists, who have contributed their research works for the realization of the book. This book has materialized in the wake of emerging advancements and innovations in this field. Therefore, the need of the hour was to compile all the required researches and disseminate the knowledge to a broad spectrum of people comprising of students, researchers and specialists of the field.

Iridoviruses refer to a type of double-stranded DNA viruses with an icosahedral structure that belong to the iridoviridae family. The members of this family can infect invertebrates, protozoa, vertebrates and algae. The host range of iridoviruses differs, such as some have a broad range of hosts, while others are limited to a small number of hosts. Non-lethal and concealed infections are common in various hosts. Horizontal transmission of iridoviruses generally occurs through predation or cannibalism of infected invertebrate hosts. They produce iridescence in the infected tissues and reproduce in the cytoplasm. Iridoviruses infect a variety of animals including reptiles, molluscs, fish, insects, amphibians and crustaceans. These viruses have a distinctive trait which is terminally redundant and circularly permuted organization of the genome. This book aims to shed light on iridoviruses. The readers would gain knowledge that would broaden their perspective about the molecular virology and biology of these viruses. The extensive content of this book provides the readers with a thorough understanding of the subject.

At the end of the preface, I would like to thank the authors for their brilliant chapters and the publisher for guiding us all-through the making of the book till its final stage. Also, I would like to thank my family for providing the support and encouragement throughout my academic career and research projects.

Editor

1

Invertebrate Iridoviruses: A Glance over the Last Decade

İkbal Agah İnce [1,2,*], Orhan Özcan [1,2,3], Ayca Zeynep Ilter-Akulke [1], Erin D. Scully [4] and Arzu Özgen [1,5]

1 Department of Medical Microbiology, School of Medicine, Acıbadem Mehmet Ali Aydınlar University, Atasehir, Istanbul 34752, Turkey; orhn.ozcn@hotmail.com (O.Ö.); aycazeynep@gmail.com (A.Z.I.-A.); aozgen@gelisim.edu.tr (A.Ö.)
2 Department of Biostatistics and Medical Informatics, Acıbadem Mehmet Ali Aydınlar University, Atasehir, Istanbul 34752, Turkey
3 Evolutionary Genetics, Groningen Institute for Evolutionary Life Sciences, University of Groningen, P.O. Box 11103, 9700 CC Groningen, The Netherlands
4 Stored Product Insect and Engineering Research Unit, USDA-Agricultural Research Service, Center for Grain and Animal Health Research, Manhattan, KS 66502, USA; erin.scully@ARS.USDA.GOV
5 Vocational School of Health, Istanbul Gelisim University, Avcılar, Istanbul 34310, Turkey
* Correspondence: ikbal.agah.ince@gmail.com

Abstract: Members of the family Iridoviridae (iridovirids) are large dsDNA viruses that infect both invertebrate and vertebrate ectotherms and whose symptoms range in severity from minor reductions in host fitness to systemic disease and large-scale mortality. Several characteristics have been useful for classifying iridoviruses; however, novel strains are continuously being discovered and, in many cases, reliable classification has been challenging. Further impeding classification, invertebrate iridoviruses (IIVs) can occasionally infect vertebrates; thus, host range is often not a useful criterion for classification. In this review, we discuss the current classification of iridovirids, focusing on genomic and structural features that distinguish vertebrate and invertebrate iridovirids and viral factors linked to host interactions in IIV6 (Invertebrate iridescent virus 6). In addition, we show for the first time how complete genome sequences of viral isolates can be leveraged to improve classification of new iridovirid isolates and resolve ambiguous relations. Improved classification of the iridoviruses may facilitate the identification of genus-specific virulence factors linked with diverse host phenotypes and host interactions.

Keywords: invertebrate iridoviruses; genomics; proteomics; classification

---

1. Introduction

Iridovirids are nucleocytoplasmic large dsDNA viruses (NCLDVs) that can be divided into five genera: *Ranavirus, Lymphocystivirus, Megalocytivirus, Iridovirus,* and *Chloriridovirus* (Figures 1 and 2; Tables 1 and 2) under two sub-families defined recently as *Alpha-* and *Betairidovirinae* by ICTV (International Committee on Taxonomy of Viruses) [1,2]. Recently, the term "iridovirids" was implemented rather than using general term iridoviruses to family members to avoid confusion between members of the genus *Iridovirus* and members of the family with the same name. In this review, all members of the family will be referred to as "iridovirids", members of the genus will be referred to as "iridoviruses", and the abbreviation IIV will refer to both genera of invertebrate iridoviruses (*Iridovirus* and *Chloriridovirus*). The iridovirids were initially distinguished from one another according to their virion particle size, host preference, presence of a DNA methyltransferase gene, GC content and phylogeny based on the amino acid sequence of the major capsid protein [3,4];

however, phylogenetic analysis based on complete genome sequences now provides a superior method of differentiation and classification (Figures S1 and S2) and possibly using pan-genomic data will advance the analysis of viral phylogenies, which may ultimately lead to the development of new classification criteria on virus classification in general.

New iridovirids are continuously being discovered; however, limited number of IIVs have been sequenced, and thus, it is hard to precisely quantify or estimate the genetic heterogeneity and variability within each genus. In addition, it is still not clear what type of information/criteria would be sufficient to determine whether newly discovered iridovirids represent new species or whether these new isolates are just variants or strains derived from an existing species. A comprehensive list of identified iridovirids with completed genome sequences is shown in Table 2. Within the genus *Iridovirus*, two species have been identified: Invertebrate iridescent virus 1 (IIV1), which was the first IIV species isolated and infects the soil-dwelling European crane fly *Tipula paludosa* (Diptera) [5] and Invertebrate iridescent virus 6 (IIV-6), also referred to as Chilo iridescent virus (CIV), which was isolated from diseased larvae of the rice stem borer, *Chilo suppressalis* (Lepidoptera; Pyralidae) in Japan [6]. At least two strains of IIV6 have been identified and are routinely used in laboratories around the world and an IIV-6 isolate from Germany was completely sequenced [7]. However, the genomic differences between this isolate and the isolates commonly used in New Zealand, Australia, and the USA laboratories have not yet been investigated. The objective of the review is to discuss the status of current research on this interesting but poorly understood family of isometric DNA viruses and will focus primarily on describing molecular studies on genome organization, gene expression strategies, virion proteins, and relatedness of the IIVs using the best-studied virus, IIV-6, as the primary model.

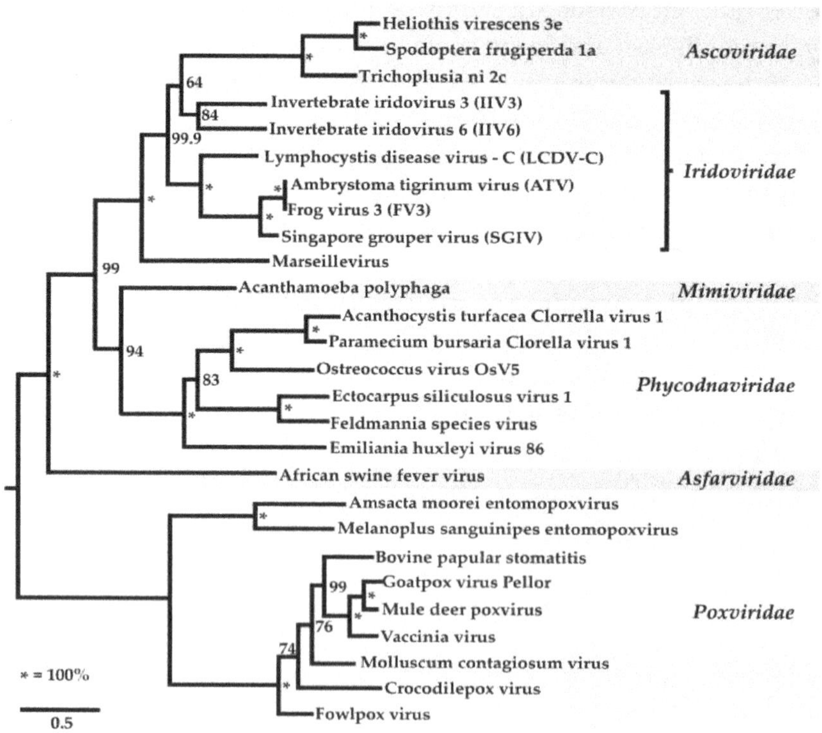

Figure 1. Phylogenetic relationships of iridovirids with other nucleocytoplasmic large DNA viruses A maximum-likelihood tree constructed based on concatenated alignments of 1849 positions of five NCLDV core proteins: A1L/VLTF2 transcription factor, A32 ATPase, D5 type ATPase, DNA polymerase B, and major capsid protein using the TreeFinder program with the estimated site rates heterogeneity and with the WAG (Whelan and Goldman) substitution model. The expected-likelihood weights of 1000 local rearrangements were used as confidence values of TreeFinder tree branches. Scale bar represents amino acid substitutions per site (This figure was used with permission from PNAS [8] and ICTV [4] where it was presented earlier and slightly modified its presentation).

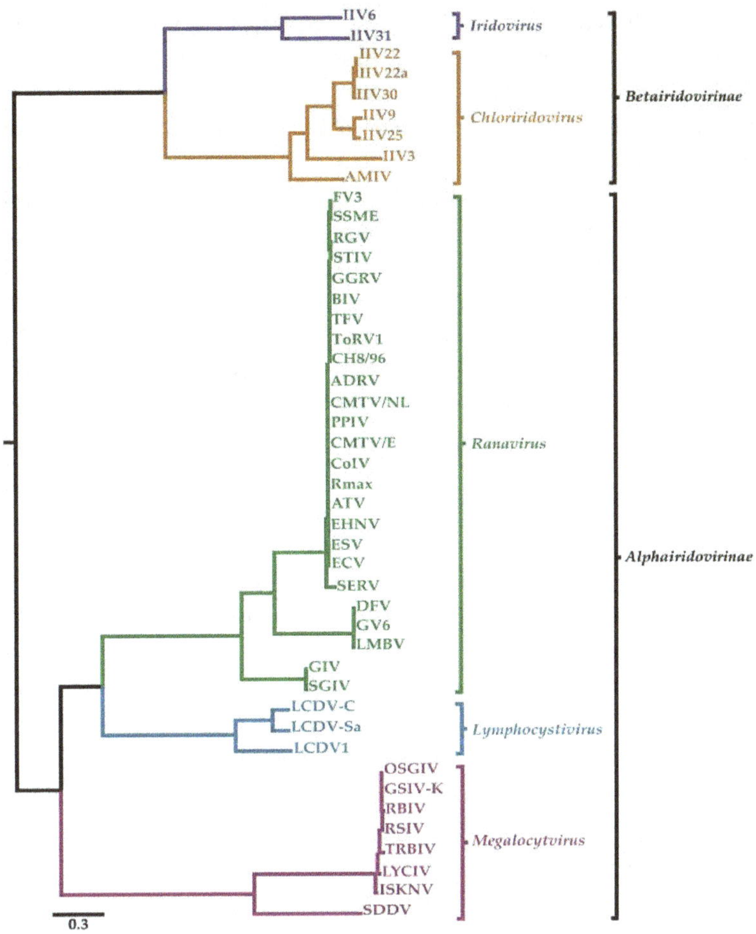

Figure 2. ML analysis based on 26 core genes conserved among all members of the family *Iridoviridae* (iridovirids). The phylogram shows a clear division of genera within the *Alphairidovirinae* (*Ranavirus*, green; *Lymphocystivirus*, cyan; *Megalocytivirus*, purple) and *Betairidovirinae* (*Iridovirus*, blue; *Chloriridovirus*, brown) subfamilies. Virus species and isolates are indicated by abbreviations, listed in Table 2. Members of accepted species are shown in bold typeface. The tree was constructed using maximum likelihood analysis using IQ-TREE open source software [9] under optimum substitution model LG+I+G4 according to Bayesian Information Criterion (BIC) [10].and the concatenated amino acid (aa) sequences of 26 core genes (19,773 aa characters including gaps) from 45 completely sequenced iridovirus genomes. The tree was midpoint rooted and branch lengths are based on the number of inferred substitutions, as indicated by the scale bar. The branch points that separate the iridovirid genera are supported by bootstrap values greater than 99%. For other branch points all bootstrap values are >70% except for those displaying high levels of amino acid similarity, e.g., TFV vs. BIV/GGRV, 66%; PPIV vs. CMTV/2013/NL et al., 49%; TRBIV vs. RSIV et al., 57%; and LYCIV vs. TRBIV et al., 56%. Scale bar represents amino acid substitutions per site. This figure was published as an ICTV report [1,2].

Table 1. Comparison of features of iridovirids.

Sub-Qualification	*Genus*	Virion Size	Hosts	GC Content	DNA Methylation
Betairidovirinae	*Iridovirus*	120–130 nm	arthropods, particularly insects	29–32%	absent
Betairidovirinae	*Chloriridovirus*	180 nm	Diptera with aquatic larval stages, mainly mosquitoes	48%	absent
Alphairidovirinae	*Ranavirus*	150 nm	Reptilia, Amphibia and Osteichthyes	54%	present
Alphairidovirinae	*Lymphocystivirus*	198–227 nm	flounder, plaice, and dab	29.1–33%	present
Alphairidovirinae	*Megalocytivirus*	140–200 nm	sea fish	54.8%	present

Table 2. Identified iridovirids with complete genome sequences. IIV-6—Invertebrate iridescent virus 6; IIV-31—Invertebrate iridescent virus 31; IIV-3—Invertebrate iridescent virus 3; IIV-9—Invertebrate iridescent virus 9; IIV-22—Invertebrate iridescent virus 22; IIV-22A—Invertebrate iridescent virus 22A; IIV-25—Invertebrate iridescent virus 25; IIV-30—Invertebrate iridescent virus 30; ADRV—Andrias davidianus ranavirus; ATV—Ambystoma tigrinum virus; BIV-ME—Bohle iridovirus isolate ME 93/35; CGSI-HN1104—Chinese giant salamander iridovirus isolate HN1104; CH8/96—Testudo hermanni ranavirus; CMTV/2008/E—Common midwife toad virus—2008/E; CMTV/2013/NL—Common midwife toad virus—2013/NL; CodV—Cod ranavirus; DFV—Doctor fish virus; ECV-13051/2012—European catfish virus; ECV-14612/12—European catfish virus; EHNV—Epizootic haematopoietic necrosis virus; ESV—European sheatfish virus; FV3—Frog virus; GGRV—German gecko ranavirus; GIV—Giant sea perch iridovirus; ATV-GUFFY—Ambystoma tigrinum virus isolate GUFFY; GV6—Guppy virus 6; LMBV—largemouth bass virus; PPIV—Pike-perch iridovirus; RCV-Z—Rana catesbeiana virus Z; REV 282/102—Rana esculenta virus isolate 282/102; RGV—Rana gryliovirus; RMax—Ranavirus maxima; RRV—Regina ranavirus; SERV—Short-finned eel ranavirus; SGIV—Singapore grouper iridovirus; SSME—Spotted salamander population in Maine; STIV—Soft-shelled turtle iridovirus; TFV—Tiger frog virus; ToRV-1—Tortoise ranavirus 1; LCDV-1—Lymphocystis disease virus 1; LCDV-C—Lymphocystis disease virus—China; LCDV-Sa—Lymphocystis disease virus—Sparus aurata; GSIV-K1—Giant sea perch iridovirus; ISKNV—Infectious spleen and kidney necrosis virus; LYCIV—Large yellow croaker iridovirus; OSGIV—Orange spotted grouper iridovirus; RBIV—Rock bream iridovirus; RSIV—Red seabream iridovirus; SDDV—Scale drop disease virus; TRBIV—Turbot reddish body iridovirus; AMIV—Anopheles minimus iridovirus; SHIV—shrimp hemocyte iridescent virus.

Genus (Subfamily)	Species, Strains, and Isolates	Size (bp)	ORFs	G + C %	GenBank Accession Number	References
Iridovirus	IIV-6	212,482	215 *	28.63	AF303741	[7]
	IIV-31	220,222	203	35.09	HF920637	[11]
Chloriridovirus	IIV-3	191,132	126	48	DQ643392	[12]
	IIV-9	205,791	191	31	GQ918152	[13]
	IIV-22	197,693	167	28.05	HF920633	[14]
	IIV-22A	196,456	174	28	HF920634	[15]
	IIV-25	204,815	177	30.32	HF920635	[16]
	IIV-30	198,533	177	28.1	HF920636	[17]
	IIV-3	191,132	126	48	DQ643392	[12]
Ranavirus	ADRV	106,734	101	55	KC865735	[18]
	ATV	106,332	96	54	AY150217	[19]
	BIV-ME	103,531	100	55.2	KX185156	[20]
	CGSI–HN1104	105,375	111	55.2	KF512820	Unpublished
	CH8/96	105,811	75	55	KP266741	[21]
	CMTV/2008/E	106,878	104	55.3	JQ231222	[22]
	CMTV/2013/NL	107,772	104	55.3	KP056312	[23]
	CodV	114,865	98	54.9	KX574342	[24]
	DFV	ND	ND	ND	ND	Unpublished
	ECV-13051/2012	127,751	135	54.2	KT989884	Unpublished
	ECV-14612/12	127,549	136	54	KT989885	Unpublished
	EHNV	127,011	100	54.05	NC028461	[25]
	ESV	127,732	136	54.23	JQ724856	[26]
	FV3	105,903	98	55	AY548484	[27]
	GGRV	103,681	73	55	KP266742	[21]
	GIV	139,793	120	49	AY666015	[28]
	ATV-GUFFY	106,437	99	54	KR075882	[29]
	GV6	ND	ND	ND	ND	Unpublished
	LMBV	ND	ND	ND	ND	Unpublished
	PPIV	108,041	109	55.3	KX574341	[30]
	RCV-Z	106,890	98	55	MF187210	[31]
	REV 282/102	107,444	101	55.2	MF538628	[32]
	RGV	105,791	106	55	JQ654586	[33]
	Rmax	115,510	100	54.9	KX574343	[24]
	ATV-RRV	106,971	102	54.1	KR075879	[34]
	SERV	126,965	111	54.7	KX353311	[35]
	SGIV	140,131	162	48.64	AY521625	[36]

Table 2. *Cont.*

Genus (Subfamily)	Species, Strains, and Isolates	Size (bp)	ORFs	G + C %	GenBank Accession Number	References
Ranavirus	SSME	105,070	95	55	KJ175144	[37]
	STIV	105,890	105	55.1	EU627010	[38]
	TFV	105,057	105	55.01	AF389451	[39]
	ToRV-1	103,876	76	55	KP266743	[21]
Lymphocystivirus	LCDV-1	102,653	108	29.1	NC_00182	[40]
	LCDV-C	186,250	239	27.2	AY380826	[41]
	LCDV-Sa	208,501	183	33	KX643370	[42]
Megalocytivirus	GSIV-K1	112,565	135	53.02	KT804738	[43]
	ISKNV	111,362	125	54.8	AF371960	[44]
	LYCIV	111,760	ND	53.92	AY779031	[45]
	OSGIV	112,636	126	54	AY894343	[46]
	RBIV-KOR-TY1	112,080	116	53	AY532606	[47]
	RSIV	112,415	93	54	AB104413	[48]
	SDDV	124,244	129	37	KR139659	[49]
	TRBIV	110,104	115	55	GQ273492	[50]
Unclassified	AMIV	163,023	148	39	KF938901	[51]
	SHIV	165,809	170	34.6	MF599468	[52]

* The IIV-6 genome contains a total of 468 ORFs when both overlapping and non-overlapping protein coding ORFs are considered [7] and 215 ORFs when only non-overlapping coding regions are considered. The revised boundaries for the 215 non-overlapping proteins are listed in Uniprot database [53–55].

2. Classification of Iridovirids

Using a maximum likelihood-based phylogenetic analysis on a concatenated sequence of 26 core genes present in all 45 iridovirid genome sequences, four major clades were apparent, with members of the genera *Ranavirus, Lymphocystivirus* and *Megalocytivirus* (vertebrate iridoviruses) forming three strongly supported monophyletic clades. The genera *Iridovirus* and *Chloriridovirus* (IIVs) formed a monophyletic clade, but the two genera diverged significantly from one another, as indicated by the relatively long branch lengths on this node (Figure 2). Due to the independent monophyletic origins of vertebrate (*Ranavirus, Lymphocystivirus,* and *Megalocytivirus*) and invertebrate iridoviruses (*Iridovirus* and *Chloriridovirus*), two new subfamilies have been proposed. The first subfamily (*Alphairidovirinae*) encompasses all iridoviruses whose hosts are ectothermic vertebrates (fish, reptiles, and amphibians) and the second family (*Betairidovirinae*) contains iridoviruses that infect invertebrates (e.g., insects, crustaceans, etc.). This latter subfamily contains the *Iridovirus* and *Chloriridovirus* genera. These new classifications were discussed in the recent ICTV study group on family *Iridoviridae* (Figure 2) [1,2].

Previously, three major clades were defined within the genus *Iridovirus* using a molecular analysis of restriction endonuclease (REN) profiles of 16 iridovirid genomes as well as *mcp* amplicons combined with DNA hybridization and dot blots by Williams and Cory [56]. Later Webby and Kalmakoff performed molecular analysis of *mcp* genes (formerly gene named as *msp*) [57], terminal redundancies and DNA-DNA homologies [58]. As a result of both studies, three major groups in genus *Iridovirus* were identified: custaceoiridovirus (group 1), oligoiridovirus (group 2), and polyiridovirus (group 3). Group 1 comprised IIV-31 and Popillia japonica IV [56] whereas group 2 included IIV-6 and IIV-21. Group 3 was the largest group, which contained nine different species; IIV-1, IIV-21, IIV-22, IIV-23, IIV-24, IIV-25, IIV-29, IIV-30 and Anticarsia gemmatalis IV.

However, a more recent study performed by Wong et al. [13] using 26 iridovirid core genes indicated that IIV-3 was more closely related to the IIVs than previously estimated. IIV-3 showed a closer relationship to the group 3 polyiridovirus complex when compared to IIV-6, which belongs to the oligoiridovirus complex, group 2 [13]. In addition, recent MCP protein sequence-based phylogenetic information from mosquito iridescent viruses from *Culex pipiens* [59] and *Anopheles minimus* [51] also supported the phylogenetic proximity of the group 3 polyiridovirus viruses to IIV-3.

The iridovirid proteins D5 helicase primase, DNA packaging ATPase (A32), viral late transcription factor 3 (A2L/VLTF3), major capsid protein (MCP) and subfamily B viral DNA polymerase show high amino acid similarities when compared to other nucleocytoplasmic large DNA viruses (NCLDV), especially in the families *Ascoviridae*, *Phycodnaviridae*, *Marseilleviridae*, *Mimiviridae*, *Poxviridae*, Iridoviridae and Asfarviridae [4,8]. Based on a phylogenetic analysis using genes conserved among NCLDVs, the proposals of Colson et al. [60] were supported: (1) all iridovirids share a common ancestor and (2) NCLDVs should be assigned to a new viral order, the Megavirales [60,61].

This analysis also supported the close relationship between ascoviruses and IIVs, suggesting that ascoviruses emerged recently and share a common ancestor with IIV-6 and IIV-3 [62]. In addition, the study indicated that IIV-6 and IIV-31 and IIV-3 and IIV-9 each formed monophyletic groupings, suggesting that IIV-6 and IIV-31 and IIV-3 and IIV-9 are more closely related than previous studies indicated [62]. In light of these recent findings, the current IIV classification needs to be revised.

Phylogenetic analysis based on MCP sequences [63] is outdated and the more current and accepted approach is to use the conserved iridovirid genes (26 genes) in phylogenetic reconstructions [1,2,13,64]. Whole genome sequences of iridovirids are becoming more common and phylogenetic reconstructions based on whole genome sequences may provide additional resolution in comparison to approaches that use only a subset of the genes. For example, phylogenetic analysis based on complete genome sequences of 50 members of the family *Iridoviridae* (for virus abbreviation see Table 2) was performed using a pipeline created in Geneious R11. Two marseillevirus (Cannes 8 virus; KF261120 and Lausannevirus; NC015326) and three ascovirus (DpAV4a; NC011335, SfAV1a; NC004361, TnAV6a; NC008518) genomes were included as outgroups. First, the sequence directions were rearranged so that all genes were oriented in the same direction. Next, the alignment was computed using parameters for clustering closely related sequences and refined using realignment parameters optimized for distant organisms. Finally, tree construction was performed using Phylogenetic Analysis Using Parsimony (PAUP) or RAxML for Neigbour Joining (NJ) and maximum likelihood approaches, respectively. SGIV/GIV fall into a distinct cluster from the rest of the ranaviruses, a large group of viruses that infect fish, reptiles, and amphibians. Six megalocytiviruses formed a single well-separated clade. Furthermore, three fully-sequenced lymphocystivirus isolates appear to be more closely related to the two IIV genera than other vertebrate iridoviruses (VIVs). However, LCDV-1 was quite distant from LCDV-C/LCDV-Sa9. Based on this information, there is clear incongruence of the phylogenetic status of LCDVs with previous phylogenies constructed based on gene subsets. Our analysis also suggests the presence of two clades in *Chloriridovirus* including IIV-22, IIV-22a, IIV-25, IIV-30, and IIV-9/IIV-3. The AMIV virus [51] has not been classified yet; however, this analysis indicates that it possibly represents a new genus for IIVs. SHIV, a new iridovirid isolate (unclassified), also showed relatedness to both IIV genera (*Iridovirus* and *Chloriridovirus*) (Figures S1 and S2). Recent phylogenetic studies of SHIV based on different subsets (27 or 16 proteins) of concatenated iridovirid protein sequences conserved with SHIV orthologs suggested that SHIV should be considered a member of the proposed new genus "Xiairidovirus" [52,65] which is supported by our analysis presented in this review (Figures S1 and S2). Phylogenetic analysis also indicates that there is considerable diversity among members of both genera of invertebrate iridescent viruses.

3. Morphology and Composition

As we mentioned earlier IIV-6 is the most studied member of IIVs with a large dsDNA genome packed into an icosahedral capsid with a T = 147 lattice. The virus particle is comprised of three concentric domains including the outer proteinaceous capsid layer, an intermediate lipid membrane with associated polypeptides and nucleocytoplasmic dsDNA genome (Figure 3) [3,66–68]. The size of IIV-6 virions was determined to be between 120-130nm in ultrathin sections. However, using CryoEM and 3D image reconstruction analyses on naked virions, [66] showed that IIV-6 can reach a maximum diameter of 185 nm when fibrils emanating from the surface of the virion were included in these measurements [66,67].

The lipid membrane of IIV-6 is comprised predominantly of phosphatidylinositol and diglycerides, which differs from the composition of the lipid membrane of its host cells, indicating that the viral lipid membrane is not likely derived from the host cell membrane [69]. Specifically, the major constituent of the viral lipid membrane is glycophosphatidylinositol (GPI), which acts as a lipid anchor to bind proteins via their C-terminus. The GPI-anchored proteins function as extracellular receptors or cell surface antigens/cell adhesion molecules, but their most important role is to provide a stable anchoring for extracellular proteases and lipases [70]. Other lipid membranes of invertebrate iridescent viruses (IIV-1, 2, 9, 21, 22, 30, 31, Anticarsia IV) showed similar lipid compositions to IIV-6 [71], suggesting that high GPI content might be a common feature of iridovirids. Since levels of glycerophospholipids are typically much higher in the endoplasmic reticulum (ER) than in other cellular compartments in mammalian and yeast cells [72], the ER may function as the cellular region where iridovirids gain their lipid components. Therefore, the ER of iridovirid-infected host cell should be analyzed.

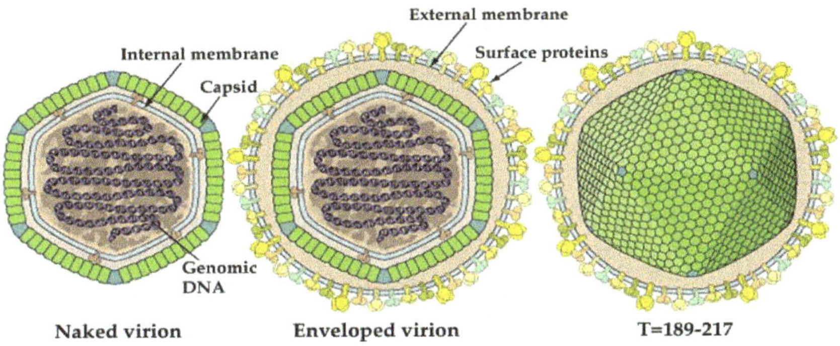

Figure 3. Graphical representation of the iridovirid structure [73].

4. IIV-6 Persistence and Sensitivity to External Factors

Some IIVs habitually can infect mosquitoes in aquatic habitats [74]. Owing to their structures, IIVs are highly stable in water as indicated by only small, 10-fold decreases in titres following 50 days either at 4 °C or at room temperature in aqueous suspensions [75]. To identify other environmental variables that influence virus stability, studies were performed investigating correlations between soil moisture levels and microorganisms on the stability of IIV-6. Although the virus was active across a range of soil moisture levels, dry soil rapidly inactivated the infectivity of IIV-6. Additionally, the half-life of IIV-6 propagated in non-sterile and sterile soil was recorded as 4.9 and 6.3 days, respectively, when compared to a control virus suspension (12.9 days) [76], indicating that extracellular enzymes or other environmental conditions influenced by the presence of microbes can impact the durability of IIV-6 in soil. However, the mechanisms by which microorganisms contribute to iridoviral persistence are not yet known.

IIV-6 is thermolabile and can be rapidly inactivated when the temperature is above 55 °C [77]. For example, IIV-6 viral activity decreased by approximately four logs when the virus was heated at 70 °C for 60 min or 80 °C for 30 min [78]. Furthermore, both solar UV light and ultraviolet radiation also reduced IIV-6 infectivity especially in aqueous habitats. Solar UV light caused a complete loss of infectivity (99.99%) after 36 h of exposure [79].

Exposure to various solvents and enzymes can also impact IIV-6 infectivity and sensitivity both in cell culture (i.e., *Spodoptera frugiperda* (Sf9) cells) and in whole insects (i.e., *Galleria mellonella* larvae) [80]. In addition, differences in sensitivity to the suspensions were observed depending on whether the assay was performed in cell culture or in whole insects. For example, while iridoviruses from both vertebrates and invertebrates were sensitive to ether, chloroform, sodium deoxycholate, only VIVs were sensitive to phospholipase A2 (PLA2) and only IIV-6 was sensitive to 1% Triton-x-100, 70% ethanol, and methanol (both in cell culture and in whole insects). Furthermore, no differences were detected in the infectivity of IIV-6 when the virus was treated with Tween-80, lipase, proteinase K, trypsin, magnesium chloride ($MgCl_2$), ethylenediaminetetraacetic acid (EDTA) and other proteinases either in insect cell culture or whole insects. Notably, when the IIV-6 lipid bilayer was disrupted, the virus was still capable of infecting the cultured cells, but not insect larvae when the inoculum was injected into the haemocoel [78]. These observations highlight the need for studies on the role of the virus lipid component in the process of entry into invertebrate cells as well as host specificity and may also explain the observed differences in sensitivity to organic solvents and enzymes documented between vertebrate and invertebrate viruses. The roles of the lipid component in invertebrate IVs have not yet been reported; however, their role is well-documented in vertebrate IVs [81].

5. Host Range and Pathology

In the class Insecta, the host range of IIV-6 includes more than 100 species belonging to six orders; Coleoptera, Diptera, Hemiptera, Hymenoptera, Lepidoptera and Orthoptera, the natural hosts were reviewed by T. Williams [82]. In addition, several reports suggested that IIV-31 can naturally infect several different isopod species (Crustacea) [83–85] as well as *Drosophilidae* (Diptera) [86]. Iridovirid-like particles similar to IIVs have also been detected in marine invertebrates such as *Nautilus* spp. [87], the marine worm *Nereis diversicolor* [88] and from shrimp hemocytes [52,65] although the limited availability of genomic information or related viruses has hampered classification of these viruses.

Blue iridescent coloring is typically associated with patent infections, which occur due to the presence of crystalline arrays of particles in the cells of almost all tissues, especially the fat body [89]. This iridescent coloring of heavily infected invertebrates is the most obvious sign of IIV disease [68]. Patent IIV infections are almost consistently lethal; however, covert (inapparent) infections are more common, which may reduce the reproductive capacity and longevity of infected adult insects and arthropods [68,74,89–95]. In some cases, the link between viral presence and mortality is obscure. For example, Bromenshenk et al. [96] used mass-spectrometry-based proteomics to suggest a link between co-infection of *Nosema* species and IIV-6 with insect mortality; however, two subsequent studies failed to find a significant correlation between IIV-6 and colony collapse disorder (CCD) in the USA [97,98].

In addition to their natural hosts, IIV-6 can be propagated in twelve different insect cell lines [80], including *Drosophila* line 2 (DL2), *Drosophila* line 1 (DR1), *Aedes albopictus*, *Aedes aegypti*, *Anticarsia gemmatalis* (BCIRL-AG-AM), *Helicoverpa zea* (BCIRL-Hz-AM1), *Heliothis virescens* (BCIRL-HV-AM1), *Pieris rapae* (PR-5), *Plutella xylostella* (BCIRL-PX2-HNV3), *Spodoptera frugiperda* line 21 (Sf21), *Spodoptera frugiperda* line 9 (Sf9) and *Trichoplusia ni* (TN-CL1). Although all these cell lines are susceptible to infection, assays have indicated that the two *Drosophila* lines, DL2 and DR1, had the highest susceptibility to IIV-6 whereas cell lines from *Aedes albopictus* and *Plutella xylostella* were four orders of magnitude less susceptible compared to the others. Similar studies performed on cell lines derived from a hemipteran whitefly *Bemisia tabaci* [99], a leafhopper *Circulifer tenellus*, a lacewing, *Ceraeochrysa cubana* [100], a root weevil *Diaprepes abbreviatus* [89] and boll weevil *Anthonomus grandis* [101] showed that these cell lines were also susceptible to IIV-6 infection. Although it was reported that, IIV-6 can infect European corn borer *Ostrinia nubilalis* (Lepidoptera) larvae in several tissues and cells, including the fat body and hemocytes, a cell line derived from hemocytes of *Ostrinia nubilalis* could not be productively infected with IIV-6 [102], indicating that in some insects, extracellular host factors may influence viral infectivity and that not all cell lines derived from susceptible hosts can be infected with IIV-6.

Interestingly, reptiles and amphibians fed IIV-infected insects appear to become infected [103–107] and invertebrate iridovirus (IIV-6) propagation has also been achieved in some poikilothermic vertebrate cell lines e.g., a viper spleen cell line [108]. Furthermore, intraperitoneal injections of large doses of native or ultraviolet-irradiated IIV-6 were lethal to frogs and mice, while heat- or antiserum-inactivated IIV-6 had no lethal toxicity [109,110]. Although vertebrates can become infected with IIVs and lethality was observed under laboratory conditions and at high doses, natural infections are rarely symptomatic or lethal, suggesting that IIV replication is more limited in vertebrate hosts and that the host immune system is able to inhibit viral replication. Supporting the hypothesis of limited replication in vertebrates, IIVs tested in reptilian (VH2) and insect cell lines (Sf9) showed cytopathogenic effects similar to those of other reported iridoviruses [83], but viral titres were 2 logs higher in Sf9 compared to VH2. IIVs have been isolated from various stocks of crickets, suggesting that mass produced feeder insects could serve as a source of viral diseases for reptiles and amphibians kept as pets [103,111–114]. However, large-scale IIV outbreaks from ingestion of infected feeder insects have not yet been observed in reptiles or amphibians.

IIV-6 infection causes different immune responses in vertebrate and invertebrate hosts. Double stranded RNAs (ds-RNA) are the main activator of the innate immune system in insects and are produced by numerous RNA and DNA viruses during infection. dsRNAs are converted into viral small interfering RNAs (vsiRNA) by *Dicer* (*Dcr-2*) and then integrated with Argonaute-2 (AGO2) in the RNA-induced silencing complex (RISC) to facilitate the cleavage of the viral target RNAs [115]. Confirming the involvement of RNAi in IIV-6 immune responses, previous studies reported *Drosophila melanogaster Dcr2* or *AGO2* mutants were more sensitive to IIV-6 infection compared to wild type flies, which ultimately led to slight increases in viral titres [116]. Later, Kemp et al. [117] showed more dramatic increases in IIV-6 replication when different mutants of *Dcr2* and *AGO2* were used. Similar increases in viral replication were also observed in *rd2d* mutants. Furthermore, a subsequent study confirmed that the viral protein 340R suppressed RNAi responses in *D. melanogaster*, indicating that the virus encodes proteins that can interfere with insect RNAi immune responses. This viral protein binds to long dsRNAs and impedes the formation of vsiRNAs via Dicer-2 [115].

The contribution of the JAK-STAT pathway in antiviral immunity in *Drosophila* was tested by using loss-of-function mutants of the JAK kinase Hopscotch with IIV-6 and several RNA viruses [117]. Although the JAK-STAT pathway in *Drosophila* confers protection against some RNA viruses (e.g., DCV and CrPV, but not FHV, SINV, and VSV; [118], it does not provide immunity against IIV-6 infection [117]. In addition, epigenetic modifications associated with regulation of the JAK-STAT pathway could not be linked to IIV-6 viral responses in *Drosophila*. For example, H3 lysine 9 methyltransferase G9a, which is the one of the three H3K9 methyltransferases coded by *Drosophila* [119], regulates the JAK-STAT pathway and is required to regulate resistance to RNA viruses (e.g., nodaviruses) except DCV that shows a tolerance phenotype, but not to DNA viruses (e.g., IIV-6) [120]. Another study sought to address the involvement phagocytosis and autophagy in the control of viral infections in insects using a panel of six viruses including IIV-6. However, hemocytes did not respond to IIV-6 infection in *Drosophila* and infection did not trigger autophagy or apoptosis in virus-infected hemocytes, suggesting that autophagy and/or apoptosis may not be typical host responses to IIV-6 in *Drosophila* [121].

Vertebrates respond to IIV-6 using different immune responses from those of insects. Even though IIV-6 is a DNA virus, it can induce the type I IFN-dependent antiviral response in mammalian cells and is likely triggered by a cytosolic sensor, Retinoic acid inducible gene-I (RIG-I), upon viral dsRNA recognition. Transcription of viral DNA into RNA by RNA Polymerase III is required to activate the RLR pathway to obtain maximum IFN secretion [122]. In addition, activation of RLR-driven innate immune response against IIV-6 protects cells (Primary wild-type mouse embryonic fibroblasts) from possible infection with other viruses, such as Vesicular Stomattis and Kunjin arboviruses [122].

Overall, studies investigating the interactions between IIVs and invertebrate and vertebrate hosts suggest that IIVs can occasionally cross infect certain vertebrates, but that their immune systems may be able to inhibit replication under natural conditions, reducing or preventing symptoms.

6. Genomic Organization and Codon Usage

The IIV-6 genome consists of a single linear 212,482 bp dsDNA molecule with 28.63% G + C content [7]. Relative to other animal viruses, the IIV-6 genome has a unique arrangement with circular permutations and terminal redundancies, which causes variations in the number of direct terminal repeats during DNA replication and genome packaging [123,124]. While the coding capacity of the genome was originally estimated at 215 non-overlapping open reading frames (ORFs), the total number of ORFs rises to 468 when both non-overlapping and overlapping ORFs are considered (Figure 4). The revised coding boundaries of the 215 non-overlapping proteins are listed in the Uniprot database [53–55]. Importantly, gene transcripts of iridovirids have no evidence of introns [36,53,54] and viral mRNAs lack poly(A) tails [125]. However, biochemical and in silico evidence suggest the existence of viral microRNAs (miRNA) that may modulate viral gene expression [116]. In addition, important differences in methylation patterns have been observed in iridoviruses that infect invertebrates and vertebrates. For example, in contrast to iridoviruses that infect vertebrates (*Ranavirus*, *Lymphocystivirus* and *Megalovirus*), IIVs do not have a high frequency of methylation in their genomes, which is generally directed by viral DNA methyltransferases [126]. IIVs do not encode a DNA methyltransferase; however, VIVs have high levels of 5′-methylcytosine and code for a DMTase, except for SGIV [36]. We have performed in silico prediction of methylation sites of 50 iridovirid genomes and have demonstrated that there is obvious distinction of methylation patterns between IIVs and VIVs, except for IIV3 from IIVs and SDDV and LCDV-1 from VIVs that have yet unclear phylogenetic status. IIV3 showed predicted methylation sites in its genome while LCDV-1 had no obvious methylation sites predicted (Figure 5).

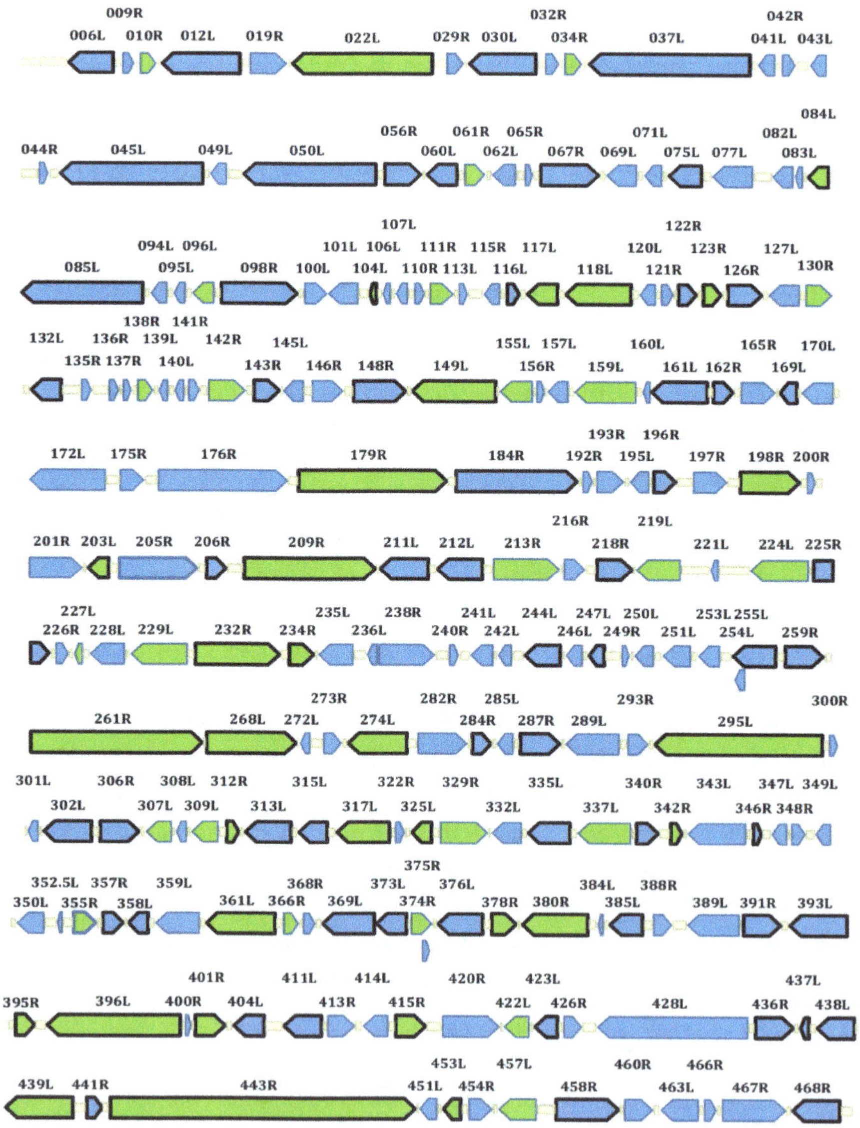

Figure 4. IIV-6 linear genomic map presenting virion and viral proteins identified in infected cells using mass spectrometry (genome size 212,482 bp). Arrows represent the transcriptional direction of the ORFs. All arrows with a black frame represent ORFs for which the corresponding protein was detected in the infected-cell proteomic analysis [127]. Green arrows represent known virion proteins [128]. This figure was published previously [127].

Figure 5. In silico prediction of methylation sites of 50 iridovirid genomes. The Geneious CpG Plugin (Biomatters Ltd., Auckland, New Zealand) predicts likely methylation sites according to the Hidden Markov models described by Richard Durbin. Green indicates potential methylation sites and yellow arrows indicates viral coding region of a genes. Methylation patterns IIVs versus VIVs could be distinguished, with the exceptions of IIV3 from IIVs and SDDV from SDDV as well as LCDV-1.

To date, 26 core genes have been identified in both vertebrate and invertebrate viruses within the family *Iridoviridae*, which have been linked to virus replication, gene transcription, protein structure and nucleocapsid assembly. The rest of the encoded proteins may have roles in virus-host interactions, hijacking of host cell machinery and establishment of viral infection. For example, core genes, including 022L, 037L, 142R, 143R, 176R/343L, 184R, 282R, 349L, 355R, 369L, 376L, 428L, and 436L, have key roles in viral transcription/DNA replication while 098R, 179R/439L and 380R have roles in protein processing, such as removal of signal sequences and modifications [129].

The core genes, 022L (nucleoside-triphosphatase) and 376L (ribonucleoside-diphosphate reductase) have roles in nucleic acid metabolism in the host cell. Other genes such as 037L (DNA polymerase), 142R (ribonuclease), 143R (putative protein), 184R (putative protein), 282R (putative protein), 176R/343L, 349L (transcription elongation factor), 428L (DNA-dependent RNA polymerase), 436L (signal localization protein), 355R (transcription regulator protein), and 369L (endonuclease homolog) are required for viral transcription/DNA replication while 098R (Serine-threonine protein kinase), 179R/439L (putative tyrosine kinase/lipopolysaccharide modifying enzyme) and 380R (Serine-threonine protein kinase), are required for protein processing such as removal of signal sequences and modifications [7,53,130]. The remaining ORFs related to core genes are identified as 274R (major capsid protein), 393L (immediate early protein), 75L (ATPase-like protein), 118L/458R and 337L (myristoylated membrane protein), 307L (a hypothetical protein of *Clostridium tetani*), 067R (helicase family protein), and 347L (ErvI/AIr family protein) whereas the functions of two core genes 295L and 287R are not yet clear [53].

7. Virion Proteins

IIV virion proteins have been previously analyzed in IIV-1 [131] (Watson and Seligy, 1997), IIV-6 [128,132,133] and IIV-9 [13]. Additionally, one- or two-dimensional SDS-PAGE methods were carried out to characterize the polypeptides in IIV-6 virions, which revealed 21–28 polypeptides and 35 polypeptides, respectively [132–134]. An additional study performed by Kelly and Tinsley [134] detected 20 polypeptides from IIV-2. Elliot et al. [135] identified and compared the polypeptides from IIV-2, IIV-22 and IIV-25 and demonstrated that IIV-22 and IIV-25 have 25 polypeptides that differ markedly in terms of molecular weight from the 20 polypeptides identified from IIV-2 [134]. However, in these earlier studies, we have observed that different experimental settings can affect the resolution of the gel analyses and could have thus influenced previous results and comparisons.

More recently, a comprehensive proteomic analysis using 1- and 2-dimensional SDS-PAGE combined with LC-MS/MS was performed to identify IIV-6 virion proteins and 54 proteins were detected [128]. Possible functions of fifteen of the putative IIV-6 ORFs were inferred by blast searches and included serine-threonine kinase, 209R, 380R, 439R; dual specificity protein phosphatase, 123R; DNA polymerase (viral) N terminal domain, 232R; carboxy terminal domain phosphatase, 355R; nucleodide triphophatase (NTP) I, 22L; fasciclin, 96L; ribonuclease III, 142R; tyrosine protein kinase, 179R; cathepsin, 361L; DNA-binding protein, 401R; protein disulfide isomerase, 453L; lysosome associated membrane glycoprotein, 061R and a ranavirus enveloped protein homolog, 118L. The remaining 39 proteins lacked similarity to any other annotated viral proteins and their functions should be investigated further. In addition, the molecular masses of the finger and zip proteins, the anchor protein, and the monomer of the pentameric complex were estimated as 19.7, 11.9, 32.4 and 39.3 kDa, respectively [67]. Based on these size approximations, seven candidate genes coding for zip proteins were identified, including; 010R, 138R and 321R, which had estimated sizes ranging from 10.5 to 13.3 kDa. The predicted size of the protein products of ORFs 329R and 219L were similar to the monomer of the pentameric complex and 457L and 142R represent candidates for anchor proteins [128]. Genomic analysis of Wiseana iridescent virus (IIV-9) identified 191 predicted genes, with 20% of its repeated sequences mainly located within the coding regions. 97 of 211 IIV-6 genes have detectable orthologs in IIV-9, whereas 108 out of 191 IIV-9 genes have orthologs in IIV-3.

Additionally, proteomic analysis of IIV-9 revealed 64 virion proteins and, when SF21 cells were infected by IIV-9, the presence of 94 viral proteins in infected cells were confirmed [13].

Phosphorylation reactions are important post-translational protein modifications that can modify or regulate the function of a protein. They also represent common host responses to iridovirid infection. During IIV-6 infection, increased phosphorylation of the ribosome associated proteins during the early phase of the infection cycle requires the expression of the viral genome [136]. However, phosphorylation events were observed in ribosomal proteins when permissive cells were infected with UV-irradiated IIV-6, suggesting that increasing phosphorylation reactions were likely a host cell response rather than being related to an expression of the viral genome.

Iridovirid genomes contain multiple genes coding for kinases and phosphatases that perform important post-translational modifications of proteins to help regulate protein activities over the course of the infection [127]. Active expression of these virus-encoded phosphatases and kinases involved in dephosphorylation and phosphorylation reactions was observed by proteomic analysis of infected cells. Four protein kinases, three putative serine/threonine kinases (ORFs 209R, 380R and 439R) and one tyrosine protein kinase (179R) were identified from the proteome of IIV-6 virions, while an additional protein kinase (098R) was identified from the viral proteome of infected cells. Two phosphatases were identified from virions, which included a dual specificity protein phosphatase, 123R and carboxy terminal domain phosphatase, 355R.

Although kinases and phosphatases are among the most abundant proteins, it is challenging to assign functions to these enzymes without expressing them individually in cell culture, making knock-out viruses, unravelling their subcellular localization, or identifying the protein targets that are specifically phosphorylated. However, the roles of several kinases have been elucidated. One putative serine/threonine kinase (380R) was present in high abundances in IIV-6 virions [127]. A purified serine/threonine protein kinase 389L (also called iridoptin) induced apoptosis in insect cells and was initially thought to be associated with the IIV-6 virion structure [137]. However, no peptides were identified from this protein in either the virion proteome or the proteome from infected cells [127,128]. Thus, it remains uncertain whether 389L is a virion associated protein or not.

8. Viral Entry, Replication, and Release Strategy

Frog virus 3 (FV3) is the best-characterized member of the genus (*Ranavirus*) and the family (*Iridoviridae*) and has served as a model to elucidate iridoviral transcription, genome replication and virus-mediated host-shutoff [138] (Figure 6). Iridovirids have nucleocytoplasmic replication. Initially, the virus attaches to cell surface receptors, and the virus is engulfed by the host cell via clathrin-mediated endocytosis or macropinocytosis in a pH-dependent manner [139]. ORF 096L of IIV-6, annotated as a fasciclin domain, is proposed to act as an insect cell adhesion molecule (CAM) [140] which may facilitate "fusion" between the virus shell and the cellular membranes [141].

Following the entry into the host cell, immediate early and delayed early transcripts are synthesized using the virion DNA as a template [142–144]. These transcripts encode proteins that are crucial for DNA replication and expression of late genes [133]. Then, newly synthesized viral DNA is translocated from the nucleus to the cytoplasm where a second stage of viral DNA replication occurs by the formation of DNA concatemers [145] (Figure 6). Late viral transcripts, including transcripts of the structural proteins for virion formation, are likely synthesized by a host RNA polymerase or a viral encoded RNA polymerase that has not yet been identified in IIVs. Evidence from VIVs seems to suggest that late viral messages are synthesized by a virus-encoded transcriptase. This hypothesis is supported because knocking down the large subunit of the viral transcriptase of SGIV with asMOs (antisense morpholino oligonucleotides) eliminates late protein synthesis [146]. Once the replicated DNA has reached the cytoplasm, it forms large concatemers, which are eventually packaged into virions. Progeny virions accumulate in large paracrystalline arrays or egress from the cell by either budding or cell lysis, as mentioned previously (Figure 6).

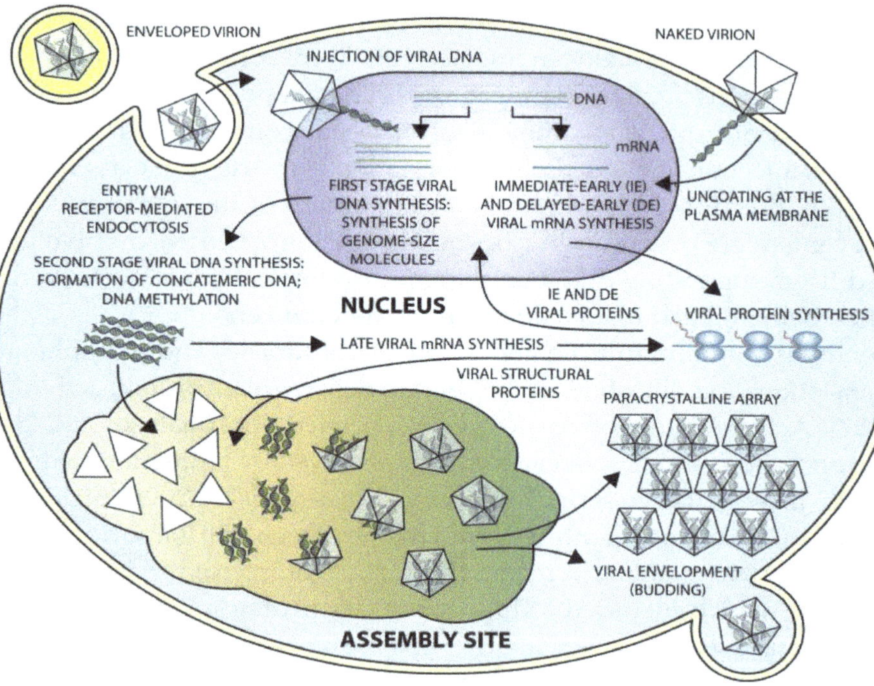

Figure 6. Schematic representation of iridoviral replication. This figure was obtained from the 10th report of ICTV with permission [1,2].

According to the mechanism by which viral particles are released from host cells, viruses can be either be characterized as "extracellular viruses", which are released via budding from the host cell membrane or "intracellular (naked) viruses", which are released along with mature virions from infected cells via lysis (Figure 6). The mechanism of release from host cells is unknown for IIVs. However, the virion assembly and budding process have recently been illustrated for a vertebrate iridovirus (SGIV) using a high pressure freezing (HPR) method. The mitochondria surrounding the viral assembly site (VAS) became enlarged and eventually lost their inner cristae, suggesting that they may eventually become empty vesicles, vulnerable to viral invasion. Mature virions can bud into those vesicles and these virus-filled vesicles may fuse with one another to form larger vacuoles. Eventually, cell lysis will release the mature virions from the cell. This new budding method may be an immune system evasion strategy of SGIV [147].

The specific classes of lipids in the virion structure can influence virus infectivity [148]. In addition, changes in host cellular lipid composition associated with cell proliferation, apoptosis, cellular stress and viral infection can also occur during viral replication and may also influence virus-host interactions [81] and impact the virus' ability to infect its hosts. For example, infection with IIV-6 can induce modifications in the phospholipids of invertebrate cells by decreasing the incorporation of phosphorous into cell phospholipids and stimulating phosphatidylcholine synthesis [149]. To understand viral lipid–protein interactions, entry process, intracellular trafficking, viral assembly and releasing strategy, the lipid composition of the vertebrate iridovirus SGIV was quantitatively analyzed using a combined approach of liquid chromatography and mass spectrometry. The study revealed that selective losses of some of the lipid classes can significantly diminish viral infectivity. For example, enzymatic digestion of viral lipids with phospholipases and sphingomyelinase decreased viral infectivity [81], but did not otherwise compromise the integrity of the viral capsid proteins. As expected, recovery of the viral lipids with liposomes obtained from a grouper embryonic cell line (GEC) restored the infectivity of the virus [147]. In addition, as discussed earlier, disruption of the lipid bilayer in IIVs had a profound effect on their ability to develop in vivo infections [78].

9. Transcriptional Regulation

In IIV-6 infected cells, viral transcription occurs, in a regulated temporal gene expression cascade that can be subdivided into three temporal classes: immediate-early (IE or α), delayed-early (DE, β), and late (L, γ), with the majority of the genes expressed as IE genes [125,142–144]. Both *cis*-acting and *trans*-acting regulatory factors initiate the transcription that coincide with various stages of viral infection. However, some IE and DE transcripts are still present during the late phase of infection and no meaningful correlation has yet been identified between transcription and the prevalence of particular proteins during infection [127,143]. Overall, a previous study performed by northern blot analyses using putative gene-specific probes on total cellular RNA from IIV-6 infected IPRI-CF-124T cells confirmed the expression of 137 viral transcripts including, 38 IE, 34 DE and 65 L gene transcripts. Proteins associated with virions are likely key activators of immediate early genes. For example, purified IIV-6 DNA is inadequate to achieve infection; however, infectivity can be recovered by adding virion proteins, suggesting that these proteins are required to trigger transcription of early genes [150,151]. Furthermore, IE class transcripts appeared around 0.5 h post infection (h.p.i) while the transcription of the delayed-early class genes was detected around 3 h.p.i. The expression of DE genes was impeded when protein synthesis inhibitors were applied which clearly indicates that DE genes require at least one earlier gene product to initiate transcription. In the late class genes, 65 transcripts were detected around at 6 h.p.i. [143].

A more recent study provided detailed information about the kinetics of IIV-6 viral protein levels in infected *Drosophila* S2 cells using a label-free quantitative proteomic approach. The transcripts belonging to the IE class included 022L, 104L, 118L, 123L, 155L, 179R, 209R, 232R, 261R, 268L, 295L, 312R, 361L, 380R, 415R, 439L, and 453L. These gene products are mainly involved in blocking host cell apoptosis, nucleoside metabolism, post-translational protein modifications and transcriptional activation of DE genes [125,143]. DE genes included 117L, 149L, 198R, 229L, 378R, 396L, 457L, which function as viral DNA polymerases, protein kinases, and transcriptional activators of late genes. The major capsid protein MCP (274L), one of the most abundant proteins in IIV-6, was expressed with other late class transcripts, including 247L, 234R, 317L, 342R and 401R [127].

10. Promoter Elements and Transcription Initiation Sites

It is difficult to conclusively identify promoter elements that regulate the temporal expression patterns of iridovirid genes [53]. However, several studies have led to the identification of motifs in putative promoter regions that may be important for transcriptional regulation. For example, in IIV-6, the transcription profiles and promoter structures of the DNA polymerase (*DNApol*; 037L) and major capsid protein genes (*mcp*; 274L) have been analyzed in detail [152]. IIV-6 infection of *Bombyx mori* SPC-BM-36 cells with using DNA or protein synthesis inhibitors demonstrated that *DNApol* belongs to the IE gene class whereas *mcp* is transcribed as a late gene. An AAAAT motif located between the −19 and −2 region was determined to be responsible for the promoter activity of *DNApol*.

A second IE promotor study clearly showed that sequences between −20 and −10 relative to the transcription start site have key promoter activity for IIV-6 012L. Although similar motifs were not found in the upstream regions of any other potential IIV-6 IE genes, the promoter regions of IE and DE genes appear to differ in their organization in IIV-6. The major differences in promoter organization were in the 20 nt region upstream of the transcription initiation site [151,153]. IIV-6 virion proteins are likely able to either directly or indirectly activate the promoters to trigger transcription of some IE genes. Supporting this hypothesis, reporter assays confirmed that without virion proteins the promoters were inactive even though the reporter plasmid constructs carrying these promoters were transfected.

Interestingly, *DNApol*, *helicase*, and *mcp* transcripts of IIV-6 lack polyA tails which was discovered when oligo-DT primers in reverse transcriptase reactions did not lead to any amplification. Furthermore, clear polyadenylation signals downstream of these two IIV-6 ORFs were not found [152]. To identify the transcription termination signals of IIV-6 genes, the LACE (Ligation-based amplification of cDNA ends) method was implemented to amplify the 3′ UTR regions of viral transcripts [125]. LACE showed that about half of all IIV-6 genes contained complementary TAATG and CATTA motifs in the 3′ regions of their mRNAs. These CATTA motifs may be conserved features that enable the formation of hairpins (22–56 nt in length) in IIVs as they are missing in many vertebrate viruses. In the absence of polyadenylation signals, these hairpins may serve as transcriptional terminators [154].

11. Induction/Inhibition of Apoptosis in Infections

Although the signaling pathways and recruited proteins linked to apoptosis can vary between vertebrates and invertebrates, the principles and the dynamics of apoptosis in response to iridovirid infection are similar. During the early stages of iridovirid infection, both vertebrate and invertebrate host cells trigger apoptotic machinery to impede the cell-to-cell movement of progeny viruses [155,156]; however, many viruses have evolved mechanisms to protect the progeny viruses from the adaptive immune system and enhance the likelihood of spreading to neighboring cells by producing anti-apoptotic proteins [157,158]. The expression of these anti-apoptotic genes prolongs survival of infected cells, which increases the production of progeny virus and/or may lead to virus persistence. Furthermore, viruses contain genes that may stimulate apoptosis (protein tyrosine phosphate or viral kinases), which may contribute to viral dissemination to neighboring cells. Fish iridoviruses, for instance, appear to exploit apoptosis for their dissemination.

Initially, expression of early viral gene/s (IE) inhibited apoptosis, which suggests that one or more IE genes likely functions as anti-apoptotic factors (e.g., *iap*). Among members of the genus *Iridovirus*, IIV-6 was the first virus identified to contain three ORFs (157L, 193R, and 332L) with homology to the *iap* (*inhibitor of apoptosis*) gene of a baculovirus [158]. Among these genes, only 193R, which was expressed as an IE gene, was confirmed as a functional IAP, [158]. The other two *iap* candidate ORFs, 157L and 332L, contained only RING finger domains and lacked the BIR domains, suggesting that they likely do not directly inhibit apoptosis. A subsequent study also identified a putative *iap* gene in a recently sequenced *Armadillidium vulgare* iridescent virus (IIV-31) [9] which shares homology with 193R of IIV-6 but its functionality has not yet been tested. IIVs also code for pro-apoptotic factors, which trigger apoptosis at later stages during the infection to facilitate viral dissemination. For example, IIV-6 infected SPC-Bm-36 cells formed apoptotic-like vesicles after 18 h of infection [159] and further evidence exists that pro-apoptotic factors may be associated with virions. For example, a comprehensive study showed that high doses of IIV-6 can initiate apoptotic responses, yet at lower doses, apoptosis was inhibited [156]. Further supporting the hypothesis of virion associated pro-apoptotic factors, de novo synthesis of viral proteins was not required for induction of apoptosis [156] and IIV-6 virion protein extract induced apoptosis in spruce budworm and boll weevil cell lines, as detected by blebbing, DNA fragmentation, TUNEL assay and protein kinase activity [160]. More recently, an *istk* gene product and a virion protein of IIV-6, 389L, which is a 49 kDa polypeptide, was identified as apoptosis inducing factor. Additionally, a 35-kDa cleavage product of ISTK (iridoptin) also functions as a potent inducer of apoptosis in host cells [137] however it is unknown when it is produced in infection and whether it is a virion or ICP protein [127,128].

12. Concluding Remarks

Recent genomic, transcriptomic, proteomic, and molecular studies have elucidated the routes of infection, mechanisms of replication, classification, and the identities of viral proteins that are associated with virulence or modulate host cellular responses to IIV infection. Despite these recent advances, several important knowledge gaps exist. For example, it is not known which host and/or viral factors trigger viral entry and replication into host cells and the regulatory mechanisms of genes expressed at various stages during infection remain uncharacterized. While transcriptomic and proteomic studies may address some of these short-comings, other important questions regarding iridovirid-host interactions cannot be addressed without the use of functional genomics. Previously, functional genomics studies in IIVs have been hampered by the lack of genetic recombination systems for producing genetic (knockout) mutants of the virus; however, the first recombinant IIV using a homologous recombination site was recently established [161,162]. The identification of viral proteins responsible for viral infectivity will be an important next step towards the development of a genetic recombination system for iridovirids (possibly similar to a bacmid system) as iridovirid DNA by itself is not infectious. The development of a recombination system will expand the utility of IIVs as a model for identifying and characterizing iridovirid genetic factors that drive virus-host interactions, impact host fitness, and influence immune system response, which could ultimately help limit or prevent disease in aquatic ecosystems or lead to the development of IIVs as biocontrol agents against medically important vectors.

Next, IIV-6 has attracted little attention as a potential biopesticide [163] due to its broad host range and limited mortality, although epizootics have been reported occasionally [82,90,164]. However, transcriptomics and proteomics studies have identified virally coded genes that may be linked to host toxicity and could be exploited as biocontrol compounds. For example, serine/threonine kinases can induce toxicity in several insect groups, such as weevils and whiteflies. Specifically, the serine/threonine kinase "iridoptin" has been identified as a potential viral toxin against non-lepidopteran insects, a group of insects for which there are limited options for biological control [137]. Additional functional genomics studies along with an expanding repertoire of genomic resources may lead to the identification of new virulence factors and toxins that could be exploited and engineered for targeted control against certain groups of insects, for example, insects of agricultural importance or that transmit plant pathogens and/or disease-causing parasites, such as mosquitoes, whiteflies and grasshoppers [89,100,113,165,166]. Furthermore, virally coded kinases could represent potential antiviral drug targets, as they play crucial roles to guarantee successful viral infection [130]. Understanding the roles of these kinases in IIV-6 may help to develop therapeutics against devastating iridovirus outbreaks in fish farming, for example.

Finally, current phylogenetic approaches for iridovirids have a limited ability to conclusively resolve species level differences; however, the availability of additional genomic resources could help to overcome some of the limitations of current phylogenetic methods and also improve support for proposed classifications. Specifically, we propose to establish a phylogenetic approach using raw sequencing data to incorporate pan-genomic information into the analysis of viral phylogenies, which may ultimately lead to the development of new classification criteria. This approach would allow researchers to leverage sequences of viral quasispecies in infected cells and/or host organisms [167], to extract neglected sequence data due to infectome complexity and gauge the genetic diversity and complexity of a viral species, e.g., gene variants, fusions and recombination products. Ultimately, pan-genomic information may expand the capacity to resolve phylogenetic relationships beyond the capabilities of current methods based on using gene subsets or complete genome information from one representative isolate.

Acknowledgments: We thank to Erica Ras (Pragmatika Biotech R&D LTD) for designing graphical illustrations from genomic analysis and phylogeny for supplementary documentation and V. Gregory Chinchar (University of Mississippi Medical Center) for his thoughtful and constructive comments on our review.

References

1. Chinchar, V.G.; Hick, P.; Ince, I.A.; Jancovich, J.K.; Marschang, R.; Qin, Q.; Subramaniam, K.; Waltzek, T.B.; Whittington, R.; Williams, T.; et al. ICTV Virus Taxonomy Profile: *Iridoviridae*. *J. Gen. Virol.* **2017**, *98*, 890–891. [CrossRef] [PubMed]
2. Chinchar, V.G.; Waltzek, T.B.; Subramaniam, K. Ranaviruses and other members of the family *Iridoviridae*: Their place in the virosphere. *Virology* **2017**, *511*, 259–271. [CrossRef] [PubMed]

3. Chinchar, V.; Essbauer, S.; He, J.; Hyatt, A.; Miyazaki, T.; Seligy, V.; Williams, T. Family *Iridoviridae*. In *Virus Taxonomy Eighth Report of the International Committee on Taxonomy of Viruses*; Elsevier Academic Press: San Diego, CA, USA, 2005; pp. 145–162.
4. King, A.M.; Lefkowitz, E.; Adams, M.J.; Carstens, E.B. *Virus Taxonomy: Ninth Report of the International Committee on Taxonomy of Viruses*; Elsevier: Amsterdam, The Netherlands, 2011.
5. Xeros, N. A second virus disease of the leatherjacket, *Tipula paludosa*. *Nature* **1954**, *174*, 562–563. [CrossRef]
6. Fukaya, M.; Nasu, S. An iridescent virus (CIV) from the rice stem borer, *Chilo suppressalis* Walker (Lepidoptera: Pyralidae). *Appl. Entomol. Zool.* **1966**, *1*, 69–72. [CrossRef]
7. Jakob, N.J.; Muller, K.; Bahr, U.; Darai, G. Analysis of the first complete DNA sequence of an invertebrate iridovirus: Coding strategy of the genome of *Chilo* iridescent virus. *Virology* **2001**, *286*, 182–196. [CrossRef] [PubMed]
8. Boyer, M.; Yutin, N.; Pagnier, I.; Barrassi, L.; Fournous, G.; Espinosa, L.; Robert, C.; Azza, S.; Sun, S.; Rossmann, M.G.; et al. Giant Marseillevirus highlights the role of amoebae as a melting pot in emergence of chimeric microorganisms. *Proc. Natl. Acad. Sci. USA* **2009**, *106*, 21848–21853. [CrossRef] [PubMed]
9. Kalyaanamoorthy, S.; Minh, B.Q.; Wong, T.K.F.; von Haeseler, A.; Jermiin, L.S. ModelFinder: Fast Model Selection for Accurate Phylogenetic Estimates. *Nat. Methods* **2017**, *14*, 587–589. [CrossRef] [PubMed]
10. Nguyen, L.-T.; Schmidt, H.A.; von Haeseler, A.; Minh, B.Q. IQ-TREE: A fast and effective stochastic algorithm for estimating maximum likelihood phylogenies. *Mol. Biol. Evol.* **2015**, *32*, 268–274. [CrossRef] [PubMed]
11. Piégu, B.; Guizard, S.; Yeping, T.; Cruaud, C.; Asgari, S.; Bideshi, D.K.; Federici, B.A.; Bigot, Y. Genome sequence of a crustacean iridovirus, IIV31, isolated from the pill bug, *Armadillidium vulgare*. *J. Gen. Virol.* **2014**, *95*, 1585–1590. [CrossRef] [PubMed]
12. Delhon, G.; Tulman, E.R.; Afonso, C.L.; Lu, Z.; Becnel, J.J.; Moser, B.A.; Kutish, G.F.; Rock, D.L. Genome of Invertebrate iridescent virus type 3 (mosquito iridescent virus). *J. Virol.* **2006**, *80*, 8439–8449. [CrossRef] [PubMed]
13. Wong, C.K.; Young, V.L.; Kleffmann, T.; Ward, V.K. Genomic and proteomic analysis of Invertebrate iridovirus type 9. *J. Virol.* **2011**, *85*, 7900–7911. [CrossRef] [PubMed]
14. Piégu, B.; Guizard, S.; Spears, T.; Cruaud, C.; Couloux, A.; Bideshi, D.K.; Federici, B.A.; Bigot, Y. Complete genome sequence of invertebrate iridescent virus 22 isolated from a blackfly larva. *J. Gen. Virol.* **2013**, *94*, 2112–2116. [CrossRef] [PubMed]
15. Piégu, B.; Guizard, S.; Yeping, T.; Cruaud, C.; Couloux, A.; Bideshi, D.K.; Frederici, B.A.; Bigot, Y. Complete genome sequence of invertebrate iridovirus IIV22A, a variant of IIV22, isolated originally from a blackfly larva. *Stand Genom. Sci.* **2014**, *9*, 940–947. [CrossRef] [PubMed]
16. Piégu, B.; Guizard, S.; Spears, T.; Cruaud, C.; Couloux, A.; Bideshi, D.K.; Federici, B.A.; Bigot, Y. Complete genome sequence of invertebrate iridovirus IIV25 isolated from a blackfly larva. *Arch. Virol.* **2014**, *159*, 1181–1185. [CrossRef] [PubMed]
17. Piégu, B.; Guizard, S.; Spears, T.; Cruaud, C.; Couloux, A.; Bideshi, D.K.; Federici, B.A.; Bigot, Y. Complete genome sequence of invertebrate iridovirus IIV30 isolated from the corn earworm, *Helicoverpa zea*. *J. Invertebr. Pathol.* **2014**, *116*, 43–47. [CrossRef] [PubMed]
18. Chen, Z.; Gui, J.; Gao, X.; Pei, C.; Hong, Y.; Zhang, Q. Genome architecture changes and major gene variations of Andrias davidianus ranavirus (ADRV). *BMC Vet. Res.* **2013**, *44*, 101. [CrossRef] [PubMed]
19. Jancovich, J.K.; Mao, J.; Chinchar, V.G.; Wyatt, C.; Case, S.T.; Kumar, S.; Valente, G.; Subramanian, S.; Davidson, E.W.; Collins, J.P.; et al. Genomic sequence of a ranavirus (family *Iridoviridae*) associated with salamander mortalities in North America. *Virology* **2003**, *316*, 90–103. [CrossRef] [PubMed]
20. Hick, P.M.; Subramaniam, K.; Thompson, P.; Whittington, R.J.; Waltzek, T.B. Complete genome sequence of a Bohle iridovirus isolate from ornate burrowing frogs (*Limnodynastes ornatus*) in Australia. *Genome Announc.* **2016**, *4*, e00632-16. [CrossRef] [PubMed]
21. Stöhr, A.C.; López-Bueno, A.; Blahak, S.; Caeiro, M.F.; Rosa, G.M.; de Matos, A.P.A.; Martel, A.; Alejo, A.; Marschang, R.E. Phylogeny and differentiation of reptilian and amphibian ranaviruses detected in Europe. *PLoS ONE* **2015**, *10*, e0118633. [CrossRef] [PubMed]
22. Mavian, C.; López-Bueno, A.; Balseiro, A.; Casais, R.; Alcamí, A.; Alejo, A. The genome sequence of the emerging common midwife toad virus identifies an evolutionary intermediate within ranaviruses. *J. Virol.* **2012**, *86*, 3617–3625. [CrossRef] [PubMed]

23. Van Beurden, S.J.; Hughes, J.; Saucedo, B.; Rijks, J.; Kik, M.; Haenen, O.L.; Engelsma, M.Y.; Gröne, A.; Verheije, M.H.; Wilkie, G. Complete genome sequence of a common midwife toad virus-like ranavirus associated with mass mortalities in wild amphibians in The Netherlands. *Genome Announc.* **2014**, *2*, e01293-14. [CrossRef] [PubMed]
24. Ariel, E.; Steckler, N.K.; Subramaniam, K.; Olesen, N.J.; Waltzek, T.B. Genomic sequencing of ranaviruses isolated from turbot (*Scophthalmus maximus*) and Atlantic cod (*Gadus morhua*). *Genome Announc.* **2016**, *4*, e01393-16. [CrossRef] [PubMed]
25. Jancovich, J.K.; Bremont, M.; Touchman, J.W.; Jacobs, B.L. Evidence for multiple recent host species shifts among the ranaviruses (Family *Iridoviridae*). *J. Virol.* **2010**, *84*, 2636–2647. [CrossRef] [PubMed]
26. Mavian, C.; López-Bueno, A.; Somalo, M.P.F.; Alcamí, A.; Alejo, A. Complete genome sequence of the European sheatfish virus. *J. Virol.* **2012**, *86*, 6365–6366. [CrossRef] [PubMed]
27. Tan, W.G.; Barkman, T.J.; Chinchar, V.G.; Essani, K. Comparative genomic analyses of Frog virus 3, type species of the genus *Ranavirus* (family *Iridoviridae*). *Virology* **2004**, *323*, 70–84. [CrossRef] [PubMed]
28. Tsai, C.T.; Ting, J.W.; Wu, M.H.; Wu, M.F.; Guo, C.; Chang, C.Y. Complete genome sequence of the grouper iridovirus and comparison of genomic organization with those of other iridoviruses. *J. Virol.* **2005**, *79*, 2010–2023. [CrossRef] [PubMed]
29. Jancovich, J.K.; Davidson, E.W.; Parameswaran, N.; Mao, J.; Chinchar, V.G.; Collins, J.P.; Jacobs, B.L.; Storfer, A. Evidence for emergence of an amphibian iridoviral disease because of human-enhanced spread. *Mol. Ecol.* **2005**, *14*, 213–224. [CrossRef] [PubMed]
30. Holopainen, R.; Subramaniam, K.; Steckler, N.K.; Claytor, S.C.; Ariel, E.; Waltzek, T.B. Genome sequence of a *Ranavirus* isolated from pike-perch *Sander lucioperca*. *Genome Announc.* **2016**, *4*, e01295-16. [CrossRef] [PubMed]
31. Claytor, S.C.; Subramaniam, K.; Landrau-Giovannetti, N.; Chinchar, V.G.; Gray, M.J.; Miller, D.L.; Mavian, C.; Salemi, M.; Wisely, S.; Waltzek, T.B. Ranavirus phylogenomics: Signatures of recombination and inversions among bullfrog ranaculture isolates. *Virology* **2017**, *511*, 330–343. [CrossRef] [PubMed]
32. Ariel, E.; Subramaniam, K.; Imnoi, K.; Sriwanayos, P.; Ahasan, M.S.; Olesen, N.J.; Amedeo, M.; Toffan, A.; Waltzek, T.B. Genomic sequencing of ranaviruses isolated from edible frogs (*Pelophylax esculentus*). *Genome Announc.* **2017**, *5*, e01015-17. [CrossRef] [PubMed]
33. Lei, X.Y.; Ou, T.; Zhu, R.L.; Zhang, Q.Y. Sequencing and analysis of the complete genome of Rana grylio virus (RGV). *Arch. Virol.* **2012**, *157*, 1559–1564. [CrossRef] [PubMed]
34. Epstein, B.; Storfer, A. Comparative genomics of an emerging amphibian virus. *G3 Genes Genomes Genet.* **2016**, *6*, 15–27. [CrossRef] [PubMed]
35. Subramaniam, K.; Toffan, A.; Cappellozza, E.; Steckler, N.K.; Olesen, N.J.; Ariel, E.; Waltzek, T.B. Genomic sequence of a ranavirus isolated from short-finned eel (*Anguilla australis*). *Genome Announc.* **2016**, *4*, e00843-16. [CrossRef] [PubMed]
36. Song, W.J.; Qin, Q.W.; Qiu, J.; Huang, C.H.; Wang, F.; Hew, C.L. Functional genomics analysis of Singapore grouper iridovirus: Complete sequence determination and proteomic analysis. *J. Virol.* **2004**, *78*, 12576–12590. [CrossRef] [PubMed]
37. Morrison, E.A.; Garner, S.; Echaubard, P.; Lesbarrères, D.; Kyle, C.J.; Brunetti, C.R. Complete genome analysis of a Frog virus 3 (FV3) isolate and sequence comparison with isolates of differing levels of virulence. *Virol. J.* **2014**, *11*, 46. [CrossRef] [PubMed]
38. Huang, Y.; Huang, X.; Liu, H.; Gong, J.; Ouyang, Z.; Cui, H.; Cao, J.; Zhao, Y.; Wang, X.; Jiang, Y. Complete sequence determination of a novel reptile iridovirus isolated from soft-shelled turtle and evolutionary analysis of *Iridoviridae*. *BMC Genom.* **2009**, *10*, 224. [CrossRef] [PubMed]
39. He, J.G.; Lü, L.; Deng, M.; He, H.H.; Weng, S.P.; Wang, X.H.; Zhou, S.Y.; Long, Q.X.; Wang, X.Z.; Chan, S.M. Sequence analysis of the complete genome of an iridovirus isolated from the tiger frog. *Virology* **2002**, *292*, 185–197. [CrossRef] [PubMed]
40. Tidona, C.A.; Darai, G. The complete DNA sequence of lymphocystis disease virus. *Virology* **1997**, *230*, 207–216. [CrossRef] [PubMed]
41. Zhang, Q.Y.; Xiao, F.; Xie, J.; Li, Z.Q.; Gui, J.F. Complete genome sequence of Lymphocystis disease virus isolated from China. *J. Virol.* **2004**, *78*, 6982–6994. [CrossRef] [PubMed]

42. López-Bueno, A.; Mavian, C.; Labella, A.M.; Castro, D.; Borrego, J.J.; Alcami, A.; Alejo, A. Concurrence of iridovirus, polyomavirus, and a unique member of a new group of fish papillomaviruses in lymphocystis disease-affected gilthead Sea Bream. *J. Virol.* **2016**, *90*, 8768–8779. [CrossRef] [PubMed]
43. Chen, X.Y.; Wen, C.M.; Wu, J.L.; Su, Y.C.; Hong, J.R. Giant seaperch iridovirus (GSIV) induces mitochondria-mediated cell death that is suppressed by bongkrekic acid and cycloheximide in a fish cell line. *Virus Res.* **2016**, *213*, 37–45. [CrossRef] [PubMed]
44. He, J.G.; Deng, M.; Weng, S.P.; Li, Z.; Zhou, S.Y.; Long, Q.X.; Wang, X.Z.; Chan, S.M. Complete genome analysis of the mandarin fish infectious spleen and kidney necrosis iridovirus. *Virology* **2001**, *291*, 126–139. [CrossRef] [PubMed]
45. Kurita, J.; Nakajima, K. Megalocytiviruses. *Viruses* **2012**, *4*, 521–538. [CrossRef] [PubMed]
46. Lü, L.; Zhou, S.Y.; Chen, C.; Weng, S.P.; Chan, S.M.; He, J.G. Complete genome sequence analysis of an iridovirus isolated from the orange-spotted grouper, *Epinephelus coioides*. *Virology* **2005**, *339*, 81–100. [CrossRef] [PubMed]
47. Do, J.W.; Moon, C.H.; Kim, H.J.; Ko, M.S.; Kim, S.B.; Son, J.H.; Kim, J.S.; An, E.J.; Kim, M.K.; Lee, S.K.; et al. Complete genomic DNA sequence of Rock bream iridovirus. *Virology* **2004**, *325*, 351–363. [CrossRef] [PubMed]
48. Nakajima, K.; Kunita, J. Red sea bream iridoviral disease. *Uirusu* **2005**, *55*, 115–125. [CrossRef] [PubMed]
49. De Groof, A.; Guelen, L.; Deijs, M.; van der Wal, Y.; Miyata, M.; Ng, K.S.; van Grinsven, L.; Simmelink, B.; Biermann, Y.; Grisez, L.; et al. A novel virus causes scale drop disease in *Lates calcarifer*. *PLoS Pathog.* **2015**, *11*, e1005074. [CrossRef] [PubMed]
50. Shi, C.Y.; Jia, K.T.; Yang, B.; Huang, J. Complete genome sequence of a *Megalocytivirus* (family *Iridoviridae*) associated with turbot mortality in China. *Virol. J.* **2010**, *7*, 159. [CrossRef] [PubMed]
51. Huang, Y.; Li, S.; Zhao, Q.; Pei, G.; An, X.; Guo, X.; Zhou, H.; Zhang, Z.; Zhang, J.; Tong, Y. Isolation and characterization of a novel invertebrate iridovirus from adult *Anopheles minimus* (AMIV) in China. *J. Invertebr. Pathol.* **2015**, *127*, 1–5. [CrossRef] [PubMed]
52. Qiu, L.; Chen, M.M.; Wang, R.Y.; Wan, X.Y.; Li, C.; Zhang, Q.L.; Dong, X.; Yang, B.; Xiang, J.H.; Huang, J. Complete genome sequence of shrimp hemocyte iridescent virus (SHIV) isolated from white leg shrimp, *Litopenaeus vannamei*. *Arch. Virol.* **2018**, *163*, 781–785. [CrossRef] [PubMed]
53. Eaton, H.E.; Metcalf, J.; Penny, E.; Tcherepanov, V.; Upton, C.; Brunetti, C.R. Comparative genomic analysis of the family *Iridoviridae*: Re-annotating and defining the core set of iridovirus genes. *Virol. J.* **2007**, *4*, 11. [CrossRef] [PubMed]
54. Tsai, C.T.; Lin, C.H.; Chang, C.Y. Analysis of codon usage bias and base compositional constraints in iridovirus genomes. *Virus Res.* **2007**, *126*, 196–206. [CrossRef] [PubMed]
55. Magrane, M.; Consortium, U. UniProt Knowledgebase: A hub of integrated protein data. *Database J. Biol. Databases Curation Database* **2011**, *2011*, Bar009. [CrossRef] [PubMed]
56. Williams, T.; Cory, J.S. Proposals for a new classification of iridescent viruses. *J. Gen. Virol.* **1994**, *75*, 1291–1301. [CrossRef] [PubMed]
57. Webby, R.; Kalmakoff, J. Sequence comparison of the major capsid protein gene from 18 diverse iridoviruses. *Arch. Virol.* **1998**, *143*, 1949–1966. [CrossRef] [PubMed]
58. Webby, R.J.; Kalmakoff, J. Comparison of the major capsid protein genes, terminal redundancies, and DNA-DNA homologies of two New Zealand iridoviruses. *Virus Res.* **1999**, *59*, 179–189. [CrossRef]
59. Muttis, E.; Miele, S.A.; Belaich, M.N.; Micieli, M.V.; Becnel, J.J.; Ghiringhelli, P.D.; Garcia, J.J. First record of a mosquito iridescent virus in *Culex pipiens* L. (Diptera: Culicidae). *Arch. Virol.* **2012**, *157*, 1569–1571. [CrossRef] [PubMed]
60. Colson, P.; De Lamballerie, X.; Yutin, N.; Asgari, S.; Bigot, Y.; Bideshi, D.K.; Cheng, X.W.; Federici, B.A.; Van Etten, J.L.; Koonin, E.V. "Megavirales", a proposed new order for eukaryotic nucleocytoplasmic large DNA viruses. *Arch. Virol.* **2013**, *158*, 2517–2521. [CrossRef] [PubMed]
61. Jancovich, J.K.; Qin, Q.; Zhang, Q.Y.; Chinchar, V.G. Ranavirus replication: Molecular, Cellular, and Immunological events. In *Ranaviruses: Lethal Pathogens of Ectothermic Vertebrates*; Gray, M.J., Chinchar, V.G., Eds.; Springer: New York, NY, USA, 2015; pp. 105–139.
62. Piégu, B.; Asgari, S.; Bideshi, D.; Federici, B.A.; Bigot, Y. Evolutionary relationships of iridoviruses and divergence of ascoviruses from invertebrate iridoviruses in the superfamily Megavirales. *Mol. Phylogenet. Evol.* **2015**, *84*, 44–52. [CrossRef] [PubMed]

63. Tidona, C.A.; Schnitzler, P.; Kehm, R.; Darai, G. Is the major capsid protein of iridoviruses a suitable target for the study of viral evolution? *Virus Genes* **1998**, *16*, 59–66. [CrossRef] [PubMed]
64. Eaton, H.E.; Ring, B.A.; Brunetti, C.R. The genomic diversity and phylogenetic relationship in the family *Iridoviridae*. *Viruses* **2010**, *2*, 1458–1475. [CrossRef] [PubMed]
65. Qiu, L.; Chen, M.M.; Wan, X.Y.; Li, C.; Zhang, Q.L.; Wang, R.Y.; Cheng, D.Y.; Dong, X.; Yang, B.; Wang, X.H.; et al. Characterization of a new member of Iridoviridae, Shrimp hemocyte iridescent virus (SHIV), found in white leg shrimp (*Litopenaeus vannamei*). *Sci. Rep.* **2017**, *7*, 11834. [CrossRef] [PubMed]
66. Yan, X.; Olson, N.H.; Van Etten, J.L.; Bergoin, M.; Rossmann, M.G.; Baker, T.S. Structure and assembly of large lipid-containing dsDNA viruses. *Nat. Struct. Mol. Biol.* **2000**, *7*, 101–103.
67. Yan, X.; Yu, Z.; Zhang, P.; Battisti, A.J.; Holdaway, H.A.; Chipman, P.R.; Bajaj, C.; Bergoin, M.; Rossmann, M.G.; Baker, T.S. The capsid proteins of a large, icosahedral dsDNA virus. *J. Mol. Biol.* **2009**, *385*, 1287–1299. [CrossRef] [PubMed]
68. Williams, T. Invertebrate iridescent viruses. In *The Insect Viruses*; Miller, L., Ball, A., Eds.; Plenum Press: New York, NY, USA, 1998; pp. 31–68.
69. Balange-Orange, N.; Devauchelle, G. Lipid composition of an Iridescent virus type 6 (CIV). *Arch. Virol.* **1982**, *73*, 363–367. [CrossRef] [PubMed]
70. Low, M.G. Glycosyl-phosphatidylinositol: A versatile anchor for cell surface proteins. *FASEB J.* **1989**, *3*, 1600–1608. [CrossRef] [PubMed]
71. Williams, T.; Thompson, I. Fatty acid profiles of invertebrate iridescent viruses. *Arch. Virol.* **1995**, *140*, 975–981. [CrossRef] [PubMed]
72. Van Meer, G.; Voelker, D.R.; Feigenson, G.W. Membrane lipids: Where they are and how they behave. *Nat. Rev. Mol. Cell Biol.* **2008**, *9*, 112–124. [CrossRef] [PubMed]
73. ViralZone. SIB Swiss Institute of Bioinformatics. Available online: www.expasy.org/viralzone (accessed on 28 March 2018).
74. Tonka, T.; Weiser, J. Iridovirus infection in mayfly larvae. *J. Invertebr. Pathol.* **2000**, *76*, 229–231. [CrossRef] [PubMed]
75. Marina, C.F.; Feliciano, J.M.; Valle, J.; Williams, T. Effect of temperature, pH, ion concentration, and chloroform treatment on the stability of Invertebrate iridescent virus 6. *J. Invertebr. Pathol.* **2000**, *75*, 91–94. [CrossRef] [PubMed]
76. Reyes, A.; Christian, P.; Valle, J.; Williams, T. Persistence of Invertebrate iridescent virus 6 in soil. *BioControl* **2004**, *49*, 433–440. [CrossRef]
77. Day, M.; Mercer, E. Properties of an iridescent virus from the beetle, *Sericesthis pruinosa*. *Aust. J. Biol. Sci.* **1964**, *17*, 892–902. [CrossRef]
78. Martinez, G.; Christian, P.; Marina, C.; Williams, T. Sensitivity of Invertebrate iridescent virus 6 to organic solvents, detergents, enzymes and temperature treatment. *Virus Res.* **2003**, *91*, 249–254. [CrossRef]
79. Hernandez, A.; Marina, C.F.; Valle, J.; Williams, T. Persistence of Invertebrate iridescent virus 6 in tropical artificial aquatic environments. *Arch. Virol.* **2005**, *150*, 2357–2363. [CrossRef] [PubMed]
80. Constantino, M.; Christian, P.; Marina, C.F.; Williams, T. A comparison of techniques for detecting Invertebrate iridescent virus 6. *J. Virol. Methods* **2001**, *98*, 109–118. [CrossRef]
81. Wu, J.; Chan, R.; Wenk, M.R.; Hew, C.L. Lipidomic study of intracellular *Singapore grouper iridovirus*. *Virology* **2010**, *399*, 248–256. [CrossRef] [PubMed]
82. Williams, T. Natural invertebrate hosts of iridoviruses (Iridoviridae). *Neotrop. Entomol.* **2008**, *37*, 615–632. [CrossRef] [PubMed]
83. Federici, B. Isolation of an iridovirus from two terrestrial isopods, the pill bug, *Armadillidium vulgare* and the sow bug, *Porcellio dilatatus*. *J. Invertebr. Pathol.* **1980**, *36*, 373–381. [CrossRef]
84. Karasawa, S.; Takatsuka, J.; Kato, J. Report on iridovirus IIV31 (*Iridoviridae, Iridovirus*) infecting terrestrial isopods (Isopoda, Oniscidea) in Japan. *Crustaceana* **2012**, *85*, 1269–1278. [CrossRef]
85. Lupetti, P.; Montesanto, G.; Ciolfi, S.; Marri, L.; Gentile, M.; Paccagnini, E.; Lombardo, B.M. Iridovirus infection in terrestrial isopods from Sicily (Italy). *Tissue Cell* **2013**, *45*, 321–327. [CrossRef] [PubMed]
86. Webster, C.L.; Longdon, B.; Lewis, S.H.; Obbard, D.J. Twenty-five new viruses associated with the *Drosophilidae* (Diptera). *Evol. Bioinform. Online* **2016**, *12*, 13–25. [CrossRef] [PubMed]
87. Gregory, C.R.; Latimer, K.S.; Pennick, K.E.; Benson, K.; Moore, T. Novel iridovirus in a nautilus (*Nautilus* spp.). *J. Vet. Diagn. Investig.* **2006**, *18*, 208–211. [CrossRef] [PubMed]

88. Devauchelle, G. Ultrastructural characterization of an iridovirus from the marine worm *Nereis diversicolor* (O. F. Müller) Architecture of the virion and virus morphogenesis. *Virology* **1977**, *81*, 237–246. [CrossRef]
89. Hunter, W.; Lapointe, S.; Sinisterra, X.; Achor, D.; Funk, C. Iridovirus in the root weevil *Diaprepes abbreviatus. J. Insect Sci.* **2003**, *3*, 1–6. [CrossRef]
90. Hernandez, O.; Maldonado, G.; Williams, T. An epizootic of patent iridescent virus disease in multiple species of blackflies in Chiapas, Mexico. *Med. Vet. Entomol.* **2000**, *14*, 458–462. [CrossRef] [PubMed]
91. Marina, C.F.; Arredondo-Jimenez, J.; Castillo, A.; Williams, T. Sublethal effects of iridovirus disease in a mosquito. *Oecologia* **1999**, *119*, 383–388. [CrossRef] [PubMed]
92. Marina, C.F.; Ibarra, J.E.; Arredondo-Jimenez, J.I.; Fernandez-Salas, I.; Valle, J.; Williams, T. Sublethal iridovirus disease of the mosquito *Aedes aegypti* is due to viral replication not cytotoxicity. *Med. Vet. Entomol.* **2003**, *17*, 187–194. [CrossRef] [PubMed]
93. Williams, T. Covert iridovirus infection of blackfly larvae. *Proc. R. Soc. Lond. B* **1993**, *251*, 225–230. [CrossRef]
94. Williams, T. Patterns of covert infection by invertebrate pathogens: Iridescent viruses of blackflies. *Mol. Ecol.* **1995**, *4*, 447–458. [CrossRef]
95. Williams, T.; Valle, J.; Viñuela, E. Is the naturally-derived insecticide Spinosad compatible with insect natural enemies? *Biocontrol. Sci. Technol.* **2003**, *13*, 459–475. [CrossRef]
96. Bromenshenk, J.J.; Henderson, C.B.; Wick, C.H.; Stanford, M.F.; Zulich, A.W.; Jabbour, R.E.; Deshpande, S.V.; McCubbin, P.E.; Seccomb, R.A.; Welch, P.M.; et al. Iridovirus and microsporidian linked to honey bee colony decline. *PLoS ONE* **2010**, *5*, e13181. [CrossRef] [PubMed]
97. Tokarz, R.; Firth, C.; Street, C.; Cox-Foster, D.L.; Lipkin, W.I. Lack of evidence for an association between Iridovirus and colony collapse disorder. *PLoS ONE* **2011**, *6*, e21844. [CrossRef] [PubMed]
98. Foster, L.J. Interpretation of data underlying the link between colony collapse disorder (CCD) and an invertebrate iridescent virus. *Mol. Cell Proteom.* **2011**, *10*, M110.006387. [CrossRef] [PubMed]
99. Funk, C.J.; Hunter, W.B.; Achor, D.S. Replication of insect iridescent virus 6 in a whitefly cell line. *J. Invertebr. Pathol.* **2001**, *77*, 144–146. [CrossRef] [PubMed]
100. Hunter, W.B.; Patte, C.P.; Sinisterra, X.H.; Achor, D.S.; Funk, C.J.; Polston, J.E. Discovering new insect viruses: Whitefly iridovirus (Homoptera: Aleyrodidae: *Bemisia tabaci*). *J. Invertebr. Pathol.* **2001**, *78*, 220–225. [CrossRef] [PubMed]
101. Henderson, C.; Johnson, C.; Lodhi, S.; Bilimoria, S. Replication of Chilo iridescent virus in the cotton boll weevil, *Anthonomus grandis*, and development of an infectivity assay. *Arch. Virol.* **2001**, *146*, 767–775. [CrossRef] [PubMed]
102. Belloncik, S.; Petcharawan, O.; Couillard, M.; Charpentier, G.; Larue, B.; Guardado, H.; Chareonsak, S.; Imanishi, S. Development and characterization of a continuous cell line, AFKM-On-H, from hemocytes of the European corn borer *Ostrinia nubilalis* (Hübner) (Lepidoptera, Pyralidae). *In Vitro Cell. Dev. Biol. Anim.* **2007**, *43*, 245–254. [CrossRef] [PubMed]
103. Just, F.; Essbauer, S.; Ahne, W.; Blahak, S. Occurrence of an invertebrate iridescent-like virus *(Iridoviridae)* in reptiles. *J. Vet. Med. B* **2001**, *48*, 685–694. [CrossRef]
104. Weinmann, N.; Papp, T.; Pedro Alves de Matos, A.; Teifke, J.P.; Marschang, R.E. Experimental infection of crickets (*Gryllus bimaculatus)* with an invertebrate iridovirus isolated from a high-casqued chameleon (*Chamaeleo hoehnelii*). *J. Vet. Diagn. Investig.* **2007**, *19*, 674–679. [CrossRef] [PubMed]
105. Papp, T.; Spann, D.; Marschang, R.E. Development and use of a real-time polymerase chain reaction for the detection of group II invertebrate iridoviruses in pet lizards and prey insects. *J. Zoo Wildl. Med.* **2014**, *45*, 219–227. [CrossRef] [PubMed]
106. Marschang, R.E. Viruses infecting reptiles. *Viruses* **2011**, *3*, 2087–2126. [CrossRef] [PubMed]
107. Stöhr, A.C.; Papp, T.; Marschang, R.E. Repeated detection of an invertebrate iridovirus in amphibians. *J. Herpetol. Med. Surg.* **2016**, *26*, 54–58. [CrossRef]
108. McIntosh, A.; Kimura, M. Replication of the insect Chilo iridescent virus (CIV) in a poikilothermic vertebrate cell line. *Intervirology* **1974**, *4*, 257–267. [CrossRef] [PubMed]
109. Ohba, M.; Aizawa, K. Lethal toxicity of arthropod iridoviruses to an amphibian *Rana limnocharis. Arch. Virol.* **1981**, *68*, 153–156. [CrossRef] [PubMed]
110. Ohba, M.; Aizawa, K. Mammalian toxicity of an insect iridovirus. *Acta Virol.* **1982**, *26*, 165–168. [PubMed]
111. Boucias, D.G.; Maruniak, J.E.; Pendland, J.C. Characterization of an iridovirus from the southern mole cricket, *Scapteriscus vicinus. J. Invertebr. Pathol.* **1987**, *50*, 238–245. [CrossRef]

112. Fowler, H.G. Natural microbial control of cricket populations (Orthoptera: *Gryllotalpidae*: *Scapteriscus borelli*): Regulation of populations aggregated in time and space. *Rev. Brasil. Biol.* **1989**, *49*, 1039–1051. [PubMed]
113. Kleespies, R.G.; Tidona, C.A.; Darai, G. Characterization of a new iridovirus isolated from crickets and investigations on the host range. *J. Invertebr. Pathol.* **1999**, *73*, 84–90. [CrossRef] [PubMed]
114. Just, F.; Essbauer, S. Characterization of an iridescent virus isolated from *Gryllus bimaculatus* (Orthoptera: Gryllidae). *J. Invertebr. Pathol.* **2001**, *77*, 51–61. [CrossRef] [PubMed]
115. Bronkhorst, A.W.; van Cleef, K.W.; Venselaar, H.; van Rij, R.P. A dsRNA-binding protein of a complex invertebrate DNA virus suppresses the *Drosophila* RNAi response. *Nucleic Acids Res.* **2014**, *42*, 12237–12248. [CrossRef] [PubMed]
116. Bronkhorst, A.W.; van Cleef, K.W.; Vodovar, N.; İnce, İ.A.; Blanc, H.; Vlak, J.M.; Saleh, M.C.; van Rij, R.P. The DNA virus Invertebrate iridescent virus 6 is a target of the *Drosophila* RNAi machinery. *Proc. Natl. Acad. Sci. USA* **2012**, *109*, E3604–E3613. [CrossRef] [PubMed]
117. Kemp, C.; Mueller, S.; Goto, A.; Barbier, V.; Paro, S.; Bonnay, F.; Dostert, C.; Troxler, L.; Hetru, C.; Meignin, C.; et al. Broad RNA interference-mediated antiviral immunity and virus-specific inducible responses in *Drosophila*. *J. Immunol.* **2013**, *190*, 650–658. [CrossRef] [PubMed]
118. Dostert, C.; Jouanguy, E.; Irving, P.; Troxler, L.; Galiana-Arnoux, D.; Hetru, C.; Hoffmann, J.A.; Imler, J.L. The Jak-STAT signaling pathway is required but not sufficient for the antiviral response of drosophila. *Nat. Immunol.* **2005**, *6*, 946–953. [CrossRef] [PubMed]
119. Kramer, J.M.; Kochinke, K.; Oortveld, M.A.; Marks, H.; Kramer, D.; de Jong, E.K.; Asztalos, Z.; Westwood, J.T.; Stunnenberg, H.G.; Sokolowski, M.B.; et al. Epigenetic regulation of learning and memory by *Drosophila* EHMT/G9a. *PLoS Biol.* **2011**, *9*, e1000569. [CrossRef] [PubMed]
120. Merkling, S.H.; Bronkhorst, A.W.; Kramer, J.M.; Overheul, G.J.; Schenck, A.; Van Rij, R.P. The epigenetic regulator G9a mediates tolerance to RNA virus infection in Drosophila. *PLoS Pathog.* **2015**, *11*, e1004692. [CrossRef] [PubMed]
121. Lamiable, O.; Arnold, J.; de Faria, I.J.D.S.; Olmo, R.P.; Bergami, F.; Meignin, C.; Hoffmann, J.A.; Marques, J.T.; Imler, J.L. Analysis of the contribution of hemocytes and autophagy to *Drosophila* antiviral immunity. *J. Virol.* **2016**, *90*, 5415–5426. [CrossRef] [PubMed]
122. Ahlers, L.R.; Bastos, R.G.; Hiroyasu, A.; Goodman, A.G. Invertebrate iridescent virus 6, a DNA virus, stimulates a mammalian innate immune response through RIG-I-like receptors. *PLoS ONE* **2016**, *11*, e0166088. [CrossRef] [PubMed]
123. Delius, H.; Darai, G.; Flügel, R. DNA analysis of Insect iridescent virus 6: Evidence for circular permutation and terminal redundancy. *J. Virol.* **1984**, *49*, 609–614. [PubMed]
124. Fischer, M.; Schnitzler, P.; Delius, H.; Rösen-Wolff, A.; Darai, G. Molecular biology of insect iridescent virus type 6. In *Molecular Biology of Iridoviruses*; Springer: Boston, MA, USA, 1990; pp. 47–80.
125. İnce, İ.A.; Özcan, K.; Vlak, J.M.; van Oers, M.M. Temporal classification and mapping of non-polyadenylated transcripts of an invertebrate iridovirus. *J. Gen. Virol.* **2013**, *94*, 187–192. [CrossRef] [PubMed]
126. Willis, D.B.; Granoff, A. Frog virus 3 DNA is heavily methylated at CpG sequences. *Virology* **1980**, *107*, 250–257. [CrossRef]
127. İnce, İ.A.; Boeren, S.; van Oers, M.M.; Vlak, J.M. Temporal proteomic analysis and label-free quantification of viral proteins of an invertebrate iridovirus. *J. Gen. Virol.* **2015**, *96*, 196–205. [CrossRef] [PubMed]
128. İnce, İ.A.; Boeren, S.A.; van Oers, M.M.; Vervoort, J.J.; Vlak, J.M. Proteomic analysis of Chilo iridescent virus. *Virology* **2010**, *405*, 253–258. [CrossRef] [PubMed]
129. Nalçacıoğlu, R.; Ince, I.A.; Demirbağ, Z. The biology of Chilo iridescent virus. *Virol. Sin.* **2009**, *24*, 285–294. [CrossRef]
130. Jacob, T.; Van den Broeke, C.; Favoreel, H.W. Viral serine/threonine protein kinases. *J. Virol.* **2011**, *85*, 1158–1173. [CrossRef] [PubMed]
131. Watson, D.; Seligy, V.L. Characterization of iridovirus IV1 polypeptides: Mapping by surface labelling. *Res. Virol.* **1997**, *148*, 239–250. [CrossRef]
132. Barray, S.; Devauchelle, G. Study of the structural polypeptides of Chilo suppressalis iridescent virus (Iridovirus type 6). *Can. J. Microbiol.* **1979**, *25*, 841–849. [CrossRef] [PubMed]
133. Barray, S.; Devauchelle, G. Protein synthesis in cells infected by Chilo iridescent virus (iridovirus, type 6). *Arch. Virol.* **1985**, *86*, 315–326. [CrossRef] [PubMed]

134. Kelly, D.C.; Tinsley, T.W. The proteins of Iridescent virus type 2 and 6. *J. Invertebr. Pathol.* **1972**, *19*, 273–274. [CrossRef]
135. Elliott, R.M.; Lescott, T.; Kelly, D.C. Serological relationships of an iridescent virus (type 25) recently isolated from *Tipula* sp. with two other iridescent viruses (types 2 and 22). *Virology* **1977**, *81*, 309–316. [CrossRef]
136. Petit, F.; Devauchelle, G. Modifications in the phosphorylation of ribosomal proteins and ribosome-associated proteins in invertebrate cells infected with iridovirus type 6. *Ann. Inst. Pasteur Virol.* **1986**, *137*, 205–214. [CrossRef]
137. Chitnis, N.S.; Paul, E.R.; Lawrence, P.K.; Henderson, C.W.; Ganapathy, S.; Taylor, P.V.; Virdi, K.S.; D'Costa, S.M.; May, A.R.; Bilimoria, S.L. A virion associated protein kinase induces apoptosis. *J. Virol.* **2011**, *85*, 13144–13152. [CrossRef] [PubMed]
138. Granoff, A. Frog virus 3: A DNA virus with an unusual life-style. *Prog. Med. Virol.* **1984**, *30*, 187–198. [PubMed]
139. Wang, S.; Huang, X.; Huang, Y.; Hao, X.; Xu, H.; Cai, M.; Wang, H.; Qin, Q. Entry of a novel marine DNA virus, Singapore grouper iridovirus, into host cells occurs via clathrin-mediated endocytosis and macropinocytosis in a pH-dependent manner. *J. Virol.* **2014**, *88*, 13047–13063. [CrossRef] [PubMed]
140. Clout, N.J.; Tisi, D.; Hohenester, E. Novel fold revealed by the structure of a FAS1 domain pair from the insect cell adhesion molecule fasciclin I. *Structure* **2003**, *11*, 197–203. [CrossRef]
141. Braunwald, J.; Nonnenmacher, H.; Tripier-Darcy, F. Ultrastructural and biochemical study of frog virus 3 uptake by BHK-21 cells. *J. Gen. Virol.* **1985**, *66*, 283–293. [CrossRef] [PubMed]
142. Barray, S.; Devauchelle, G. Protein synthesis in cells infected by Chilo iridescent virus: Evidence for temporal control of three classes of induced polypeptides. *Ann. l'Inst. Pasteur/Virol.* **1987**, *138*, 253–261. [CrossRef]
143. D'Costa, S.M.; Yao, H.J.; Bilimoria, S.L. Transcription and temporal cascade in Chilo iridescent virus infected cells. *Arch. Virol.* **2001**, *146*, 2165–2178. [CrossRef] [PubMed]
144. D'Costa, S.M.; Yao, H.J.; Bilimoria, S.L. Transcriptional mapping in Chilo iridescent virus infections. *Arch. Virol.* **2004**, *149*, 723–742. [CrossRef] [PubMed]
145. Goorha, R. Frog virus 3 DNA replication occurs in two stages. *J. Virol.* **1982**, *43*, 519–528. [PubMed]
146. Wang, F.; Bi, X.; Chen, L.M.; Hew, C.L. ORF018R, a highly abundant virion protein from Singapore grouper iridovirus, is involved in serine/threonine phosphorylation and virion assembly. *J. Gen. Virol.* **2008**, *89*, 1169–1178. [CrossRef] [PubMed]
147. Liu, Y.; Tran, B.N.; Wang, F.; Ounjai, P.; Wu, J.; Hew, C.L. Visualization of assembly intermediates and budding vacuoles of Singapore grouper iridovirus in Grouper embryonic cells. *Sci. Rep.* **2016**, *6*, 18696. [CrossRef] [PubMed]
148. Mercer, J.; Helenius, A. Vaccinia virus uses macropinocytosis and apoptotic mimicry to enter host cells. *Science* **2008**, *320*, 531–535. [CrossRef] [PubMed]
149. Orange-Balangé, N.; Devauchelle, G. Effect of infection with iridovirus type 6 on invertebrate cell phospholipids. *Ann. Inst. Pasteur Virol.* **1982**, *133*, 205–214. [CrossRef]
150. Cerutti, M.; Cerutti, P.; Devauchelle, G. Infectivity of vesicles prepared from Chilo iridescent virus inner membrane: Evidence for recombination between associated DNA fragments. *Virus Res.* **1989**, *12*, 299–313. [CrossRef]
151. Nalcacioglu, R.; Ince, I.A.; Vlak, J.M.; Demirbag, Z.; van Oers, M.M. The Chilo iridescent virus DNA polymerase promoter contains an essential AAAAT motif. *J. Gen. Virol.* **2007**, *88*, 2488–2494. [CrossRef] [PubMed]
152. Nalçacioğlu, R.; Marks, H.; Vlak, J.M.; Demirbağ, Z.; van Oers, M.M. Promoter analysis of the Chilo iridescent virus DNA polymerase and major capsid protein genes. *Virology* **2003**, *317*, 321–329. [CrossRef] [PubMed]
153. Dizman, Y.A.; Demirbag, Z.; Ince, I.A.; Nalcacioglu, R. Transcriptomic analysis of Chilo iridescent virus immediate early promoter. *Virus Res.* **2012**, *167*, 353–357. [CrossRef] [PubMed]
154. İnce, İ.A.; Pijlman, G.P.; Vlak, J.M.; van Oers, M.M. Hairpin structures with conserved sequence motifs determine the 3' ends of non-polyadenylated invertebrate iridovirus transcripts. *Virology* **2017**, *511*, 344–353. [CrossRef] [PubMed]
155. Barber, G.N. Host defense, viruses and apoptosis. *Cell Death Differ.* **2001**, *8*, 113–126. [CrossRef] [PubMed]
156. Chitnis, N.S.; D'Costa, S.M.; Paul, E.R.; Bilimoria, S.L. Modulation of iridovirus-induced apoptosis by endocytosis, early expression, JNK, and apical caspase. *Virology* **2008**, *370*, 333–342. [CrossRef] [PubMed]

157. Everett, H.; McFadden, G. Viruses and apoptosis: Meddling with mitochondria. *Virology* **2001**, *288*, 1–7. [CrossRef] [PubMed]
158. Razvi, E.S.; Welsh, R.M. Apoptosis in viral infections. *Adv. Virus Res.* **1995**, *45*, 1–60. [PubMed]
159. İnce, İ.A.; Westenberg, M.; Vlak, J.M.; Demirbağ, Z.; Nalçacioğlu, R.; van Oers, M.M. Open reading frame 193R of Chilo iridescent virus encodes a functional inhibitor of apoptosis (IAP). *Virology* **2008**, *376*, 124–131. [CrossRef] [PubMed]
160. Paul, E.R.; Chitnis, N.S.; Henderson, C.W.; Kaul, R.J.; D'Costa, S.M.; Bilimoria, S.L. Induction of apoptosis by iridovirus virion protein extract. *Arch. Virol.* **2007**, *152*, 1353–1364. [CrossRef] [PubMed]
161. Ozgen, A.; Muratoglu, H.; Demirbag, Z.; Vlak, J.M.; van Oers, M.M.; Nalcacioglu, R. Construction and characterization of a recombinant invertebrate iridovirus. *Virus Res.* **2014**, *189*, 286–292. [CrossRef] [PubMed]
162. Nalcacioğlu, R.; Muratoglu, H.; Yeşilyurt, A.; van Oers, M.; Vlak, J.M.; Demirbağ, Z. Enhanced insecticidal activity of Chilo iridescent virus expressing an insect specific neurotoxin. *J. Invertebr. Pathol.* **2016**, *138*, 104–111. [CrossRef] [PubMed]
163. Williams, T.; Barbosa-Solomieu, V.; Chinchar, V.G. A decade of advances in iridovirus research. *Adv. Virus Res.* **2005**, *65*, 173–248. [PubMed]
164. Ricou, G. Production de *Tipula paludosa* Meig. en prarie en function de l'humiditié du sol. *Revue d'Ecologie et de Biologie du Sol.* **1975**, *12*, 69–89.
165. Fukuda, T. Per os transmission of Chilo iridescent virus to Mosquitoes. *J. Invertebr. Pathol.* **1971**, *18*, 152–153. [CrossRef]
166. Ohba, M.; Aizawa, K. Multiplication of Chilo iridescent virus in noninsect arthropods. *J. Invertebr. Pathol.* **1979**, *33*, 278–283. [CrossRef]
167. Andino, R.; Domingo, E. Viral quasispecies. *Virology* **2015**, *479–480*, 46–51. [CrossRef] [PubMed]

2

Cloning of the Major Capsid Protein (MCP) of Grouper Iridovirus of Taiwan (TGIV) and Preliminary Evaluation of a Recombinant MCP Vaccine against TGIV

Hsin-I Liu [1], Pinwen Peter Chiou [1,2], Hong-Yi Gong [1,2] and Hsin-Yiu Chou [1,2,*]

Academic Editors: Jun Li and Li Lin

1 Department of Aquaculture, National Taiwan Ocean University, Keelung 20224, Taiwan; duo.liu@msa.hinet.net (H.-I.L.); ppchiou@mail.ntou.edu.tw (P.P.C.); hygong@mail.ntou.edu.tw (H.-Y.G.)

2 Center of Excellence for the Oceans, National Taiwan Ocean University, No. 2, Pei-Ning Rd., Keelung 20224, Taiwan

* Correspondence: hychou@mail.ntou.edu.tw

Abstract: Fish iridoviruses cause systemic diseases with high mortality in various species of wild and farm-raised fish, resulting in severe economic losses. In 1998, we isolated a new epizootic iridovirus in cultured grouper (*Epinephelus* sp.) in Taiwan, thus named as grouper iridovirus of Taiwan (TGIV). We report here the cloning of TGIV major capsid protein (MCP). Phylogenetic analysis of the iridoviral MCPs confirmed the classification of TGIV into the *Megalocytivirus* genus. Recombinant TGIV MCP and GIV MCP were then generated to produce polyclonal antibodies. Western blot analysis revealed that the two antisera were species-specific, indicating no common epitope shared by the MCPs of the two viruses. We further assayed the potency of a subunit vaccine containing recombinant TGIV MCP. The vaccine effectively protected grouper from TGIV infection. The result demonstrated that MCP is a suitable antigen for anti-TGIV vaccines.

Keywords: grouper iridovirus of Taiwan (TGIV); major capsid protein (MCP); recombinant vaccine

1. Introduction

Grouper (*Epinephelus coioides*) is a commercially important culture species in many Asian countries. In Taiwan, the yield of grouper in 2013 reached 26,000 tons and was worth of 6.825 billion NT dollars. However, the production of grouper fry has dropped from 282.48 million tails in 1999 to only 65.14 million tails in 2013 [1]. Viral diseases are a major cause to the decline of grouper fry production. Among the viral pathogens, iridoviruses have been reported to cause severe infection in various stages of the life cycle of grouper.

Iridovirus are icosahedral cytoplasmic DNA virus that can infect invertebrates and poikilothermic vertebrates, for example insects, fishes, amphibians, and reptiles [2,3]. Piscine iridoviruses infect a wide range of fish and are classified into three genera: *Lymphocystivirus*, *Ranavirus* and *Megalocytivirus* [4]. While lymphocystiviruses generally instigate non-fatal, superficial dermal infections, ranaviruses and megalocytiviruses are notoriously known for causing high mortality in many economic fish species. Ranaviruseses that are known to infect marine fish include Singapore grouper iridovirus (SGIV) [5] and grouper iridovirus (GIV) [6]. Megalocytiviruses that infect marine fish include red seabream iridovirus (RSIV) [3,7], sleepy grouper disease iridovirus (GSDIV) [8], and infectious spleen and kidney necrosis virus (ISKNV) [9–11]. In addition, we have reported the first outbreak of megalocytivirus in cultured grouper in Taiwan, and named the pathogen grouper iridovirus of Taiwan (TGIV) [12].

TGIV could cause up to 60% mortality in the infected grouper fry. Diseased fish show clinical symptoms including swimming in circles and darkening of the body color as a result of anemia. By electron microscopy, abundant number of icosahedral virus particles, of about 230 ± 10 nm in size, are observed in the spleen of diseased fish [12]. Since its discovery in 1998, TGIV has been threatening the grouper fry culture industry in Taiwan [12].

TGIV is equipped with a major capsid protein (MCP) that is of approximately 50 kDa in mass. The MCP is the predominant structural protein in an iridovirus particle and is estimated to account for up to 45% of all virion proteins in the infected cells [13,14]. Virus structural proteins often serve as a key antigen capable of stimulating potent immune response against the viral infection [15]. Hence, MCP has been considered as an important candidate antigen for vaccine against iridoviral infection [16,17]. However, TGIV MCP has not been cloned and characterized up to this moment. In this study, we report the cloning and characterization of TGIV MCP. Furthermore, we tested the potency of a recombinant MCP subunit vaccine against TGIV infection in grouper. The data showed that the vaccine could provide protection with 86% of relative percent survival (RPS) in the infected grouper.

2. Results

2.1. *Sequence Analysis of TGIV-MCP*

The TGIV-MCP gene is 1362 bp in length, encoding a putative 453-amino acid protein with a predicted molecular mass of 49.96 kDa (accession number KT989778). Compared to its counterparts in *Megalocytivirus* genus, TGIV-MCP amino acid sequence is 99.8%, 99.8%, 99.6%, 99.6% and 99.3% identical to the MCPs of orange-spotted grouper iridovirus (OSGIV, no. AAX82316.1), grouper sleepy disease iridovirus (GSDIV, no. AAP37443.1), red seabream iridovirus (RSIV, no. BAK14277.1), rock bream iridovirus (RBIV, no. AAW48183.1), and infectious spleen and kidney necrosis virus (ISKNV, no. ADU25248.1), respectively. Additionally, sequence was 61.9%, 59.6% and 59.6% identical to the homologs of *Ranavirus*: Frog virus 3 (FV3, no. ACP19256.1), Grouper iridovirus (GIV, AAV91066.1), Singapore grouper iridovirus (SGIV, no. YP_164167.1), respectively. The phylogenetic tree established on the amino acid sequences of MCPs in megalocytiviruses and ranaviruses support the classification of TGIV into the *Megalocytivirus* genus (data not shown).

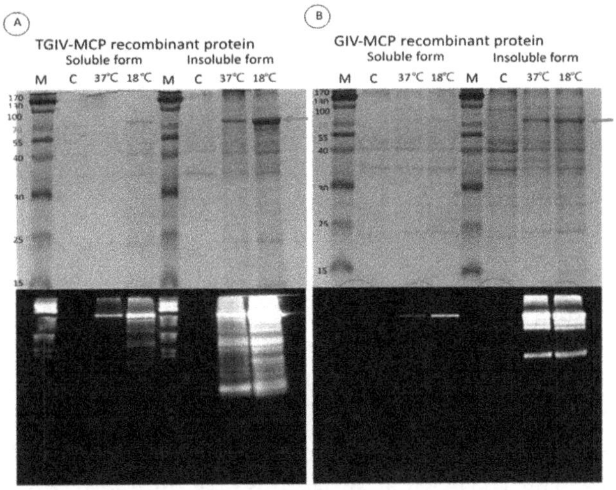

Figure 1. Induction of recombinant TGIV-MCP and GIV-MCP proteins at different temperatures. After addition of IPTG, the transformed *E. coli* cultures (**A**: TGIV-MCP, **B**: GIV-MCP) were incubated at 18 or 37 °C for 12 h. Bacterial cells were harvested and homogenized. Both soluble and insoluble proteins were subjected to SDS-PAGE (**upper** panels), followed by Western blotting with anti-His monoclonal antibody (**A**) or anti-GIV polyclonal antibody (**B**) (**lower** panels). C: sample harvested prior to IPTG induction.

2.2. Expression and Purification of Recombinant TGIV-MCP and GIV-MCP for Production of Polyclonal Antibodies

The pGS-21a-TGIV-MCP prokaryotic expression vector was employed to express recombinant His–GST–TGIV–MCP–His and His–GST–GIV–MCP–His proteins. Optimal expression of both recombinant proteins was achieved by incubation with 1 mM IPTG for 12 h at 18 °C (Figure 1, upper panel). The recombinant proteins were further verified by Western blotting with anti-His monoclonal serum (Figure 1, lower panel) and subsequently purified by Ni–NTA column (Figure 2, left panel). The purified recombinant TGIV–MCP and GIV–MCP proteins were then used to immunize rabbit to generate anti-TGIV–MCP and anti-GIV–MCP polyclonal antibodies, respectively. The avidity of the two polyclonal antibodies was evaluated by Western blotting (Figure 2, right panel). Both antisera could be diluted up to 1:10,000 in the assay.

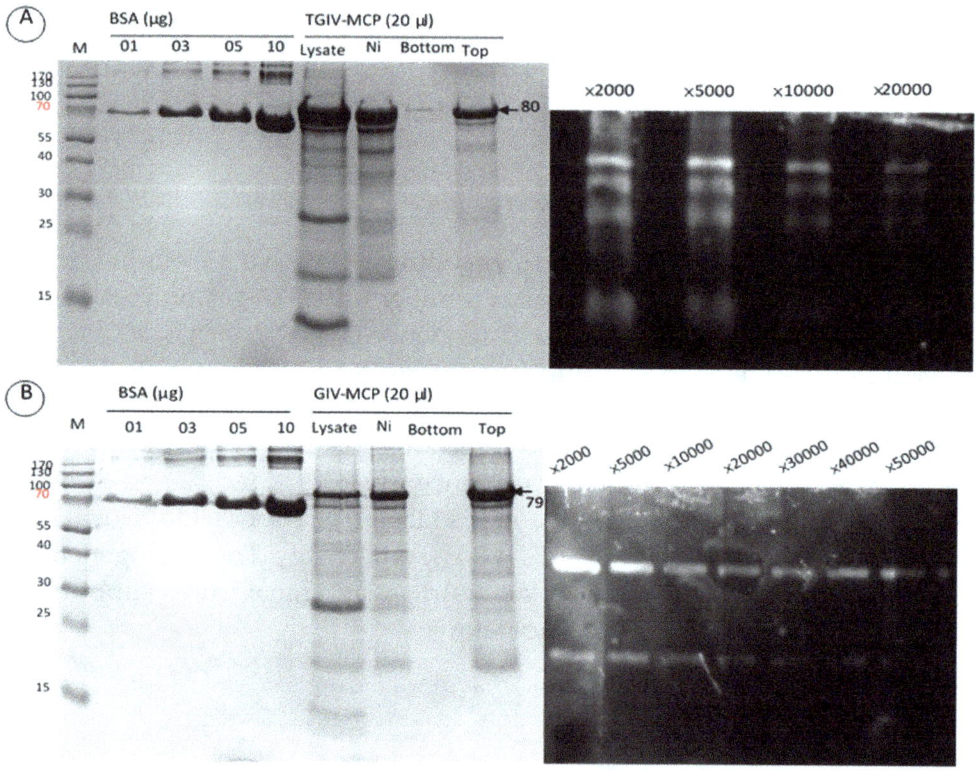

Figure 2. Generation of anti-TGIV–MCP and anti-GIV–MCP polyclonal antibodies. The purification of recombinant MCP proteins and the specificity of the polyclonal antibodies are shown in the left and right panels, respectively. **Left** panels: Recombinant TGIV–MCP (**A**) and GIV–MCP (**B**) proteins were purified by Ni–NTA column (Ni), followed by centrifugation in centricon filter to displace imidazole in the lysis buffer. The purified recombinant proteins were used to immunize rabbit to generate polyclonal antibodies against TGIV–MCP and GIV–MCP, respectively. **Top**: the purified recombinant proteins; **Bottom**: the flow-through waste. **Right** panels: Western blotting was carried out to verify the specificity of the polyclonal antibodies generated from the recombinant proteins. Recombinant TGIV–MCP and GIV–MCP proteins were subjected to SDS-PAGE and transferred to a PVDF membrane, respectively. Western blotting was conducted with anti-TGIV-MCP (**A**) or anti-GIV–MCP (**B**) antisera. The TGIV–MCP antiserum was diluted in PBS at 2000, 5000, 10,000 and 20,000 folds, and the GIV–MCP antiserum was diluted in PBS at 2000, 5000, 10,000, 20,000, 30,000, 40,000 and 50,000 folds.

2.3. Specificity of Anti-TGIV-MCP and Anti-GIV-MCP Sera

Grouper fry (0.4 g) were challenged with TGIV and GIV by intraperitoneal injection. Spleen tissue samples were collected at the appearance of typical clinical symptoms and were subjected to

Western blot analysis with anti TGIV-MCP and GIV-MCP antibodies. As shown in Figure 3, both TGIV-MCP and GIV-MCP antibodies successfully identified their own target MCP in the infected tissues. Interestingly, there appeared no cross-reactivity between the two antisera as indicated by that each antiserum specifically recognize its own target MCP.

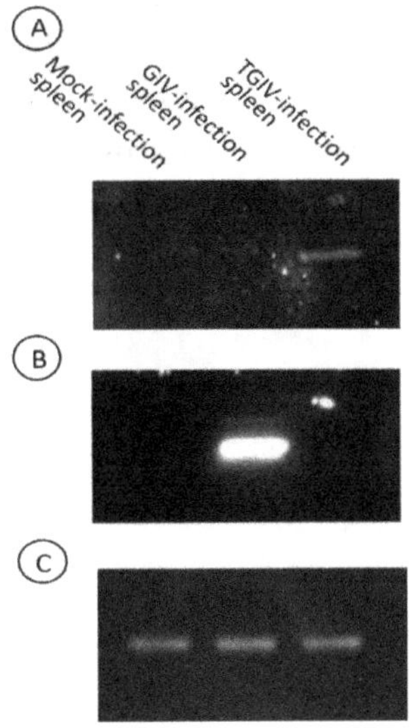

Figure 3. Detection of MCPs in TGIV and GIV infected tissues with specific polyclonal antibodies. The spleen tissues were collected from morbid fish infected with TGIV or GIV. Cell lysates were subjected SDS-PAGE, transferred to a PDVF membrane, and detected by TGIV-MCP (**A**) and GIV MCP (**B**) polyclonal antibodies (5000-fold dilution in PBS), respectively. β-Actin detected with specific monoclonal antibody (**C**) was included as a control.

2.4. *Immunogenicity of Recombinant TGIV-MCP*

To verify the immunogenicity of recombinant TGIV-MCP in grouper, sera collected from grouper immunized with the recombinant protein were assessed by ELISA on days 7 and 14 post-immunization. As shown in Figure 4, the titer of TGIV-MCP-specific antibodies was elevated in the immunized individuals as compared with the controls. The result indicates that the recombinant TGIV-MCP protein could be a candidate vaccine against TGIV infection.

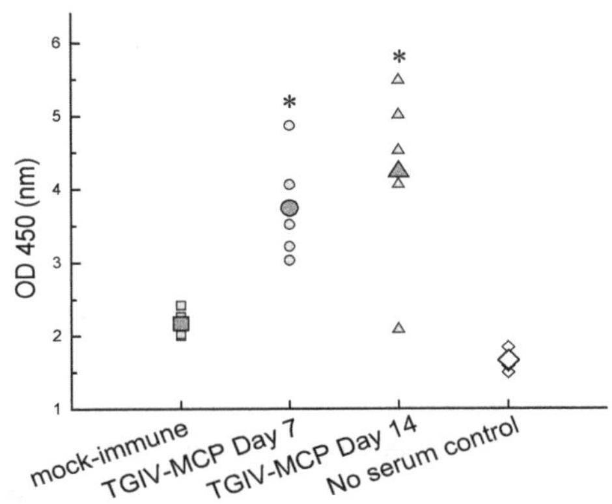

Figure 4. Induction of TGIV-MCP specific antibodies in immunized grouper. Serum samples were taken at 7 and 14 days post-immunization from 5 grouper immunized with recombinant TGIV-MCP protein. The serum samples were assayed by ELISA with anti-TGIV-MCP polyclonal antibodies. The large symbols represent average titer of the five samples in each treatment. * $p < 0.05$.

2.5. *Efficacy of Recombinant TGIV-MCP Vaccine*

To evaluate the protection efficacy of the recombinant TGIV-MCP, grouper were immunized via intraperitoneal injection of the recombinant protein, followed by TGIV infection. As shown in Figure 5A, the vaccine enhanced the survival rate in the infected vaccinated fish with a survival rate 84.4% (RPS = 86%) on day 21 post-infection. An autopsy was performed on the TGIV-infected or non-infected control fish. As shown in Figure 5B, enlargement of spleen, a clinical sign of TGIV infection, was observed in the infected fish but was absent in the non-infected fish. We have also investigated the evidence of TGIV replication in the infected fish. As shown in Figure 5C, the MCP transcript was detected by RT-PCR in the head kidney and spleen of the infected fish but absent in the non-infected control. Overall, these data indicate that the mortality was due to TGIV and that the recombinant TGIV-MCP could function as an effective vaccine against TGIV infection in grouper.

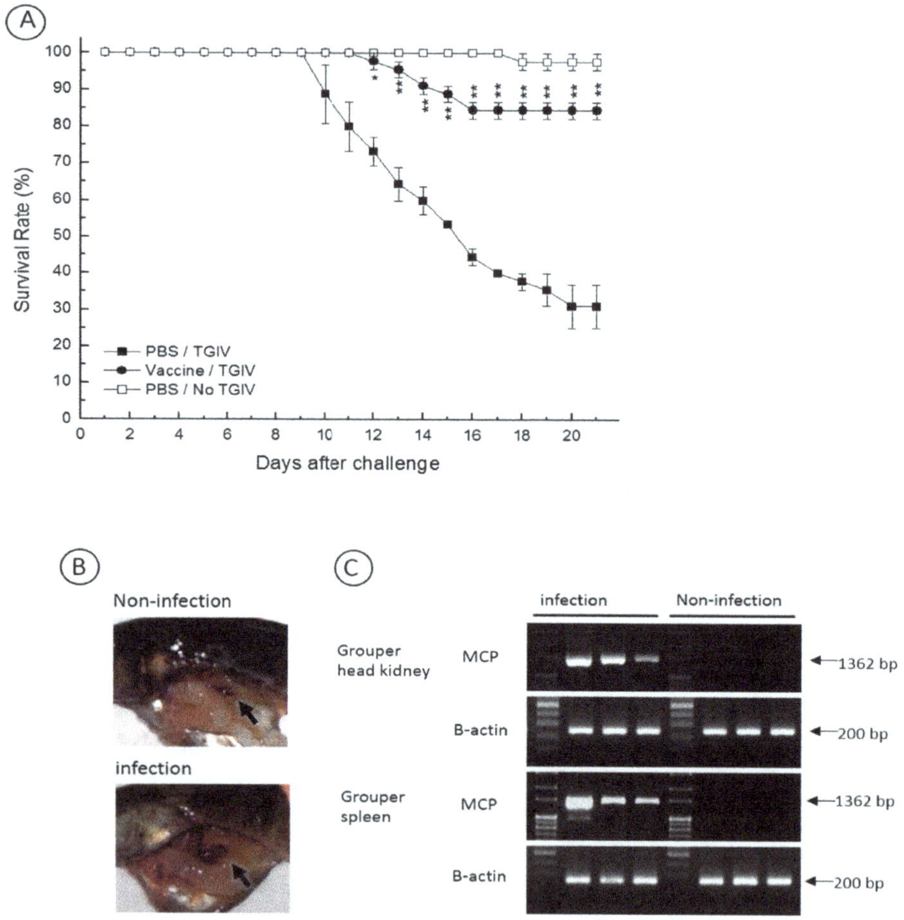

Figure 5. Survival rate of vaccinated and non-vaccinated grouper at different times after challenge with TGIV. (**A**) To test the vaccine efficacy, three treatments were analyzed: (1) vaccination via intraperitoneal injection of recombinant MCP protein, followed by TGIV infection; (2) mock-vaccination via intraperitoneal injection of PBS, followed by TGIV infection; and (3) mock-immunization with PBS only. All groups were carried out in triplicates of 15 fish. The figure shows the average survival rate of the three different groups. The vaccine apparently enhanced the survival rate against TGIV infection with a RPS of about 86% on day 21 post-infection; (**B**) An autopsy was performed on the TGIV-infected and non-infected fish of the mock-vaccinated treatment. Enlargement of spleen, a clinical sign of TGIV infection, was observed in the infected fish but was absent in the non-infected fish. Arrows indicate the spleens; (**C**) The evidence of TGIV replication in the infected fish was assayed. The MCP transcript was detected by RT-PCR in the head kidney and spleen of the infected fish but absent in the non-infected control. * $p < 0.05$–0.01, ** $p < 0.01$.

3. Discussion

Despite the devastating impact of TGIV on the grouper industry, the genome sequence of TGIV remains largely unknown. Being the most predominant structural protein, the iridoviral MCPs have been used to serve as a marker for classification as well as an ideal target for vaccine against the viral infection. Unlike its counterparts in GIV and SGIV of the *Ranavirus* genus, the MCP sequence of TGIV has not yet been determined. We showed here the first report of the nucleotide sequence of TGIV MCP. Further phylogenetic analysis placed TGIV-MCP into the clad of megalocytiviruses. The result provides the first molecular evidence of the phylogenetic position of TGIV in the *Megalocytivirus* genus of the Iridoviridae family.

Antibodies are a powerful tool for virus detection and characterization. For example, Hill and Way have classified aquatic birnavirus infectious pancreatic necrosis (IPNV) into 2 serogroups by serum neutralization assay. Serogroup A contains 9 serotypes, *i.e.*, West Buxton, SP, Ab, He, Te, Canada 1, Canada 2, Canada 3 and Jasper, whereas serogroup B contains only TV-1 serotype [18]. In this study, we have developed polyclonal antibodies against TGIV-MCP and GIV-MCP, respectively. Based on the Western blot analysis, both TGIV-MCP and GIV-MCP effectively interacted with their target protein, respectively. Interestingly, it appeared that there was no cross-reactivity between the two antisera despite a 59.6% similarity in amino sequences between the MCPs of the two viruses. The results suggest that there is likely no common epitope shared by the MCPs of GIV and TGIV. The results further indicate difficulty in developing MCP-based vaccine capable of recognizing both GIV and TGIV. In Taiwan, incidents of GIV and TGIV have been reported annually. Based on the results of this study, future vaccination strategy needs to focus on the development of bi-valent vaccine targeting at both GIV-MCP and TGIV-MCP.

Conventional inactivated vaccines have shown great value in protecting fish from viral infections [19,20]. Successful examples of inactivated vaccines include vaccines for IPNV [21] and Infectious hematopoietic necrosis virus (IHNV) [22] in salmonids, NNV in grouper [23], and RSIV in red seabream [24]. However, the manufacture of inactivated vaccines is labor-intensive and expensive due to the requirement of cultured cells to propagate viruses. Biotechnology vaccines such as subunit vaccines provide an alternate approach. For megalocytiviruses, the viral MCPs have been successfully used to develop subunit vaccine against RSIV, ISKNV and RBIV. Shin and colleagues have developed an oral vaccine containing recombinant RBIV MCP expressed in rice callus [25]. Seo and colleagues have developed another form of RBIV oral vaccine containing recombinant RBIV MCP expressed in yeast [26]. Furthermore, a DNA vaccine targeting at the RSIV MCP has been successfully developed by Caipang and colleagues [27]. Nonetheless, no vaccine against TGIV has been reported so far. In this study, we have developed a subunit vaccine containing recombinant GST-TGIV-MCP fusion protein. The vaccine was capable of assisting grouper against TGIV challenge with a RPS up to 86%. This preliminary test sets the first step to future development of commercial vaccine against TGIV infection in grouper.

4. Experimental Section

4.1. Cell Culture

Leibovitz's L-15 medium (Flow Laboratories, Ayrshire, UK) supplemented with fetal bovine serum (FBS) and antibiotics (50 IU/mL penicillin, 50 μg/mL streptomycin, and 1.75 μg/mL fungizone) was used to culture GSB cells derived from grouper swim bladder [28]. For routine passage and virus titration, 10% and 2% FBS was added to the medium, respectively. The cells were incubated at 25 °C.

4.2. Grouper

Diseased grouper (2.8 ± 0.5 cm in total length, 0.4 ± 0.15 g in weight) were collected from a culture farm at Kaohsiung Prefecture in southern Taiwan. The healthy grouper were obtained from another hatchery (886 Apex Aquaculture) in southern Taiwan and were verified to be virus-free by polymerase chain reaction (PCR). All of the groupers were maintained in 25–28 °C aquaria with aeration and fed commercially obtained artificial food twice daily.

4.3. TGIV Major Capsid Protein Gene Clone

The full-length open reading frame (ORF) of the MCP gene was amplified from infected grouper head kidney and spleen DNA using PCR. TGIV-infected grouper organs were added in a digestion buffer (20 mM NaCl, 200 mM Tris-HCl, 200 mM EDTA, 4 M urea, pH = 8.0) and homogenized by TissueLyser II (Qiagen, Hannover, Germany). The lysate was treated with proteinase K (0.2 mg/mL) at 55 °C for 3 h, and was pelleted down by centrifugation at 2000× *g* for 10 min. The supernatant was subjected to treatment of phenol/chloroform/IAA (25:24:1), and was precipitated with ethyl alcohol. The DNA pellet was dissolved in TE buffer (10 mM Tris-HCl, 1 mM EDTA, pH = 8.0) and used as a template for PCR. The cloning primers, designed based on RSIV-MCP (accession number AB461856), were as follows: MCP-forward: 5′–ATGACTTGTACAACGGGTGC–3′; MCP-reverse: 5′–TTACAGGATAGGGAAGCCTG–3′. The PCR amplification reaction conditions were as follows: 1 cycle of 94 °C for 5 min; 35 cycles of 94 °C for 30 s, 50 °C for 30 s, and 72 °C for 2 min, followed by incubation at 72 °C for 10 min. The TGIV-MCP PCR product was cloned into yT&A vector (Yeastern Biotech, Taipei, Taiwan) for sequencing. The MCP gene sequence was determined by sequencing at least twice. Multiple sequence alignments of TGIV MCP with the counterparts of other iridoviruses were conducted with the program ClustalW. A phylogenetic tree based on the amino acid sequences of iridoviral MCPs was constructed with neighbour-joining method [29] using the MEGA 6.06 program [30].

4.4. Expression of Recombinant TGIV Major Capsid Protein

The TGIV MCP ORF was cloned into a pGS-21a *E. coli* expression vector to generate the plasmid pGS-21a-TGIV-MCP. The cloned plasmid was transformed into *E. coli* BL21 via heat shock method and cultured in the LB-plated containing ampicillin antibiotic. The pGS-21a-TGIV-MCP expressed the MCP fusion protein with glutathione *S*-transferase protein (GST) and 6-histidine tagged in *E. coli* under the induction of isopropyl-1, 1-thio-β-D-galactopyranoside (IPTG). The expression of the recombinant TGIV MCP was confirmed by sodium dodecyl sulfate polyacrylamide gel electrophoresis (SDS-PAGE). After confirming the size of the recombinant protein, the protein was solubilized with 0.5% sarcosin, and purified by affinity chromatography using a nickel–nitrilotriacetic acid (Ni–NTA) column according to the manufacturer's instructions (GE)

4.5. SDS-PAGE

The quality of recombinant TGIV-MCP protein was analyzed with 10% SDS-PAGE [31]. The protein samples were resuspended in laemmli sample buffer at 95 °C for 5 min, prior to electrophoresis at content current of 100 V. The proteins were visualized by staining with coomassie brilliant blue after electrophoresis.

4.6. Western Blot

The identity of the recombinant TGIV-MCP protein was further verified by Western blot. After electrophoresis, the separated proteins on SDS-PAGE were transferred onto a polyvinylidene difluoride (PVDF) membrane. After transfer, the PVDF membrane was soaked in blocking buffer (5% skim milk in PBS-T) at 25 °C for 2 h, followed by incubation with primary antibody for 2 h. The PVDF membrane was washed 5 times in PBS-T (0.2% tween 20 in PBS). Subsequently, the membrane was incubated in solution containing horseradish peroxidase (HRP) conjugated secondary antibody for 1 h, followed by incubation with Immobilon Western Chemiluminescent HRP Substrate (Merck Millipore, Darmstadt, Germany).

4.7. Analysis of Antibody Production in Response to Immunization

After 7 and 14 days post-immunization, the levels of anti-TGIV-MCP antibodies in fish from each treatment group were assayed by enzyme-linked immunosorbent assay (ELISA). Fifty nanograms per well of purified recombinant TGIV-MCP protein was coated onto the wells of ELISA plate and let dry in a biosafety cabinet overnight. The plates were washed with PBS, and then blocked with PBS containing 2% BSA at 37 °C for 1 h. Subsequently, the plates were washed with PBS-T three times, followed by incubation with sera from the vaccinated/control grouper at 37 °C for 2 h. The plates were washed with PBS-T three times each for 2 min. After washing, the plates were incubated with 100 μL of anti-grouper IgM monoclonal antibody for 2 h. The plates were then washed with PBS-T for 2 min three times and further incubated with 100 μL of HRP-conjugated anti-mouse IgG for 1 h. After washing with PBS-T for 2 min five times, the plates were incubated with substrate TetraMethylBenzidine (TMB) (Thermo Fisher Scientific, Waltham, MA, USA), and the optical density was measured at 450 nm using an automated ELISA reader (Thermo Fisher Scientific, Waltham, MA, USA).

4.8. Vaccination and TGIV Challenge

To evaluate the efficacy of the recombinant MCP protein as vaccine, juveniles of grouper (averaged 2.8 cm in body length and 0.4 g in body weight) were used in the vaccine trials. Groupers were divided into three groups (15 fish per group) for the vaccine trial, which was repeated three times. In group 1, groupers were injected with multiple emulsified PBS (negative control). In group 2, groupers were intraperitoneally injected with 2 μg/fish-g of multiple emulsified recombinant TGIV-MCP. In group 3 (non-vaccinated group), the groupers were intraperitoneally administrated with blank multiple emulsion PBS. After 1 week, booster was carried out with equal dose of recombinant TGIV-MCP. At 2 week post-vaccination, fish from groups 1 and 2 were intraperitoneally injected with TGIV at 1×10^4 copies· $fish^{-1}$. At 21 days post-infection, the survival rate of each group was recorded daily and autopsy was performed to assess the cause of death. Relative percent survival was calculated using the following formula: RPS = [1 − (vaccinated mortality %/control mortality)] × 100% [32]. The significance of difference within treatments was analyzed by Duncan's multiple range test. For each treatment, 3 repeats were conducted to obtain the mean values, which were considered to be significantly different at $p < 0.05$ [33].

4.9. Reverse Transcription-PCR, RT-PCR

Total RNA was extracted from grouper spleen and head kidney by using RNAzol®RT (Molecular Research Center, Inc., Cincinnati, OH, USA), according to manufacturer's recommendations. 1 μg of RNA sample was used for the generation of cDNA by using HiScript I™ Reverse Transcriptase (Bionovas, Toronto, ON, Canada) with final volume of 20 μL according to manufacturer's recommendations.

PCR reaction of TGIV-MCP and β-actin genes was performed using the Taq DNA Polymerase 2× Master Mix (Ampliqon, Odense, Denmark), with final concentrations of 1 μL of cDNA, 1× PCR buffer, 1.5 mM of $MgCl_2$, 0.2 mM of dNTP mixture, and 0.2 μM of each primer in a total reaction volume of 25 μL. PCR was performed with an initial incubation at 95 °C for 3 min, followed by 40 amplification cycles (95 °C for 30 s; 54 °C for 30s; 72 °C for 2 min (MCP) or 30s (β-actin)), and a final incubation at 72 °C for 10 min. The PCR products (TGIV-MCP gene: 1362 bp; β-actin: 200 bp) were subjected to electrophoresis in 2% agarose gel, and visualized by staining with ethidium bromide.

5. Conclusions

In conclusion, we reported the cloning of TGIV MCP gene. Comparison of the deduced amino sequences of MCPs among ranaviruses and megalocytiviruses confirms the classification of TGIV into *Megalocytivirus* genus. We further demonstrated the potency of a recombinant TGIV MCP vaccine.

The results indicate that MCP is an idea candidate antigen for vaccines against TGIV. Lastly, the species-specific TGIV MCP antiserum can be used to differentiate GIV and TGIV infections during disease outbreak.

Acknowledgments: We thank Chi-Yao Chang (Academia Sinica, Taiwan) for kindly providing the anti-GIV polyclonal antibody. This work was supported by a grant-in-aid from the Fisheries Agency of the Council of Agriculture, Executive Yuan, and the Center of Excellence for the Oceans of the National Taiwan Ocean University, and Ministry of Science and Technology in Taiwan (MOST 104-2321-B-019-004-MY3).

Author Contributions: Hsin-Yiu Chou conceived and supervised the project. Hsin-I Liu performed the experiments and wrote the main paper. Pinwen Peter Chiou gave conceptual advice and edited the manuscript. Hong-Yi Gong contributed materials. All authors discussed the results and implications and commented on the manuscript at all stages.

References

1. Council of Agriculture (COA). *Agricultural Statistics Yearbook 2014;* Council of Agriculture: Taipei, Taiwan, 2014.
2. Chinchar, V.G. Ranaviruses (family Iridoviridae): Emerging cold-blooded killers. *Arch. Virol.* **2002**, *147*, 447–470. [CrossRef] [PubMed]
3. Williams, T. The iridoviruses. *Adv. Virus Res.* **1996**, *46*, 345–412. [PubMed]
4. Wang, C.S.; Shih, H.H.; Ku, C.C.; Chen, S.N. Studies on epizootic iridovirus infection among red sea bream, *Pagrus major* (Temminck & Schlegel), cultured in Taiwan. *J. Fish. Dis.* **2003**, *26*, 127–133. [PubMed]
5. Chua, F.H.C.; Ng, M.L.; Ng, K.L.; Loo, J.J.; Wee, J.Y. Investigation of outbreaks of a novel disease, "Sleepy Grouper Disease", affecting the brown-spotted grouper, *Epinephelus tauvina* Forskal. *J. Fish. Dis.* **1994**, *17*, 417–427. [CrossRef]
6. Murali, S.; Wu, M.F.; Guo, I.C.; Chen, S.C.; Yang, H.W.; Chang, C.Y. Molecular characterization and pathogenicity of a grouper iridovirus (GIV) isolated from yellow grouper, *Epinephelus awoara* (Temminck & Schlegel). *J. Fish. Dis.* **2002**, *25*, 91–100.
7. Kawakami, H.; Nakajima, K. Cultured fish species affected by red sea bream iridoviral disease from 1996 to 2000. *Fish. Pathol.* **2002**, *37*, 45–48. [CrossRef]
8. Mahardika, K.; Zafran; Yamamoto, A.; Miyazaki, T. Susceptibility of juvenile humpback grouper Cromileptes altivelis to grouper sleepy disease iridovirus (GSDIV). *Dis. Aquat. Organ.* **2004**, *59*, 1–9. [CrossRef] [PubMed]
9. Huang, Z.; Tang, J.; Li, M.; Fu, Y.; Dong, C.; Zhong, J.F.; He, J. Immunological evaluation of Vibrio alginolyticus, Vibrio harveyi, Vibrio vulnificus and infectious spleen and kidney necrosis virus (ISKNV) combined-vaccine efficacy in *Epinephelus coioides*. *Vet. Immunol. Immunopathol.* **2012**, *150*, 61–68. [CrossRef] [PubMed]
10. He, J.G.; Weng, S.P.; Zeng, K.; Huang, Z.J.; Chan, S.M. Systemic disease caused by an iridovirus-like agent in cultured mandarinfish, *Siniperca chuatsi* (Basilewsky), in China. *J. Fish. Dis.* **2000**, *23*, 219–222. [CrossRef]
11. Wang, Y.Q.; Lü, L.; Weng, S.P.; Huang, J.N.; Chan, S.M.; He, J.G. Molecular epidemiology and phylogenetic analysis of a marine fish infectious spleen and kidney necrosis virus-like (ISKNV-like) virus. *Arch. Virol.* **2007**, *152*, 763–773. [CrossRef] [PubMed]
12. Chou, H.Y.; Hsu, C.C.; Peng, T.Y. Isolation and characterization of a pathogenic iridovirus from cultured grouper (*Epinephelus* sp.) in Taiwan. *Fish. Pathol.* **1998**, *33*, 201–206. [CrossRef]
13. Black, P.N.; Blair, C.D.; Butcher, A.; Capinera, J.L.; Happ, G.M. Biochemistry and ultrastructure of iridescent virus type 29. *J. Invertebr. Pathol.* **1981**, *38*, 12–21. [CrossRef]
14. Willis, D.B.; Goorha, R.; Miles, M.; Granoff, A. Macromolecular synthesis in cells infected by frog virus 3. VII. Transcriptional and post-transcriptional regulation of virus gene expression. *J. Virol.* **1977**, *24*, 326–342. [PubMed]
15. Sommerset, I.; Krossoey, B.; Biering, E.; Frost, P. Vaccines for fish in aquaculture. *Expert Rev. Vaccines* **2005**, *4*, 89–101. [CrossRef] [PubMed]
16. Caipang, C.M.A.; Takano, T.; Hirono, I.; Aoki, T. Genetic vaccines protect red seabream, Pagrus major, upon challenge with red seabream iridovirus (RSIV). *Fish. Shellfish Immunol.* **2006**, *21*, 130–138. [CrossRef] [PubMed]

17. Kim, T.J.; Jang, E.J.; Lee, J.I. Vaccination of rock bream, Oplegnathus fasciatus (Temminck & Schlegel), using a recombinant major capsid protein of fish iridovirus. *J. Fish. Dis.* **2008**, *31*, 547–551. [PubMed]
18. Hill, B.J.; Way, K. Serological classification of infectious pancreatic necrosis (IPN) virus and other aquatic birnaviruses. *Annu. Rev. Fish. Dis.* **1995**, *5*, 55–77. [CrossRef]
19. Nakajima, K.; Maeno, Y.; Kurita, J.; Inui, Y. Vaccination against red sea bream iridoviral disease in red sea bream. *Fish. Pathol.* **1997**, *32*, 205–209. [CrossRef]
20. Nakajima, K.; Maeno, Y.; Honda, A.; Yokoyama, K.; Tooriyama, T.; Manabe, S. Effectiveness of a vaccine against red sea bream iridoviral disease in a field trial test. *Dis. Aquat. Organ.* **1999**, *36*, 73–75. [CrossRef] [PubMed]
21. Christie, K.E. Immunization with viral antigens: Infectious pancreatic necrosis. *Dev. Biol. Stand.* **1997**, *90*, 191–199. [PubMed]
22. Anderson, E.; Clouthier, S.; Shewmaker, W.; Weighall, A.; LaPatra, S. Inactivated infectious haematopoietic necrosis virus (IHNV) vaccines. *J. Fish. Dis.* **2008**, *31*, 729–745. [CrossRef] [PubMed]
23. Kai, Y.-H.; Chi, S.-C. Efficacies of inactivated vaccines against betanodavirus in grouper larvae (*Epinephelus coioides*) by bath immunization. *Vaccine* **2008**, *26*, 1450–1457. [CrossRef] [PubMed]
24. Caipang, C.M.A.; Hirono, I.; Aoki, T. Immunogenicity, retention and protective effects of the protein derivatives of formalin-inactivated red seabream iridovirus (RSIV) vaccine in red seabream, Pagrus major. *Fish. Shellfish Immunol.* **2006**, *20*, 597–609. [CrossRef] [PubMed]
25. Shin, Y.J.; Kwon, T.H.; Seo, J.Y.; Kim, T.J. Oral immunization of fish against iridovirus infection using recombinant antigen produced from rice callus. *Vaccine* **2013**, *31*, 5210–5215. [CrossRef] [PubMed]
26. Seo, J.Y.; Chung, H.J.; Kim, T.J. Codon-optimized expression of fish iridovirus capsid protein in yeast and its application as an oral vaccine candidate. *J. Fish. Dis.* **2013**, *36*, 763–768. [CrossRef] [PubMed]
27. Fu, X.; Li, N.; Lin, Q.; Guo, H.; Zhang, D.; Liu, L.; Wu, S. Protective immunity against infectious spleen and kidney necrosis virus induced by immunization with DNA plasmid containing mcp gene in Chinese perch Siniperca chuatsi. *Fish. Shellfish Immunol.* **2014**, *40*, 259–266. [CrossRef] [PubMed]
28. Fernandez, R.D.; Yoshimizu, M.; Kimura, T.; Ezura, Y.; Inouye, K.; Takami, I. Characterization of three continuous cell lines from marine fish. *J. Aquat. Anim Health* **1993**, *5*, 127–136. [CrossRef]
29. Saitou, N.; Nei, M. The neighbor-joining method: A new method for reconstructing phylogenetic trees. *Mol. Biol. Evol.* **1987**, *4*, 406–425. [PubMed]
30. Kumar, S.; Tamura, K.; Nei, M. MEGA3: Integrated software for Molecular Evolutionary Genetics Analysis and sequence alignment. *Brief. Bioinform.* **2004**, *5*, 150–163. [CrossRef] [PubMed]
31. Laemmli, U.K. Cleavage of structural proteins during the assembly of the head of bacteriophage T4. *Nature* **1970**, *227*, 680–685. [CrossRef] [PubMed]
32. Amend, D.F. Potency testing of fish vaccines. *Dev. Biol Stand.* **1980**, 447–454.
33. Norušis, M.J. SPSS Inc. In *SPSS Advanced Statistics User's Guide*; SPSS: Chicago, IL, USA, 1990; p. 285.

3

Geographic Distribution of *Epizootic haematopoietic necrosis virus* (EHNV) in Freshwater Fish in South Eastern Australia: Lost Opportunity for a Notifiable Pathogen to Expand its Geographic Range

Joy A. Becker [1], Dean Gilligan [2], Martin Asmus [3], Alison Tweedie [1] and Richard J. Whittington [1,4,*]

1 Sydney School of Veterinary Science, The University of Sydney, Camden 2570, Australia; joy.becker@sydney.edu.au (J.A.B.); alison.tweedie@sydney.edu.au (A.T.)
2 NSW Industry and Investment, Batemans Bay Fisheries Office, Batemans Bay 2536, Australia; dean.gilligan@dpi.nsw.gov.au
3 NSW Industry and Investment, Narrandera Fisheries Centre, Narrandera 2700, Australia; martin.asmus@dpi.nsw.gov.au
4 OIE Reference Laboratory for Epizootic Haematopoietic Necrosis Virus and Ranavirus Infection of Amphibians, Sydney School of Veterinary Science, The University Sydney, Camden 2570, Australia
* Correspondence: richard.whittington@sydney.edu.au

Abstract: *Epizootic haematopoietic necrosis virus* (EHNV) was originally detected in Victoria, Australia in 1984. It spread rapidly over two decades with epidemic mortality events in wild redfin perch (*Perca fluviatilis*) and mild disease in farmed rainbow trout (*Oncorhynchus mykiss*) being documented across southeastern Australia in New South Wales (NSW), the Australian Capital Territory (ACT), Victoria, and South Australia. We conducted a survey for EHNV between July 2007 and June 2011. The disease occurred in juvenile redfin perch in ACT in December 2008, and in NSW in December 2009 and December 2010. Based on testing 3622 tissue and 492 blood samples collected from fish across southeastern Australia, it was concluded that EHNV was most likely absent from redfin perch outside the endemic area in the upper Murrumbidgee River catchment in the Murray–Darling Basin (MDB), and it was not detected in other fish species. The frequency of outbreaks in redfin perch has diminished over time, and there have been no reports since 2012. As the disease is notifiable and a range of fish species are known to be susceptible to EHNV, existing policies to reduce the likelihood of spreading out of the endemic area are justified.

Keywords: iridovirus; ranavirus; epidemiology; antibody; ELISA; virus isolation; prevalence; native-fish conservation; biosecurity; endemic disease

---

1. Introduction

Epizootic haematopoietic necrosis virus (EHNV) is an aquatic pathogen that has been of international concern since its emergence in the early 1980s and is listed by the World Organization for Animal Health [1]. Infection with EHNV is systemic and causes epizootic hematopoietic necrosis (EHN), a disease characterized by extensive visceral tissue damage leading to mortality [2,3]. The first epidemics caused by EHNV occurred in freshwater impoundments in central Victoria (VIC), including Lake Mokoan and Nillahcootie in the Broken River catchment. Several populations of wild redfin perch (*Perca fluviatilis*) were impacted in the period from 1984 to 1986 [4,5]. These first epidemics often resulted in the mortality of tens of thousands of juvenile redfin perch and, in some outbreaks, adult fish were also affected. From 1986 to the mid-1990s, outbreaks of EHN in redfin perch and farmed rainbow trout (*Oncorhynchus mykiss*) were reported across the Murray–Darling Basin (MDB) in southeastern Australia [6].

Taxonomically, EHNV is in the genus *Ranavirus* within the Alphairidovirinae subfamily, Family Iridoviridae [7]. EHNV is only known in Australia, but similar viruses have been detected in other countries, usually associated with severe diseases in fish or amphibians [8]. EHNV is a double-stranded-DNA virus that can be released from a host cell by lysis (nonenveloped virion) or budding (enveloped virion) [9]. The genome of EHNV is 125 to 127 kb, with 106 to 109 genes, which is larger than amphibian ranaviruses that are usually 105 kb [7,10]. EHNV can be detected by virus isolation in several cell lines [4,11], antigen-capture ELISA [12], and histological sections using specific anti-EHNV antibodies [2]. Further, there are several polymerase chain-reaction (PCR) assays targeting the major capsid protein [13–15] or polymerase gene [16,17].

EHNV is extremely resistant and can survive for months or years in water or frozen fish [18,19]. It may persist in sediments and equipment for lengthy periods. Transmission of EHNV between susceptible hosts is possible via water or ingestion of tissue from infected fish [18,20,21]. EHNV infects fish through the skin, gills, or mouth. Infection results in multifocal necrosis of the hematopoietic tissue in the spleen, liver, and kidney [2,5,22,23]. Most infected fish quickly succumb and die within a few weeks, but disease expression is highly dependent on water temperature [21] and host [20,24]. Experimental challenge studies have demonstrated the potential for EHNV infection in several other species of Australian native fishes, including Murray cod (*Maccullochella peelii*), Macquarie perch (*Macquaria australasica*), and Murray River rainbowfish (*Melanotaenia fluviatilis*) [20]. In addition, freshwater fish native to Europe, such as the black bullhead (*Ameiurus melas*), pike perch (*Sander lucioperca*), and northern pike (*Esox lucius*) are susceptible to EHNV [25–27]. The combined results of Langdon [18] and Becker et al. [20] suggest that susceptibility of fish species does not appear to correlate with taxonomy, so susceptibility of untested species cannot be predicted based on taxonomic relatedness.

The MDB is considered to be one of the largest catchments in the world, with the river system flowing nearly 3750 kilometers from headwaters to the sea. The basin consists of 23 river valleys covering approximately 14% of Australia's total land mass, with an area of just over one million square kilometers [28]. Mortality rates in initial EHNV epidemics in VIC and NSW were estimated to be >90% and, in some smaller water bodies, 100% mortality in redfin perch was reported [4–6]. From these outbreaks, there was evidence of both downstream and upstream spread of EHNV, as well as an instance outside of the MDB catchments [6]. Mechanisms of spread in redfin perch included fish migration, water movements, translocation of live redfin perch by recreational fishers, and mechanical vector transmission by piscivorous birds [19]. EHNV was distributed between water catchments in NSW with shipments of rainbow-trout fingerlings from two infected hatcheries in NSW in the 1990s [29], leading to biosecurity policy being implemented by the NSW government's fisheries department to prevent this.

The occurrence of EHN has been discontinuous over time and space, with discrete disease events occurring within the endemic regions without a long-term pattern of recurrence. After spreading to new locations, the incidence of disease has been infrequent [10]. There have been no reports of EHNV since an outbreak in redfin perch in NSW since 2009, ACT since 2011, and VIC in 2012 [30].

Despite its significance, very little is known about the natural distribution of EHNV in Australia, or its potential impact on native fish, as no formal surveys have ever been conducted. Impacts of disease are very difficult, time-consuming, and costly to measure in natural aquatic systems, which has deterred research investment. The objective of this study was to investigate the distribution and make inferences about the epidemiology of EHNV in wild populations of fish species in the MDB. These comprise both introduced (alien) species, such as the redfin perch, and many native species.

2. Materials and Methods

2.1. Study Design

Fish samples provided for laboratory testing were collected during established research programs by collaborating government agencies. In NSW, the Department of Primary Industries (NSW DPI) collected samples under the MDB Authority Sustainable Rivers Audit (SRA) (Figure 1) [31]. These samples were categorized according to fish species, collection year, river catchment, and altitudinal zone. In each zone, 7 fish sampling sites were chosen using a stratified random procedure, with a minimum of 18 sites per river catchment. Each was the centerpoint of a one-kilometer stream reach. This design was adopted following power and benefit–cost analyses of species-accretion data from the SRA [32]. Fish were sampled by boat-mounted or backpack electrofishing using standardized effort and methods. Sampled catchments in 2008 were Border Rivers, Broken, Darling, Loddon, Lower Murray, Mitta, Murray-Riverina, and Upper Murray. Sampled catchments in 2009 were Avoca, Goulburn, Kiewa, Lachlan, Macquarie-Bogan, Namoi, Paroo, and Warrego. Sampled catchments in 2010 were Campaspe, Castlereagh, Condamine-Culgoa, Gwydir, Murrumbidgee, Ovens, and Wimera-Avon. Other samples in NSW were collected from major impoundments that received an annual stocking of native fish and trout from several other independent research projects, and from disease investigations following notification of mortality events.

Figure 1. Maximum extent of study area and catchments of the Murray–Darling Basin.

The Arthur Rylah Institute (https://www.ari.vic.gov.au/) conducted annual fish sampling across three catchments, representing three to four altitudinal zones. Australian Capital Territory (ACT) Municipal Services undertook sampling under its Urban Lakes Monitoring Program, as well as specific collections.

Additional sites were sampled because of the presence of species that are potential hosts for EHNV, for example, Cataract Dam in the Nepean catchment, NSW (Macquarie perch); Hunter River, NSW (Australian bass), Snowy River, NSW (rainbow trout); Yarra River, VIC (redfin perch and Macquarie perch) (Figure 2 and Table 1).

Table 1. Estimated true prevalence of *Epizootic haematopoietic necrosis virus* (EHNV) in redfin perch and other species based on testing tissue and serum samples collected between July 2007 and June 2011. Where sample size was >30 per population, and prevalence, which results indicate, was not exceeded for population sizes of 10,000 and 500. A population was defined as a group of fish restricted by species, year of collection, catchment, zone, and impoundment.

Population					Total No. of Samples	No. Positive	Prevalence %			Prevalence Level Not Exceeded with $\geq$95% Confidence Population of 10,000	Prevalence Level Not Exceeded with $\geq$95% Confidence Population of 500
Species	Collection Year	Compartment	Zone	Site Restriction			Point Estimate	95% Lower Confidence Limit	95% Upper Confidence Limit		
Redfin perch	2009	Broken			74	0	0.0	0.0	4.9	5	5
Redfin perch	2009	Goulburn			112	0	0.0	0.0	3.2	5	5
Redfin perch	2010	Gwydir	Upper	Copeton dam	153	0	0.0	0.0	2.4	2	2
Redfin perch	2008	Ovens (Murray)	Lower	Lake Buffalo	214	0	0.0	0.0	1.7	2	1
Redfin perch	2010	Macquarie	Lower		115	0	0.0	0.0	3.2	5	5
Redfin perch	2008	Murray	Lower		113	0	0.0	0.0	3.2	5	5
Redfin perch	2008	Murray	Lower		42[1]	0	0.0	0.0	10.8	10	10
Redfin perch	2009	Murray	Lower		84	0	0.0	0.0	4.3	5	5
Redfin perch	2009	Murrumbidgee	Upper	Blowering Dam	109	8	7.34	3.2	14.0	na	na
Redfin perch	2007	Murrumbidgee	Upper	Burrinjuck Dam	136	0	0.0	0.0	2.7	5	2
Redfin perch	2008	Murrumbidgee	Upper	Burrinjuck Dam	97	0	0.0	0.0	3.7	5	5
Redfin perch	2009	Murrumbidgee	Upper	Burrinjuck Dam	35	0	0.0	0.0	10.0	10	10
Redfin perch	2009	Murrumbidgee	Upper	Googong Reservoir	43	0	0	0.0	8.2	5	5
Redfin perch	2008	Murrumbidgee	Upper	Lake Ginnindera	51	1	1.96	0.0	10.4	na	na
Redfin perch	2009	Murrumbidgee	Upper	Lake Ginnindera	40	0	0	0.0	8.8	5	5
Redfin perch	2010	Murrumbidgee	Upper	Lake Ginnindera	120	3	2.50	0.0	7.1	na	na
Redfin perch	2011	Murrumbidgee	Upper	Lake Ginnindera	36	0	0.0	0.0	9.7	10	10
Redfin perch	2011	Murrumbidgee	Upper	Yerrabi pond	91	0	0.0	0.0	4.0	5	5
Australian bass	2009	Hunter[2]	Upper		98	0	0.0	0.0	3.7	5	5
Blackfish	2008	Murray	Lower		108	0	0.0	0.0	3.4	5	5
Brown trout	2009	Snowy[2]	Upper		34	0	0.0	0.0	10.3	10	10
Golden perch	2009	Gwydir	Upper	Copeton dam	80	0	0.0	0.0	4.5	5	5
Macquarie perch	2009	Yarra[2]			48[1]	0	0.0	0.0	9.5	10	10
Mountain galaxias	2009	Macquarie	Upper		48	0	0.0	0.0	7.4	10	10
Mountain galaxias	2008	Murrumbidgee	Upper		60	0	0.0	0.0	6.0	5	5

Table 1. *Cont.*

Species	Population				Total No. of Samples	No. Positive	Prevalence %			Prevalence Level Not Exceeded with ≥95% Confidence Population of 10,000	Prevalence Level Not Exceeded with ≥95% Confidence Population of 500
	Collection Year	Compartment	Zone	Site Restriction			Point Estimate	95% Lower Confidence Limit	95% Upper Confidence Limit		
Mountain galaxias	2009	Lachlan	Upper		37	0	0.0	0.0	9.5	10	10
Mountain galaxias	2008	Murray	Upper		31	0	0.0	0.0	11.2	10	10
Eastern mosquitofish	2008	Murrumbidgee	Upper		419	0	0.0	0.0	0.9	1	1
Murray cod	2007	Lachlan	Lower		40[1]	0	0.0	0.0	11.4	>10	10
Murray cod	2008	Murray	Lower		102[1]	0	0.0	0.0	5.0	5	5
Murray cod	2008	Murrumbidgee	Upper	Blowering dam	31	0	0.0	0.0	11.2	10	10
Murray cod	2009	Murrumbidgee	Upper	Burrinjuck Dam	30	0	0.0	0.0	11.6	10	10
Rainbow trout	2008	Murrumbidgee	Upper		115	0	0.0	0.0	3.2	5	5
Rainbow trout	2009	Snowy[2]	Upper		31	0	0.0	0.0	11.2	10	10
Silver perch	2007	Murrumbidgee	Lower		32[1]	0	0.0	0.0	14.2	>10	>10
Silver perch	2009	Gwydir	Upper	Copeton dam	51	0	0.0	0.0	7.0	5	5
Southern pygmy perch	2009	Murray	Upper		229	0	0.0	0.0	1.6	2	1

na, not applicable; [1] serum tested by ELISA, remaining tests were by virus isolation; [2] not in the Murray–Darling Basin.

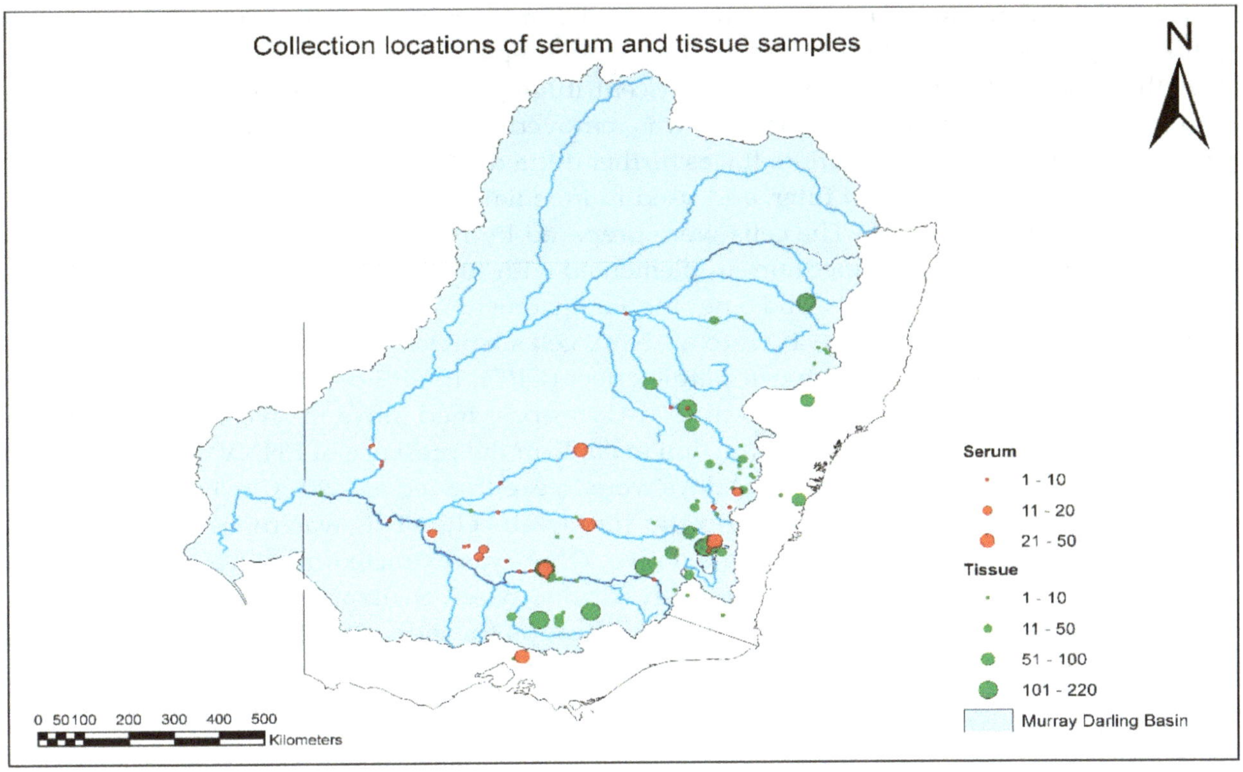

Figure 2. Distribution and number of tissue and serum samples collected from all fish species between July 2007 and June 2011.

2.2. Collection of Fish Tissue and Serum

Fish were euthanized and were kept whole or dissected in the field, individually bagged (as practicable) and labeled with collector, date, species, and location details. These samples were held on ice or in portable refrigerators until return to a research station where some were dissected. Alternatively, whole fish were frozen at −20 °C until dissection at the University of Sydney (Camden, NSW), usually within six months of collection. For transport, frozen fish or tissue were packed with ice bricks and transported to Camden by road. During dissection, the spleen, posterior kidney, and liver were removed and placed into individual sample vials. All fish tissue was kept frozen at −20 °C until it was tested for EHNV.

For serum collection, fish were placed in plastic holding tubs and anesthetized with 20 mL of AQUI-S®per 100 L water until there was deep sedation evidenced by limited movement and complete loss of equilibrium. Blood was extracted from the caudal vein using an appropriate needle for the size of the fish (25–19 gauge), and was expelled into a 1.5 mL polypropylene tube (Eppendorf, Hamburg, Germany). Tubes were labeled and placed on ice or in a portable fridge until return to a research station. Whole blood samples were allowed to clot and were kept chilled for up to 7 days in the field. Samples of clotted blood were centrifuged at 2000 to 10,000 rpm for 10 min. The separated serum was removed to a labelled tube (Eppendorf) and placed at −20 °C. Serum samples were transported to Camden at −20 °C, thawed at room temperature, diluted 1:10 in 50% v/v glycerol in 25 mM Tris, 150 mM NaCl pH 7.4 with 0.02% v/v merthiolate (TSGM), and stored at −20 °C. Moribund or dead fish were sampled for virus isolation. Kidney, liver, and spleen were the target organs, pooled either after dissecting out the viscera, or by using the whole fish with the head and tail removed (for fish <40 mm total length).

2.3. Virus Isolation and Confirmation by PCR

Tissue samples were placed in sterile microcentrifuge tubes and stored at −80 °C if not processed immediately. Each sample was weighed, and 9 × weight/volume of homogenizing medium (HM) (minimum essential medium supplemented with 200 IU/mL penicillin, 200 μg/mL streptomycin, and 5 μg/mL Fungizone) was added. Tissue samples were prepared by grinding in a chilled mortar and pestle with sterile sand and HM, then clarified by centrifuging at 900× g for 10 min in a microcentrifuge. A 200 μL aliquot of the clarified homogenate was removed for DNA extraction, and a second 500 μL aliquot was prepared for virus isolation. It was further diluted 1:4 v/v in HM, passed through a 0.22 μm low protein-binding syringe-end filter, and used to inoculate bluegill fry (BF-2) cells in suspension in 24-well tissue culture plates. The cells were prepared by resuspending 80%–90% confluent cell monolayers in minimal essential medium supplemented with 10% fetal bovine serum (FBS), 200 IU/mL penicillin, 200 μg/mL streptomycin, and 5 μg/mL Fungizone to 2×10^5 cells/mL. In duplicate, 150 μL of each sample was inoculated directly into a 1.5 mL cell suspension. Cells were incubated at 22 °C and examined for development of a cytopathic effect (CPE). If CPE developed, the infected tissue culture supernatant (TCSN) was harvested, 150 μL was passaged into a fresh cell suspension, and a 200 μL aliquot was reserved for DNA extraction to confirm the presence of EHNV by PCR. Any wells with cells that exhibited no CPE after 7–10 days were freeze-thawed at −20 °C overnight, and 150 μL of TCSN was passaged by well-to-well transfer into fresh cells. This was repeated after a further 7–10 day incubation to confirm samples as negative. DNA was extracted from 200 μL of TCSN using the HighPure Viral Nucleic Acid Extraction Kit (Roche, Basel, Switzerland), and examined using conventional or real-time PCR as described [1,13]. Results for individual samples were considered positive when either or both of the duplicates were positive.

2.4. Serology

Detection of specific anti-EHNV antibodies in fish serum was based on published methods [23]. All reagents were added to wells in volumes of 50 μL, and all incubations were at room temperature unless otherwise stated. A plain 96-well ELISA plate (Immulon®, Thermo Fisher Scientific, Waltham, MA, USA) was coated with affinity purified rabbit-anti-EHNV antibody in borate coating buffer, incubated for 90 min, then washed 5 times in reverse-osmosis purified water with 0.05% v/v Tween 20 using a plate washer (Wellwash 96-385, Thermo Electron Corporation, Waltham, MA, USA). Heat-inactivated EHNV antigen was added to pairs of wells in alternate pairs of columns on each plate to enable all sera to be tested in duplicate with and without antigen, and the plate was incubated overnight at 4 °C. After washing as above, the remaining binding sites were blocked in 1% w/v gelatin in phosphate-buffered saline pH 7.2 with 0.05% v/v Tween 20 (PBST) solution and incubated for 30 min. After washing, fish serum diluted in 0.01% w/v gelatin in PBST (PBSTG) was added and incubated for 90 min. Four positive-control sera from redfin perch were included; positive-control sera were not available for other species of fish. After washing, a sheep antifish immunoglobulin reagent appropriate for each species of fish, followed after washing by Donkey antisheep–horseradish peroxidase (HRP) conjugate (KPL) were added; both reagents were diluted in PBSTG, and incubations were for 90 min. After washing, the plate was developed with 1 mM 2,2′azino-bis(3-ethylbenzthiazoline-6-sulfonic acid) (ABTS) for 20 min before the reaction was stopped with 25 μL per well of 0.01% w/v NaN_3 in 0.1 M citric acid. Optical density (OD) was read at 405 nm with a microplate reader (Multiskan Ascent, Thermo Electron Corporation). Signal-to-noise ratio (S/N) was determined as the mean OD for duplicate wells, with EHNV antigen divided by the mean OD for duplicate wells without EHNV antigen. A positive sample was defined when OD $\geq$ 0.4 and S/N $\geq$ 2 except in Macquarie perch, in which it was defined to be when OD > 0.6 and S/N $\geq$ 1.5.

2.5. Prevalence of EHNV Infection and Likelihood of Freedom of Infection

Populations were defined as groups of fish restricted by species, year of collection, river catchment, and zone (based on altitude as above or below 400 m) with collected fish from impoundments and lakes considered to be in distinct populations from those in rivers. The prevalence of EHNV infection in populations was calculated only where there were $\geq$30 samples. Maps were created using ArcMap®(ESRI, Redlands, CA, USA).

A fish was considered positive if a sample from it was positive in either virus isolation or ELISA. The prevalence of EHNV was defined as the proportion of fish samples that tested positive in that population and 95% exact binomial confidence intervals were determined using Minitab Statistical Software. The diagnostic sensitivity and specificity of virus isolation were both assumed to be 100% [6], so apparent prevalence equaled true prevalence. For ELISA, true prevalence and Blaker's exact confidence limits were calculated assuming a test sensitivity of 70% and specificity of 100% using the calculator at http://www.ausvet.com.au.

For tissue samples tested by virus isolation where all test results were negative, the sample size required to be 95% certain that the population was free of EHNV at a specified design prevalence was determined using software at http://www.ausvet.com.au, specifically the modified hypergeometric exact calculation within the FreeCalc analysis results of freedom-testing function provided in the Survey Toolbox package [33]. Sample size was evaluated to obtain 95% confidence in detecting infection at a specified design prevalence. The following were specified as input parameters: Type I and II error levels, 0.05; sensitivity, 99.9%; specificity, 99.9%; population threshold for a binomial calculation, 10,000. For a population size of 10,000, the sample sizes required for design prevalence of 10%, 5%, 2%, and 1% were 29, 58, 145, and 270, respectively. For a population of 500, the sample sizes needed for the same design prevalence were 28, 55, 124, and 213. Where sample size was $\geq$30, these values were used to infer the prevalence that is not exceeded with 95% confidence at population sizes of 10,000 and 500.

For blood samples tested by ELISA, the same method to estimate sample size was used, but test sensitivity was assumed to be 70%. For a population of 10,000, sample sizes for 95% confidence at design prevalence of 10%, 5%, 2%, and 1% were 41, 84, 211, and 421, respectively. For a population of 500, sample sizes were 40, 78, 175, and 298.

3. Results

A total of 3622 fish-tissue samples were collected and tested over the duration of project (Figure 2, Table S1). EHNV was not isolated from any fish species other than redfin perch during this study (Tables S1 and S2). The geographic distribution and intensity of sampling is illustrated in Figure 2, and by species in Figures S1–S6. Overall MDB coverage was reasonable, but sample sizes in any year from each species and location were often small (Figure 2, Table 1).

For some populations of species such as the eastern mosquitofish (*Gambusia holbrooki*) in the upper Murrumbidgee, southern pygmy perch (*Nannoperca australisa*) in the upper Murray, and redfin perch in Lake Buffalo and the upper Gwydir (Copeton Dam), sample sizes were sufficient to suggest that EHNV is absent. That is, results showed that it is highly unlikely that, if present, EHNV exists in more than 1% to 2% of fish in each of these populations. For many other populations, this figure was <5% (Table 1).

A total of 1917 tissue and serum samples were collected from redfin perch during the project (Figure 3, Tables S1 and S2). There were several positive tissue samples from redfin perch, and all were from fish in the known endemic area in upper Murrumbidgee River catchment impoundments (Figure 4). One positive redfin perch was collected from Lake Ginninderra, ACT in December 2008 (Table 2). This sample was one of two dead redfin perch collected on 4 December 2008 near a boat ramp in Lake Ginnindera; the other dead fish and four that were electrofished at the site were negative for EHNV. It was unclear whether these fish were involved in an outbreak of EHN (Table 2). A further three positive redfin perch were from a group of 12, collected from Lake Ginninderra on 15 December 2010 (Table 2). Fish size was in the range of 30–63 mm total length and 0.3–2.6 g body weight (n = 12), and were the young of the year. In December 2009, eight positive redfin perch were identified from Blowering Dam, NSW collected during a fish kill. The estimated true prevalence was 7.3% (95% CI: 3.2%–14%; Table 1).

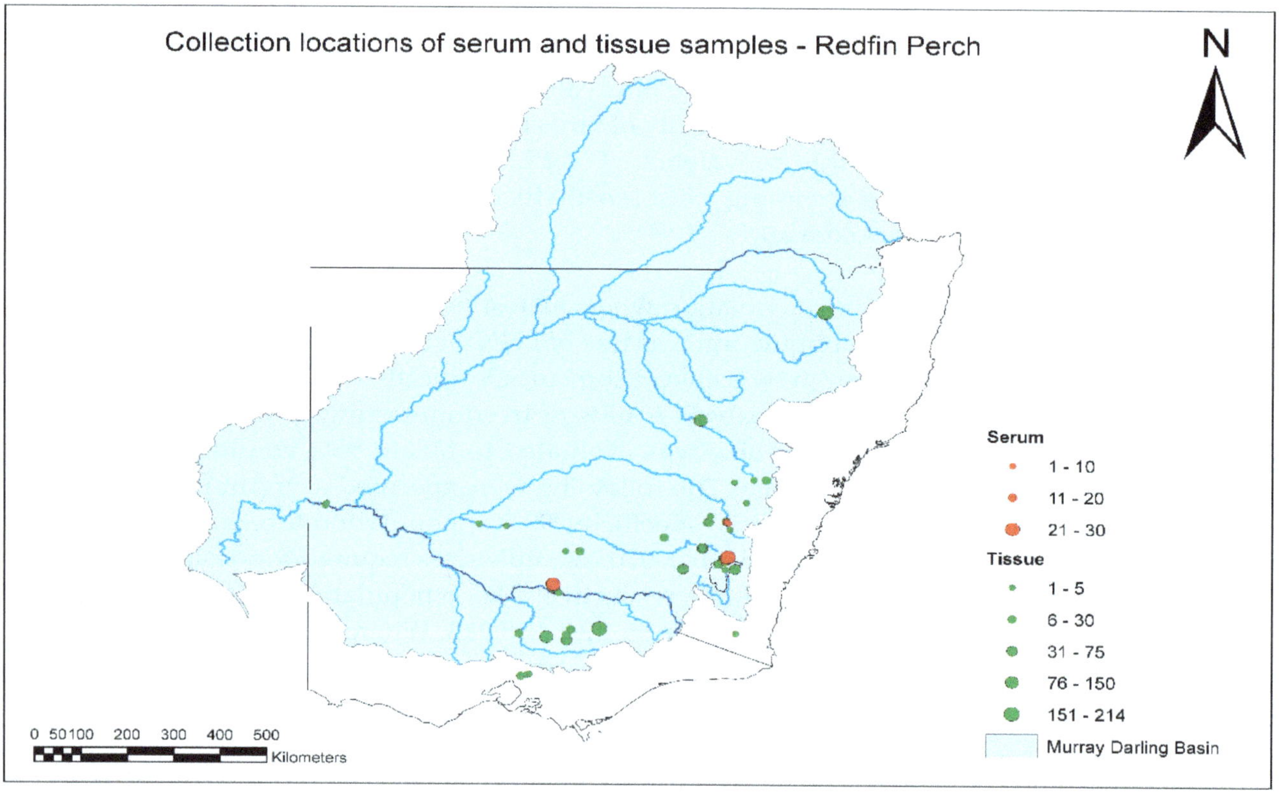

Figure 3. Distribution and number of tissue and serum samples collected from redfin perch between July 2007 and June 2011.

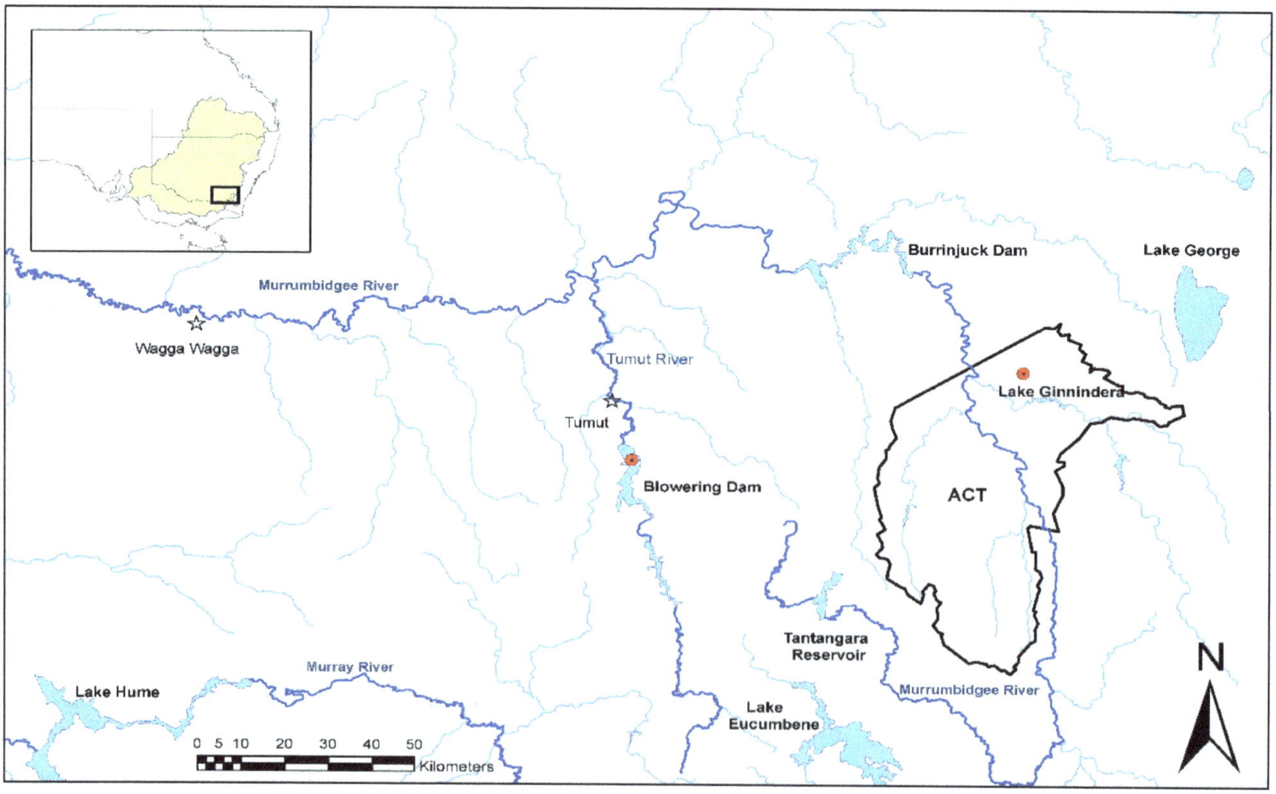

Figure 4. Moribund and dead redfin perch collected from Lake Ginnindera, ACT in December 2008 and 2010, and Blowering Dam, NSW in December 2009 were infected with EHNV.

Table 2. Summary of results of virus-isolation tests for EHNV conducted on redfin perch tissue collected from Lake Ginninderra, ACT.

Date	No. of Fish	Total Length (mm)	EHNV
15 October 2008	44	190–480	not detected
5 November 2008	1	242	not detected
4 December 2008	6	43–57	1 infected
18 March 2010	7	100–114	not detected
17 September 2010	42	92–135	not detected
15 December 2010[1]	12	30–63	3 infected
15 December 2010[2]	35	49–189	not detected
19 December 2010	2	168–172	not detected
21 December 2010	2	125–157	not detected
21 February 2011	36	90–167	not detected

[1] juvenile young of the year; [2] adult fish.

Longitudinally collected tissue samples from redfin perch were available from Lake Ginninderra (Table 2). EHNV infection was highly temporally clustered. In both 2008 and 2010, the virus emerged in the population between September/October and the end of December. Based on the sample sizes, the failure to detect EHNV in 44 and 36 adult redfin perch in October 2008 and February 2011, respectively, is consistent with <10% of individuals in those populations being infected at those times. Where EHNV was detected, the upper confidence limits for prevalence were 7.1% to 14.0% (Lake Ginninderra and Bowering Dam; Table 1).

A total of 492 blood samples were tested (Table S2). None of the blood samples from native fish was positive for anti-EHNV antibodies (Table S2; Figures S1–S3, S6). There were sufficient samples collected from Macquarie perch in the Yarra River, VIC to be 95% confident that, if present, EHNV infected <10% of fish in the population (Table 1; Figure S3). For Murray cod, there were sufficient samples collected to be 95% confident that EHNV prevalence was less than 5% in the lower Murray and <10% in the lower Lachlan rivers (Table 1 and Table S1).

A total of 142 serum samples, collected from redfin perch, were tested for anti-EHNV antibodies. The four positive control samples yielded positive results, with OD from 0.4 to >0.8, and S/N as high as about 6. There were two positive samples from the survey, both collected from adult redfin perch from Lake Ginninderra in February 2011 (Table S2). There were sufficient samples collected from redfin perch in the lower Murray river to be 95% confident that, if present, EHNV infected <10% of fish in the population (Table 1, Figure 3). As EHNV would be expected to infect a large proportion of redfin perch over time, this may be sufficient evidence to conclude that EHNV was absent from this population.

4. Discussion

Surveys to detect pathogens in wild fish in natural ecosystems are notoriously difficult, and methodological approaches are of international interest [34]. The main encountered problems include defining the population and obtaining representative samples. Consequently, it is easy to define the presence of a pathogen in a particular region when a disease is noticed, but very hard to be sure that a pathogen is absent when there are no signs of disease. A field survey to identify whether or not a pathogen is present in a population requires consideration of the likely prevalence of the pathogen if it is present (design prevalence), the sensitivity and specificity of diagnostic tests, the required degree of confidence in the results, and an ability to obtain representative samples, generally through random sampling. International standards for surveys to show that an infectious disease is not present have been proposed [35]. For diseases that are transmitted slowly or at an early stage of an outbreak, a design prevalence of 1% to 2% is recommended. For diseases that are highly transmissible, a design prevalence of 5% or more can be used. The present survey was conducted assuming that EHNV would be present in at least 10% of fish in a population. This was based on observations in redfin perch where the virus tends to affect a large proportion of the population [4,5]. It would be desirable to have much greater power of detection, down to 5%, 2%, or 1% prevalence, because, at those levels, claims about freedom from infection could be made with more certainty. An assumption was that the samples were random, but as this cannot be guaranteed, confidence limits may be wider than shown in Table 1. EHNV lacks host specificity so a range of fish species were targeted. However, it was logical to focus surveillance on the most susceptible known species [34], which is why the redfin perch was chosen to be the main target species in this survey.

Based on testing 3622 tissue and 492 blood samples, it was concluded that EHNV is endemic in some parts of the upper Murrumbidgee River catchment in the MDB. Other sites that were known to have infected redfin perch in the past were either not sampled (for example, Lake Hume, VIC) or tested negative (for example, Burrinjuck Dam, NSW, and Googong Reservoir, ACT). Although some samples were obtained from sites in Victoria near those that have previously harbored infected redfin perch (Broken River above and below Lake Nillahcootie in 2009), major potential sites of endemic infection were not sampled (for example, Lake Nillahcootie and Lake Mokoan). Similarly, lakes Albert and Alexandrina, and Mount Bold Reservoir in South Australia, all sites of past outbreaks [19], were not sampled. For these reasons, our current awareness of EHNV distribution in the known endemic region remains incomplete. However, it is reasonable to assume that it persists in places where it once occurred. This is because, during the study, an outbreak of disease due to EHNV occurred in juvenile redfin perch in Blowering Dam, NSW (December 2009), a dead redfin perch infected with EHNV was detected in Lake Ginninderra, ACT (December 2008), and there was an outbreak at this location in December 2010, consistent with prior observations of outbreaks in both locations [19]. There have been anecdotal reports of outbreaks in Blowering Dam and in the ACT since the last cases were formally confirmed in the 1990s. However, not all parts of the upper Murrumbidgee catchment were found to be infected in this study, for example, Cotter Reservoir was not. This is the location of an important population of endangered Macquarie perch, a species experimentally shown to be susceptible to EHNV [18].

The majority of samples from redfin perch came from the known EHNV-endemic area in the southeastern part of the MDB. It was not intended to obtain samples from South Australia or Queensland during this study, and fewer-than-expected samples were collected from redfin perch and any other species from northern catchments. This was partly because of the composition of the species assemblage sampled for the SRA. There are few populations of redfin perch in any of the northern catchments. Exceptions are the upper Gwydir catchment, from where a substantial number of samples of redfin perch were obtained, and a very small area of the upper Beardy River subcatchment of the Border Rivers catchment (Figure 3).

EHNV appeared to be absent from other species of fish in the MDB during the study period. Sufficient data were obtained from some regions of the MDB to be 95% confident that EHNV was present in <10% of individuals in population of the following species: river blackfish (*Gadopsis marmoratus*), brown trout (*Salmo trutta*), mountain galaxias (*Galaxias olidus*), eastern mosquitofish, murray cod, silver perch (*Bidyanus bidyanus*), southern pygmy perch, rainbow trout, and redfin perch.

The apparent absence of EHNV from most of the MDB and the demonstrated susceptibility of some fish species [18,20] should raise concern about the risk of its introduction. It is possible that the virus has not entered the middle and lower regions of southern MDB, nor its western and northern parts. However, the results of the survey only apply to the period of the survey, and sampling intensity was low for many species in most parts of the MDB.

Serology was applied in this survey, a procedure that is not routine in aquatic-animal health [36]. Two serum samples collected during this study were positive, as were many samples collected from redfin perch that survived experimental infection with EHNV, one of which was carrying the live virus in its organs [24]. The positive-control samples used in the assay had been collected from survivor redfin perch from Lake Mokoan in Victoria in 1991 to 1992, and were tested in 1991–1992 in prototype ELISA assays with similar results to those presented here (Whittington and Hyatt, unpublished).

Therefore, the antibodies were stable in frozen storage (at −20 °C). These results indicate that serology does have application in field surveys for EHNV. Surveys for EHNV, where outbreaks of disease are intermittent even in endemic regions, reinforces the important role of antibody detection. Antibody detection provides historical evidence of exposure to the virus because antibodies persist in the blood of survivors long after the resolution of an active infection.

The life history of EHNV in nature is unclear. It is unknown whether fish can become infected with EHNV and remain carriers for life, and so is whether meaningful testing could be conducted when the disease is not apparent. It is also unclear whether an environmental reservoir of the virus may exist in river sediments, fish, or, indeed, in species other than fish. As the virus has been shown in laboratory experiments to be highly resistant to drying, and persists in water, on surfaces, and in frozen fish tissue, it has been assumed that it would also be highly resistant in the environment [18]. Whether environmental sources of EHNV are available to fish is completely unknown. It may, for example, persist in river sediments, but be tightly bound to particles and unavailable to fish. The results of examination of samples collected over time from redfin perch in Lake Ginninderra showed that EHNV could only be intermittently detected there. In two separate years, 2008 and 2010, the virus was undetectable in adult fish in September/October, but had appeared by the end of December in the young of the year. It is unknown whether the virus was truly absent from the adults, or whether it was present at a very low level, sufficient to initiate an outbreak in juvenile fish when conditions were appropriate for transmission. Larger sample sizes would be required to disprove a carrier state in adult redfin perch. Alternatively, there may be another reservoir host in Lake Ginninderra.

The spread mechanisms of EHNV in the natural environment are not entirely clear. However, it is certain that trade of infected rainbow trout fingerlings has been responsible for spread between farms in different catchments, for example, the upper Snowy, upper Murrumbidgee, and Shoalhaven Rivers [29], but it is unknown whether this has led to infection of wild fish. On the contrary, it is possible that farmed rainbow trout may become infected not only from infected fingerling trout, but also from wild redfin perch—this may explain annual outbreaks in rainbow trout farmed on the Tumut river adjacent to Blowering Dam [29]. Alternatively, the virus may persist on the farm in between detectable outbreaks. Downstream movement of EHNV seeming to not have historically occurred in the highly susceptible redfin perch population [19] or during this study tends to suggest that movement of the virus with water flow in rivers is unimportant, and that other factors are necessary for spread. These factors could include the anthropogenic movement of fish (live or dead redfin perch) or movement of the virus on equipment, by birds, or other putative vectors. An alternative view is that the disease caused by EHNV is naturally restricted to specific environments, predominantly lakes and impoundments. This is possible, and suggests the involvement of an amplification host or particular risk factors, such as locally high population density to assist transmission, particular temperature gradients, and unknown mechanisms.

The lack of evidence for the downstream spread of EHNV in rivers may be explained by dilution of the virus to a level below an infectious dose, insufficient host density in rivers to initiate propagation of an outbreak, or both, but the exact reasons are not known. Although density requirements and infectious dose are uncertain, it is reasonable to assume that a young redfin perch infected with EHNV could contain 10^7 virions per gram; if the carcase is completely disaggregated in the water column, and the virus is not lost through adsorption to particulates, the concentration of virus in water in a 10 cm cube around a 10 g fish would be 10^4/L, but would decrease rapidly as this water mixes with surrounding water. Assuming fish clustering within an impoundment and large water volume, it is clear that the virus becomes rapidly diluted. It is known that EHNV remains closely associated with fish tissue, in particular, the cell cytoskeleton [37], and so it may not be efficiently released. It may require, for example, ingestion of a fish carcase to transmit efficiently to another host. The natural occurrence of EHN in juvenile redfin perch while they are in dense schools is similar to the occurrence of EHN in rainbow trout in intensive culture. Such host density may be required for transmission between fish to occur, and may explain why outbreaks in native fish have not been seen. Native species generally do not occur in high density sympatric with redfin perch. Further research is required on this topic.

5. Conclusions

The conclusions from the study are that EHNV was probably confined to the upper Murrumbidgee River and associated impoundments in NSW and the ACT in 2007 to 2011, and these observations comprise the last reports of the virus from these areas. The possibility exists that EHNV is present in other areas of the MDB but was not detected, and in the locations in VIC where it originally emerged, although these were not sampled in this study. The last detection in Australia was in VIC in 2012. The reasons for the apparent disappearance of the virus are unknown, but policies to reduce the risk of exposure to EHNV of fish in the MDB are justifiable. It is logical to prevent or reduce the likelihood of spread from the endemic area to areas that are apparently free. Not all parts of the upper Murrumbidgee catchment, for example, Cotter Reservoir, are infected, and it would be desirable to maintain this freedom to protect the resident Macquarie perch population there. All reports of fish kills in the upper Murrumbidgee catchment region should be investigated and laboratory tests for EHNV undertaken. Routine surveillance of redfin perch populations in the upper Murrumbidgee catchment should be undertaken to determine the occurrence of EHNV. Policies to exclude recreational use of these areas during EHNV outbreaks, close fisheries, or enable compulsory disinfection should be considered.

Author Contributions: All authors contributed to the design of the study. D.G. and M.A. provided a priori knowledge on fish ecology and led sample collection; R.J.W. and A.T. completed all laboratory testing; R.W. and J.A.B. completed the data analysis; J.A.B. and M.A. completed the mapping; J.A.B. and R.J.W., original draft; A.T., M.A., and D.G., review and editing.

Acknowledgments: This project was conceived by Richard Whittington and Mark Lintermans. This research was funded by the Murray–Darling Basin Authority, grant number MDBA 743. Technical assistance was provided by Rebecca Maurer, Anna Waldron, Craig Kristo, and Satoko Kawaji. We also thank the following individuals who contributed tissue and serum samples: Matthew Beitzel, Luke Pearce, Mark Lintermans, Danswell Starrs, Rhian Clear, Jarod Lyon, Pam Clunie, Joanne Kearns, Koral Hunt, Rod Cockburn, and Peter Cockburn. All research undertaken in this report was approved by the Animal Ethics Committee of University of Sydney: N00/6-2007/3/4619 and NSW DPI P01/0059.

References

1. OIE. World organisation for animal health (OIE). Chapter 10.1 epizootic haematopoietic necrosis. In aquatic animal health code. Available online: http://www.Oie.Int/en/international-standard-setting/aquatic-code/access-online/ (accessed on 28 February 2019).
2. Reddacliff, L.A.; Whittington, R.J. Pathology of epizootic haematopoietic necrosis virus (EHNV) infection in rainbow trout (oncorhynchus mykiss walbaum) and redfin perch (*Perca fluviatilis* L.). *J. Comparative Pathol.* **1996**, *115*, 103–115. [CrossRef]
3. Hick, P.; Ariel, E.; Whittington, R. Epizootic hematopoietic necrosis and european catfish virus. In *Fish Viruses and Bacteria: Pathobiology and Protection*; PTK Woo, R.C., Ed.; CABI: Oxfordshire, UK, 2017; pp. 38–50.
4. Langdon, J.S.; Humphrey, J.D.; Williams, L.M.; Hyatt, A.D.; Westbury, H.A. First virus isolation from australian fish: An iridovirus-like pathogen from redfin perch, *Perca fluviatilis* L. *J. Fish Dis.* **1986**, *9*, 263–268. [CrossRef]
5. Langdon, J.S.; Humphrey, J.D. Epizootic haematopoietic necrosis a new viral disease in redfin perch *Perca fluviatilis* L. In Australia. *J. Fish Dis.* **1987**, *10*, 289–298. [CrossRef]
6. Whittington, R.J.; Becker, J.A.; Dennis, M.M. Iridovirus infections in finfish—critical review with emphasis on ranaviruses. *J. Fish Dis.* **2010**, *33*, 95–122. [CrossRef]

7. Chinchar, V.G.; Hick, P.; Ince, I.A.; Jancovich, J.K.; Marschang, R.; Qin, Q.; Subramaniam, K.; Waltzek, T.B.; Whittington, R.; Williams, T.; et al. ICTV virus taxonomy profile: Iridoviridae. *J. Gen. Virol.* **2017**, *98*, 890–891. [CrossRef] [PubMed]
8. Mavian, C.; López-Bueno, A.; Somalo, M.P.F.; Alcamí, A.; Alejo, A. Complete genome sequence of the european sheatfish virus. *J. Virol.* **2012**, *86*, 6365–6366. [CrossRef] [PubMed]
9. Williams, T.; Barbosa-Solomieu, V.; Chinchar, V.G. A decade of advances in iridovirus research. *Adv. Virus Res.* **2005**, *65*, 173–248. [PubMed]
10. Hick, P.M.; Subramaniam, K.; Thompson, P.M.; Waltzek, T.B.; Becker, J.A.; Whittington, R.J. Molecular epidemiology of epizootic haematopoietic necrosis virus (EHNV). *Virology* **2017**, *511*, 320–329. [CrossRef] [PubMed]
11. Crane, M.S.J.; Young, J.; Williams, L.M. Epizootic haematopoietic necrosis virus (EHNV): Growth in fish cell lines at different temperatures. *Bull. Eur. Assoc. Fish Pathol.* **2005**, *25*, 228–231.
12. Whittington, R.J.; Steiner, K.A. Epozootic hematopeietic necrosis virus (EHNV)—improved ELISA for detection in fish tissues and cell cultures and an efficient method for release of antigen from tissues. *J. Virol. Methods* **1993**, *43*, 205–220. [CrossRef]
13. Jaramillo, D.; Tweedie, A.; Becker, J.A.; Hyatt, A.; Crameri, S.; Whittington, R.J. A validated quantitative polymerase chain reaction assay for the detection of ranaviruses (family iridoviridae) in fish tissue and cell cultures, using ehnv as a model. *Aquaculture* **2012**, *356–357*, 186–192. [CrossRef]
14. Marsh, I.B.; Whittington, R.J.; O»Rourke, B.; Hyatt, A.D.; Chisholm, O. Rapid differentiation of australian, european and american ranaviruses based on variation in major capsid protein gene sequence. *Mol. Cell Probes* **2002**, *16*, 137–151. [CrossRef]
15. Pallister, J.; Gould, A.; Harrison, D.; Hyatt, A.; Jancovich, J.; Heine, H. Development of real-time PCR assays for the detection and differentiation of Australian and European ranaviruses. *J. Fish Dis.* **2007**, *30*, 427–438. [CrossRef] [PubMed]
16. Stilwell, N.K.; Whittington, R.J.; Hick, P.M.; Becker, J.A.; Ariel, E.; van Beurden, S.; Vendramin, N.; Olesen, N.J.; Waltzek, T.B. Partial validation of a taqman real-time quantitative pcr for the detection of ranaviruses. *Dis. Aquat. Organ.* **2018**, *128*, 105–116. [CrossRef] [PubMed]
17. Holopainen, R.; Ohlemeyer, S.; Schutze, H.; Bergmann, S.M.; Tapiovaara, H. Ranavirus phylogeny and differentiation based on major capsid protein, DNA polymerase and neurofilament triplet h1-like protein genes. *Dis. Aquat. Organ.* **2009**, *85*, 81–91. [CrossRef] [PubMed]
18. Langdon, J.S. Experimental transmission and pathogenicity of epizootic haematopoietic necrosis virus (EHNV) in redfin perch, *Perca fluviatilis* L., and 11 other teleosts. *J. Fish Dis.* **1989**, *12*, 295–310. [CrossRef]
19. Whittington, R.J.; Kearns, C.; Hyatt, A.D.; Hengstberger, S.; Rutzou, T. Spread of epizootic haematopoietic necrosis virus (EHNV) in redfin perch (*Perca fluviatilis*) in southern Australia. *Australian Vet. J.* **1996**, *73*, 112–114. [CrossRef]
20. Becker, J.A.; Tweedie, A.; Gilligan, D.; Asmus, M.; Whittington, R.J. Experimental infection of australian freshwater fish with epizootic haematopoietic necrosis virus (EHNV). *J. Aquat. Anim. Health* **2013**, *25*, 66–76. [CrossRef] [PubMed]
21. Whittington, R.J.; Reddacliff, G.L. Influence of environmental temperature on experimental infection of redfin perch (*Perca fluviatilis*) and rainbow trout (*oncorhynchus mykiss*) with epizootic hematopoietic necrosis virus, an Australian iridovirus. *Australian Vet. J.* **1995**, *72*, 421–424. [CrossRef]
22. Langdon, J.S.; Humphrey, J.D.; Williams, L.M. Outbreaks of an EHNV-like irodovirus in cultured rainbow trout, *salmo gairdneri*, richardson, in Australia. *J. Fish Dis.* **1988**, *11*, 93–96. [CrossRef]
23. Whittington, R.J.; Philbey, A.; Reddacliff, G.L.; Macgown, A.R. Epidemiology of epizootic haematopoietic necrosis virus (EHNV) infection in farmed rainbow trout, *oncorhynchus mykiss* (walbaum): Findings based on virus isolation, antigen capture elisa and serology. *J. Fish Dis.* **1994**, *17*, 205–218. [CrossRef]
24. Becker, J.A.; Tweedie, A.; Gilligan, D.; Asmus, M.; Whittington, R.J. Susceptibility of Australian redfin perch *Perca fluviatilis* experimentally challenged with epizootic hematopoietic necrosis virus (EHNV). *J. Aquat. Anim. Health* **2016**, *28*, 122–130. [CrossRef] [PubMed]
25. Bang Jensen, B.; Holopainen, R.; Tapiovaara, H.; Ariel, E. Susceptibility of pike-perch sander lucioperca to a panel of ranavirus isolates. *Aquaculture* **2011**, *313*, 24–30. [CrossRef]
26. Bang Jensen, B.; Ersboll, A.K.; Ariel, E. Susceptibility of pike esox lucius to a panel of ranavirus isolates. *Dis. Aquat. Organ.* **2009**, *83*, 169–179. [CrossRef]

27. Gobbo, F.; Cappellozza, E.; Pastore, M.R.; Bovo, G. Susceptibility of black bullhead ameiurus melas to a panel of ranavirus isolates. *Dis. Aquat. Organ.* **2010**, *90*, 167–174. [CrossRef] [PubMed]
28. Lintermans, M. *Fishes of the Murray Darling Basin*; Murray Darling Basin Commission: Canberra, Australia, 2007.
29. Whittington, R.J.; Reddacliff, L.A.; Marsh, I.; Kearns, C.; Zupanovic, Z.; Callinan, R.B. Further observations on the epidemiology and spread of epizootic haematopoietic necrosis virus (EHNV) in farmed rainbow trout oncorhynchus mykiss in southeastern australia and a recommended sampling strategy for surveillance. *Dis. Aquat. Organ.* **1999**, *35*, 125–130. [CrossRef] [PubMed]
30. Australia, A.H. *Animal Health in Australia 2017*; Animal Health Australia: Canberra, Australia, 2018; pp. 93–97.
31. Authority, M.D.B. *Sustainable Rivers Audit 2: The Ecological Health of Rivers in the Murray–Darling Basin at the End of the Millennium Drought (2008–2010)*; Murray-Darling Basin Authority: Canberra, Australia, 2012; pp. 1–100.
32. Commision, M.D.B. *Murray–Darling Basin Rivers: Ecosystem Health Check, 2004–2007. A Summary Report Based on the Independent Sustainable Rivers Audit Group's Sra Report 1: A Report on the Ecological Health of Rivers in the Murray–Darling Basin, 2004–2007*; Murray-Darling Basin Authority: Canberra, Australia, 2008.
33. Cameron, A. *Survey Toolbox for Aquatic Animal Diseases. A Practical Manual and Software Package. Aciar Monograph*; Australian Centre for International Agricultural Research: Canberra, Australi, 2002; Volume 94, p. 375.
34. Corsin, F.; Georgiadis, M.; Hammell, K.; Hill, B. Guide for aquatic animal health surveillance. World Organisation for Animal Health (OIE): Paris, France, 2009.
35. Anon. *Aquatic Animal Health Code*; World Organisation for Animal Health (OIE): Paris, France, 2009; p. 265.
36. Jaramillo, D.; Peeler, E.J.; Laurin, E.; Gardner, I.A.; Whittington, R.J. Serology in finfish for diagnosis, surveillance, and research: A systematic review. *J. Aquat. Anim. Health* **2017**, *29*, 1–14. [CrossRef] [PubMed]
37. Eaton, B.T.; Hyatt, A.D.; Hengstberger, S. Epizootic haematopoietic necrosis virus: Purification and classification. *J. Fish Dis.* **1991**, *14*, 157–169. [CrossRef]

4

Distribution and Phylogeny of Erythrocytic Necrosis Virus (ENV) in Salmon Suggests Marine Origin

Veronica A. Pagowski [1,†], Gideon J. Mordecai [1,*,†], Kristina M. Miller [2,*], Angela D. Schulze [2], Karia H. Kaukinen [2], Tobi J. Ming [2], Shaorong Li [2], Amy K. Teffer [3], Amy Tabata [2] and Curtis A. Suttle [1,4,5,6,*]

1 Department of Earth, Ocean and Atmospheric Sciences, University of British Columbia, 2207 Main Mall, Vancouver, BC V6T 1Z4, Canada; v.pagowski@alumni.ubc.ca
2 Pacific Biological Station, Fisheries and Oceans Canada, 3190 Hammond Bay Rd, Nanaimo, BC V9T 6N7, Canada; Angela.Schulze@dfo-mpo.gc.ca (A.D.S.); Karia.Kaukinen@dfo-mpo.gc.ca (K.H.K.); Tobi.Ming@dfo-mpo.gc.ca (T.J.M.); Shaorong.Li@dfo-mpo.gc.ca (S.L.); Amy.Tabata@dfo-mpo.gc.ca (A.T.)
3 Biology Department, University of Victoria, 3800 Finnerty Rd, Victoria, BC V8P 5C2, Canada; akteffer@gmail.com
4 Department of Microbiology and Immunology, University of British Columbia, 2207 Main Mall, Vancouver, BC V6T 1Z4, Canada
5 Department of Botany, University of British Columbia, 2207 Main Mall, Vancouver, BC V6T 1Z4, Canada
6 Institute for the Oceans and Fisheries, University of British Columbia, 2207 Main Mall, Vancouver, BC V6T 1Z4, Canada
* Correspondence: gmordecai@eoas.ubc.ca (G.J.M.); Kristi.Saunders@dfo-mpo.gc.ca (K.M.M.); suttle@science.ubc.ca (C.A.S.)
† These authors contributed equally to this work.

Abstract: Viral erythrocytic necrosis (VEN) affects over 20 species of marine and anadromous fishes in the North Atlantic and North Pacific Oceans. However, the distribution and strain variation of its viral causative agent, erythrocytic necrosis virus (ENV), has not been well characterized within Pacific salmon. Here, metatranscriptomic sequencing of Chinook salmon revealed that ENV infecting salmon was closely related to ENV from Pacific herring, with inferred amino-acid sequences from Chinook salmon being 99% identical to those reported for herring. Sequence analysis also revealed 89 protein-encoding sequences attributed to ENV, greatly expanding the amount of genetic information available for this virus. High-throughput PCR of over 19,000 fish showed that ENV is widely distributed in the NE Pacific Ocean and was detected in 12 of 16 tested species, including in 27% of herring, 38% of anchovy, 17% of pollock, and 13% of sand lance. Despite frequent detection in marine fish, ENV prevalence was significantly lower in fish from freshwater (0.03%), as assessed with a generalized linear mixed effects model ($p = 5.5 \times 10^{-8}$). Thus, marine fish are likely a reservoir for the virus. High genetic similarity between ENV obtained from salmon and herring also suggests that transmission between these hosts is likely.

Keywords: erythrocytic necrosis virus (ENV); viral erythrocytic necrosis (VEN); Pacific salmon; Pacific herring; British Columbia

1. Introduction

Viral erythrocytic necrosis (VEN) is a disease associated with severe blood abnormalities in infected fish which has caused mass mortality in Pacific herring (*Clupea pallasii*) [1]. The disease is traditionally diagnosed by microscopic examination of stained blood smears for the presence of inclusion bodies within the cytoplasm of infected erythrocytes. Electron microscopy revealed that infected erythrocytes contained icosahedral virions, which were named erythrocytic necrosis virus (ENV) [2]. Although first described more than half a century ago, the virus is poorly characterized as attempts to propagate it in fish cell lines have been unsuccessful [3].

Based on the small amount of available ENV genomic sequence, the virus has been assigned to a new putative genus within the *Iridoviridae* (a family of double-stranded DNA viruses), comprising other erythrocytic viruses from ectothermic hosts [2,4]. Further genomic sequencing of ENV may provide a reference for studying genetic variation geographically, and among fish species. Currently, the only verified ENV sequences available in GenBank encode ATPase, the major capsid protein (MCP), DNA-dependent DNA polymerase, and DNA-dependent RNA polymerase.

In the NE Pacific Ocean, VEN has been described in Pacific herring [1] and in the marine phase of pink (*Oncorhynchus gorbuscha*), chum (*Oncorhynchus keta*), coho (*Oncorhynchus kisutch*), steelhead (*Oncorhynchus mykiss*), and Chinook (*Oncorhynchus tshawytscha*) salmon [5,6]. High geographic variability in VEN prevalence and disease susceptibility of chum, coho, sockeye, and Chinook salmon, as well as Pacific herring, suggests that ENV could help explain high year-to-year variability in the population dynamics of these keystone species throughout coastal regions of the NE Pacific Ocean [1,7,8]. In the laboratory, VEN has been induced after confining healthy salmon with diseased fish. Among salmonids, disease transmission has been demonstrated in pink and chum salmon [5,8,9]. Chinook, coho, and sockeye salmon appear to be more resistant to infection in challenge studies, as assessed by electron microscopy [5,9], despite natural epizootics in Chinook and coho salmon. To date, an explicit relationship between disease manifestation and viral load has not been described, and ENV isolation has not been accomplished in previous studies reporting disease transmission among species. There is substantial evidence that viruses in different genera within the family *Iridoviridae* cause different disease manifestation and severity [10–15]; thus, phylogenetic placement and genetic characterization of ENV may offer insight into factors contributing to disease outbreaks.

Considering the high prevalence of ENV in herring and salmon in the NE Pacific Ocean [1,16–20], there is a need to further characterize the virus in at-risk fish populations. Substantial herring stock declines in British Columbia since the 1970s, which have mainly been attributed to overfishing, are of concern as herring are a keystone species that underpin the coastal food web [21,22]. Similarly, Pacific salmon provide an important biological, economic, and cultural resource in British Columbia, and recent stock declines are likely to have large impacts on wildlife and fisheries in the region. The Pacific Salmon Commission, for example, reported a 60% reduction in Salish Sea Chinook abundance from 1984 to 2010 [23]. Furthermore, the Committee on the Status of Endangered Wildlife in Canada reports that half of the British Columbia Chinook salmon populations are endangered [24].

In the current study, we applied metatranscriptomic sequencing to examine the phylogenetic relationship of ENV in Chinook salmon and investigated the epidemiology and host tropism of ENV using high-throughput PCR. Our results show that ENV is widely prevalent across numerous fish species in the NE Pacific Ocean and that ENV in salmon and herring has high genetic similarity, suggesting that the virus may circulate among these species.

2. Materials and Methods

2.1. Fish Sampling

In total, 19,652 fish comprising 16 species were sampled from freshwater and marine environments as previously described [17,25]. Briefly, 3228 freshwater samples were collected at hatcheries, through beach seining, or by smolt traps for wild salmon. Marine samples were obtained mostly by purse and beach seines, and by trawl. Typically, mixed tissue samples were dissected from fish using sterile procedures in the field [26] and frozen in RNAlater before nucleic acid extraction, although some fish were flash frozen in the field and dissected in the laboratory. Both procedures have been routinely used for examining individual fish for the presence of viral nucleic acids [16,18,26–28]. Samples were collected over the course of an 11-year period from 2007–2018 in a region spanning Alaska to

Northern Washington (Figure S1) as part of a large pathogen-screening effort conducted by Fisheries and Oceans Canada.

2.2. Data Collection

The occurrence and abundance of ENV in fish was determined using the Fluidigm BioMark Platform at the Department of Fisheries and Oceans Canada [26]. Briefly, the platform provides an estimate of viral load based on copy number, as assessed by RT-qPCR. Copy numbers are calculated based on serial dilutions of artificial construct DNA standards. The calculated limit of detection (LOD) was applied to identify fish with amplifications above the 95% detection threshold [26]. ENV assays on this platform show 100% inclusivity (detection of all known strain variants of the targeted microbe) and 97.9% exclusivity (no detection of untargeted microbe species). Primers were originally obtained from a 100 bp ATPase-like protein partial gene sequence. Additional details on primer sequences as well as assay specificity and reliability are available in Miller et al. [26]. ENV monitoring was conducted alongside assays for 46 other infective agents on combined RNA and DNA extractions. In this study, ENV prevalence is reported as the proportion of fish with ENV detections, both with and without the LOD criteria applied. When not mentioned explicitly, prevalence values are reported with LOD criteria applied. Statistical analyses were done only on samples with LOD criteria applied.

2.3. Metatranscriptomic Sequencing and Bioinformatics

In order to isolate putative ENV sequences, several samples with high-load detections of this virus, as assessed using the Fluidigm BioMark, were selected for transcriptomic sequencing. In total, three aquaculture and one wild Chinook salmon were used to predict ENV proteins. Three additional fish, including an Atlantic salmon, a Chinook salmon, and a herring sample underwent an enrichment step for these predicted ENV proteins before bioinformatic analysis.

All fish except the wild Chinook salmon were sequenced on Illumina Next-Generation Sequencing (NGS) platforms using different RNA-seq protocols, depending on original targets and loads. To avoid DNA contamination from the host reducing the sequencing depth of the target virus, RNA sequencing was used to obtain transcriptomic sequences of the DNA virus. Each of these libraries was single tissue, either heart or spleen. Ribosomal RNA was removed from total RNA using the RiboMinus Invitrogen Eukaryote kit for RNA (Life Technologies, Carlsbad, CA, USA). The RNA-Seq library was prepared using the NEBNext Ultra RNA Library prep kit (New England BioLabs, Ipswich, MA, USA) with an average fragment size of 250-bp and was paired-end sequenced with 100-bp reads on the Illumina HiSeq analyzer (Illumina, San Diego, CA, USA).

For the wild Chinook sample, the ENV contigs were obtained using a similar RNAseq approach. This library was created using pooled tissue (gill, liver, heart, kidney and brain) and prepared with the ScriptSeq Complete Epidemiology NGS library kit (Illumina, San Diego, CA, USA). Briefly, ribosomal RNA was removed from Total RNA using the Epicentre ScriptSeq Complete Gold Kit (Epidemiology) (Illumina, San Diego, CA, USA) according to the manufacturer's instructions. The ScriptSeq Index reverse primers were added to the cDNA during the final amplification step which involved 14 cycles. Finally, a paired-end 125 bp sequencing run was performed on the Illumina HiSeq System.

Finally, in order to enhance our sensitivity (i.e., NGS read depth and coverage) for ENV in the medium load (~1300 copies) farmed Atlantic salmon sample, and the Chinook and herring samples (~42,670 and 214,500 copies, respectively) we employed SureSelect^{XT} enrichment technology (Agilent, Santa Clara, CA, USA). A custom set of RNA target enrichment probes (120 bp in length and staggered along the exome of interest) were designed for ENV as well as many other salmonid viruses assessed on our infectious agent monitoring platform. These sequences (497.266-kbp in total length) and subsequent bait oligonucleotides included all of the suspected ENV contigs previously assembled from high load samples. Baits which failed the SureSelect QA/QC parameters and/or significantly matched salmonid genes via BLAST searches were removed, leaving the final set of enrichment probes at 20,497. The mixed tissue (gill, atrium, ventricle, liver, pyloric caeca, spleen, head kidney, posterior kidney) RNA library, was prepared using the SureSelect^{XT} low input (NGS) target workflow (Agilent, Santa Clara, CA, USA) with a SureSelect^{XT} RNA Direct/XTHS modified protocol. Approximately 200 ng of total RNA was lyophilized and fed into the SureSelect Strand-Specific RNA library Prep kit (Agilent, Santa Clara, CA, USA) according to the manufacturer's instructions. After the 2nd strand cDNA synthesis and end repair steps, the library prep was moved to the SureSelect^{XT} low input reagent kit, starting with the end repair and A tailing step. Molecular barcoded adaptors were added using ligation and then amplified for 14 cycles according to manufacturer's instructions to create a pre-capture RNAseq library. Samples were quantified with the Qubit dsDNA HS kit (Invitrogen, Carlsbad, CA, USA), qualified with the DNA12000 chips run on the Agilent 2100 Bioanalyzer (Agilent, Santa Clara, CA, USA), and pooled into a batch of 12 prior to hybridizing 1500 ng to the bait library. Hybridizations were incubated at 65 °C, captured on streptavidin beads (Beckman Coulter, Brea, CA, USA) and washed at 70 °C according to manufacturer's instructions before a post-capture amplification of 14 cycles. These final libraries were quantified with the Qubit dsDNA HS kit (Invitrogen, Carlsbad, CA, USA) and qualified with the DNA HS chips run on the Agilent 2100 Bioanalyzer (Agilent, Santa Clara, CA, USA). Finally, a paired-end 101 bp v2 300 kit sequencing run was performed on the Illumina Miseq (Illumina, San Diego, CA, USA), which included a 5% phiX spike in.

After adapter removal, Illumina MiSeq sequencing produced between 53.3 and 59.9 M reads for each Chinook salmon used to predict ENV amino-acid sequences (quality score >28). These reads were processed as detailed below. Adapters were removed using Trimmomatic [29], and the trimmed reads aligned with genome sequence from Atlantic Salmon [30] using the Burrows-Wheeler Aligner [31]. Unmapped sequences were extracted from the dataset using Samtools [32] and assembled into contiguous sequences (contigs) using SPAdes [33]. The translated contigs were queried against the non-redundant (NR) database in GenBank using DIAMOND [34]. Contigs with top hits to members of the family *Iridoviridae* were extracted in Microsoft Excel and the Qiime script *filter_fasta.py* [35] was adapted to retain these contigs as fasta files. GeneMark [36] was used to make protein predictions for putative ENV nucleotide sequences. Predicted proteins were subject to a BLAST search against the NR database [37], the lymphocystis disease virus genome, and a set of 47 conserved genes within Nucleo-Cytoplasmic Large DNA viruses (NCLDVs) [38]. Additionally, contigs of 500 bp in length or greater were extracted and translated into all six frames using Geneious version 9.1.8 [39]. A single, most likely translation frame was selected from the BLAST results, and those with BLAST results mapping to ENV were used to create phylogenies for ATPase, DNA-dependent DNA polymerase, DNA-dependent RNA polymerase, and the MCP. For each of these proteins, phylogenetic trees were mapped using available sequences from closely related iridoviruses using ClustalW for alignments and PhyML with Le Gascuel substitution model for tree creation [40,41]. For each phylogenetic tree, *Spodoptera frugiperda ascovirus 1a* was used as an outgroup. Assembled contigs have been submitted to GenBank under the accession numbers MK638669–MK638757. To increase confidence that sequences we attributed to ENV belong to this virus, we aligned sequences obtained from the farmed Atlantic salmon (632,935 reads), Chinook salmon (658,241 reads), and herring (800,139 reads) which underwent an enrichment step for ENV viral content to our putative ENV protein-encoding sequences using the using the Burrows-Wheeler Aligner [31].

2.4. Spatial Epidemiology Analysis

All statistical analyses and plots were performed in R version 3.5.2 [42]; scripts are available in File S1. For statistical analyses, ENV detection was categorized as positive or negative with LOD criteria applied and viral loads were quantified based on estimated viral copy numbers in host fish, following Miller et al. [26]. For map plotting only, ENV load was quantified by subdividing ENV copy number data into six categorical bins based on value with no LOD criteria (0, 0–5, 5–10, 10–100, 100–10,000, >10,000 viral copies). As sample sizes for other fish were small, statistical analyses were only conducted on herring and salmon. However, ENV prevalence and load was investigated for all sampled fish species. Heat maps were produced with an inverse-distance weighting function, using R packages "ggmap" and "gstat" [43,44], while plots were constructed with "ggplot2" [45]. Differences in load among species, age class, and habitat type were assessed using Kruskal-Wallis and post-hoc pairwise Dunn tests with Benjamini-Hochberg adjusted p values for multiple comparisons. Differences in ENV prevalence among categorical variables were assessed using Chi-squared tests of independence. A post-hoc Fisher's exact test and Chi-squared tests with Bonferroni correction were conducted for pairwise comparisons to determine ENV prevalence differences among species and years, respectively, with "rcompanion" [46]. Spearman correlation was conducted to examine correlations among monthly prevalence between species. To account for residual variation in the data, a generalized linear mixed-effects model with Laplace approximation was implemented to examine differences in ENV prevalence between fresh and saltwater, with catch region, species, age class, season, population (hatchery or wild), and year considered as random effects with the R package "mlmRev" [47,48]. A similar model was used to assess differences in prevalence among age classes (smolt or adult) with habitat type used as an additional random effect instead of age class.

3. Results

3.1. Genetic Characterization

Metatranscriptomic sequencing of Chinook salmon with high ENV loads revealed high sequence identity between ENV sequences found in Chinook salmon and viral sequences from GenBank associated with Pacific herring. The available ENV sequences from Pacific herring (ATPase, DNA-dependent DNA polymerase, MCP, and DNA-dependent RNA polymerase) showed over 99% nucleic acid identity to ENV sequences from Chinook salmon in our study (Table 1). Sequences reported in Table 1 originate from heart tissue of one of the aquaculture Chinook salmon samples and from the wild Chinook salmon mixed-tissue specimen, both with a high load ENV detection (CT values of 10.7 and 13.4, respectively). The aquaculture Chinook salmon specimen had jaundice and several co-infections including *Paranucleospora theridion*, Piscine reovirus, and *Renibacterium salmoninarum*, as determined by the RT-qPCR assay. Phylogenies based on these genes (Figure 1) from ENV and its relatives (Table 2), showed that viral sequences from herring and salmon form a well-supported clade. A metatranscriptomic approach on a DNA virus only reveals virally expressed transcripts; thus, a full ENV genome could not be assembled. Within the metatranscriptomic sequences, there were 32 putative ENV proteins which consistently mapped to proteins of similar function from viruses in the family *Iridoviridae* (Table S2), as well as BLAST hits to 21 of the 47 core proteins in NCLDVs [38] (File S2). Putative ENV transcripts typically had between 20% and 60% nucleotide identity with other iridoviruses. These values are consistent with nucleotide identities reported between existing ENV sequences and other fish iridoviruses. In total 117 contigs with BLAST hits to iridoviruses were isolated from transcriptomic sequencing. Putative ENV sequences ranged in length from 152 to 4475 bp (File S3). From these, we predicted 89 protein-encoding sequences longer than 200-bp (Accession numbers MK638669-MK638757). When putative ENV protein-encoding sequences were aligned to reads obtained from the Atlantic salmon, Chinook salmon, and Pacific herring enriched for these contigs, we detected 35 of 117 putative ENV sequences. Results from this analysis are summarized in file S4.

Table 1. Basic Local Alignment Search Tool (BLAST) summary of erythrocytic necrosis virus (ENV) sequences from Chinook salmon compared with available herring derived ENV sequences on GenBank. The origin of each sequence is indicated as aquaculture (Aq.) or wild Chinook salmon. Sequences used in phylogenies are in bold.

Reference Sequence	Sequence ID	% Identity (Nucleotide)	% Identity (Amino Acid)	Sequence Alignment Length (AA)
ATPase KJ730210.1 (partial)	**80 (Aq.)**	99.9	100	849
	46 (Wild)	99.6	98.7	74
	57 (Wild)	99.5	98.5	67
	54 (Wild)	100	100	77
DNA-dependent DNA polymerase KJ756347.1 (partial)	73 (Wild)	99.9	98.3	59
	79 (Aq.)	99.8	99.7	897
	19 (Wild)	99.5	99.5	213
MCP KT211480.1 (partial, Puget Sound)	**86 (Aq.)**	99.6	100	1436
	75 (Wild)	100	100	175
DNA-dependent RNA polymerase KJ756346.1 (partial)	**78 (Aq.)**	99.9	96.8	972

Table 2. GenBank reference sequences and viral species used to create phylogenies, with colors indicating genera groups corresponding to phylogenetic trees in Figure 1.

Abbreviation	Accession Numbers				Taxon
	DNA-Dependent DNA Polymerase	DNA-Dependent RNA Polymerase	MCP	ATPase	
NA	AAC54632.1	YP_762407.1	NC_008361.1	NC_008361.1	Spodoptera ascovirus
ENV	AIQ77732.1	AIQ77731.1	KT211480.1	AIN76233.1	Erythrocytic necrosis
Megalocytivirus					
RSIV	BAA28669.1	BAK14252.1	AB109371.1	BAK14298.1	Red Sea Bream iridovirus
ISKNV	CAZ73994.1		AF370008.1	NP_612331.1	Infectious spleen and kidney necrosis virus
TRBIV	ADE34365.1	ADE34378.1	AY590687.2	ADE34443.1	Turbot reddish body iridovirus
RBIV		AAT71848.1			Rock Bream iridovirus
Ranavirus					
FV3	NC_005946.1	ASH99239.1		AHM26101.1	Frog virus 3
EHNV	ACO25234.1	YP_009182042.1	AY187045.1	YP_009182084.1	Epizootic haematopoietic necrosis virus
GIV	AY666015.1		KX284838.1		Grouper iridovirus
ECV		AMZ05024.1		YP_006347705.1	European catfish virus
ATV		ALN36639.1		YP_003852.1	Ambystoma tigrinum virus
LMRV			KM516719.1		Lacerta monticola ranavirus
LMBV			KU507317.1		Largemouth bass ranavirus
Lymphocystivirus					
LCDV(Sa)		YP_009342128.1	AY823414.1		Lymphocystis disease virus (various strains)
LCDV(China)		YP_073534.1	NC_005902.1	YP_073585.1	
LCDV1				AAX54510.1	
Iridovirus					
IIV6	NC_003038.1	AAK82288.1		NP_149647.1	Invertebrate iridescent virus 6
IIV			NC_023615.1		Invertebrate iridescent virus
AVIV			NC_024451.1	NC_024451.1	Armadillium vulgare iridescent virus
Aedes taen.			NC_008187.1		Aedes taeniorhynchus iridescent virus
Chloriridovirus					
IIV3	YP_654692.1	YP_654581.1		YP_654693.1	Invertebrate iridescent virus 3
RMIV	CAC84133.1				Regular mosquito iridescent virus

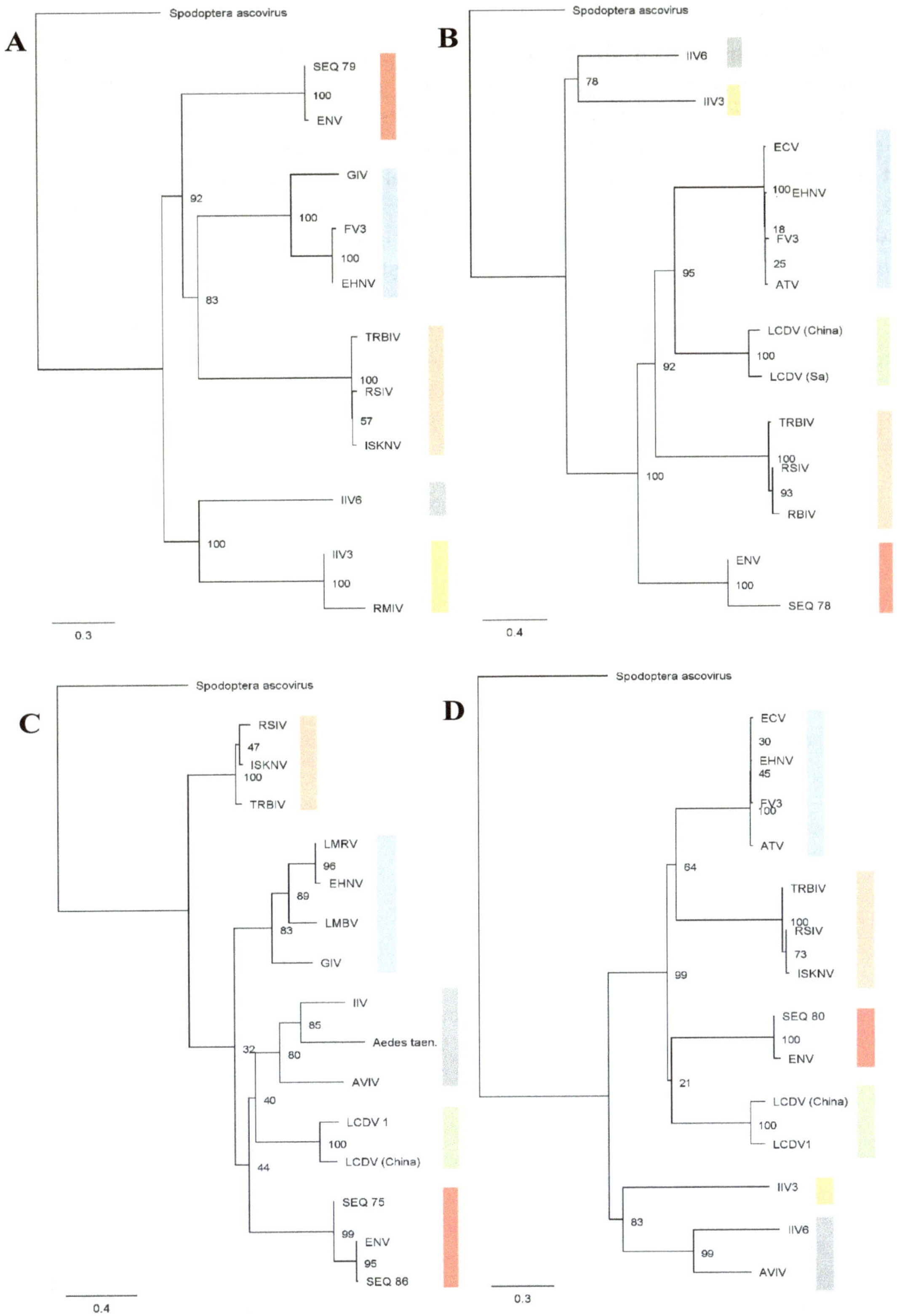

Figure 1. Substitution-rate optimized maximum likelihood phylogenetic trees of putative ENV DNA-dependent DNA polymerase (**A**), DNA-dependent RNA polymerase (**B**), major capsid protein (**C**), and ATPase (**D**) sequences (SEQ#) mapped with related iridoviruses. Alignments were based on nucleotide sequences for the major capsid protein and amino acid sequences for all other trees. Branch labels indicate bootstrap support values for 100 re-samplings and the scale bar indicates substitution rate. GenBank reference sequences and contig IDs are listed below (Table 2), with colors indicating genera groups. Putative ENV sequence lengths are listed in Table 1.

3.2. Spatial Epidemiology

ENV is widely distributed in the NE Pacific Ocean and was detected in 12 of 16 tested species (Figure 2) throughout the sampling region. High viral loads were common within the Strait of Georgia, along the west coast of Vancouver Island, and in straits and channels throughout coastal northern British Columbia and southern coastal Alaska (Figure 3). Over 19,000 fish were tested for ENV, from 16 different species collected from marine and fresh waters spanning from Washington to Alaska. ENV prevalence was highest in anchovy and herring, occurring in over 27% of all sampled fish in these species and in over 37% of smolts. In herring, the proportion of fish in which ENV was detected was significantly higher than in any salmon species tested. We also report significantly lower ENV prevalence and load within salmon smolt ($p_{prev} = 1.2 \times 10^{-17}$, $p_{load} = 1.8 \times 10^{-4}$) and greater ENV prevalence among herring smolt ($p = 4.5 \times 10^{-6}$), when compared to respective adult prevalence. Among fish with detections of ENV, viral load was highest in herring and Chinook salmon and lowest in Atlantic salmon. ENV load also appeared low in chum and pink salmon; however, these differences were not statistically significant (Figure S2, Table S3). Significant differences in ENV load were found between herring and coho, Chinook, sockeye, and Atlantic salmon (p values are given in Tables S3 and S4).

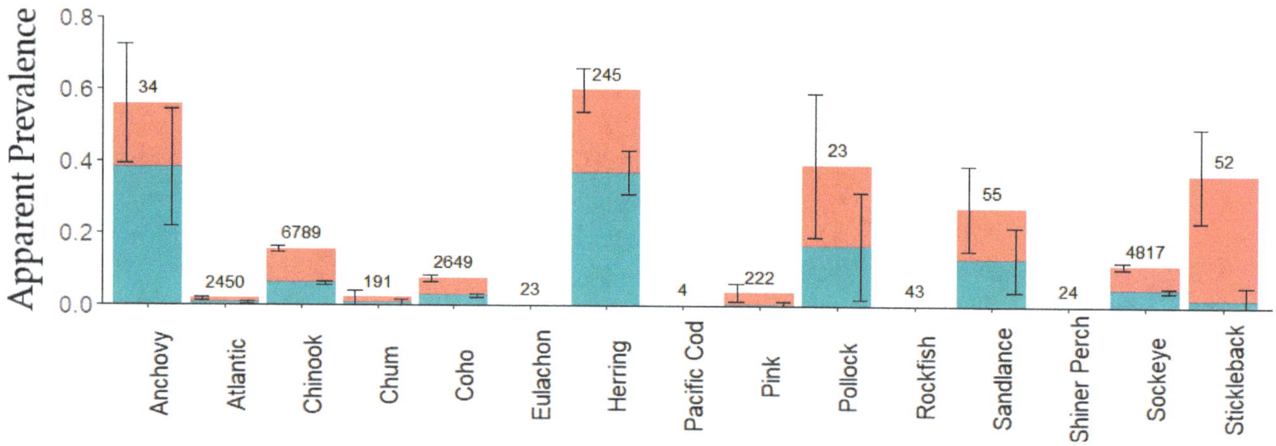

Figure 2. Apparent ENV prevalence in mixed tissue smolt samples by species. Error bars indicate 95% confidence intervals. Blue bars indicate proportions with limit of detection (LOD) criteria applied and coral bars indicate proportions without LOD criteria applied. Values indicate sample sizes for each species.

Despite high ENV prevalence and load in coastal regions, ENV was rarely detected, and percent prevalence was statistically lower in all species sampled in freshwater ($p < 2.2 \times 10^{-16}$) (Figure 4). Furthermore, after applying a generalized linear mixed-effects model with Laplace approximation to account for residual variation in the data, categorical classification of habitat type as freshwater or saltwater still had a significant effect on ENV prevalence ($p = 5.50 \times 10^{-8}$). Given that 98% of the freshwater detections were below the LOD, these all represent very low-load detections, and are unlikely biologically relevant or may represent "false" positives. As such, all statistical tests are reported for samples with LOD criteria applied only.

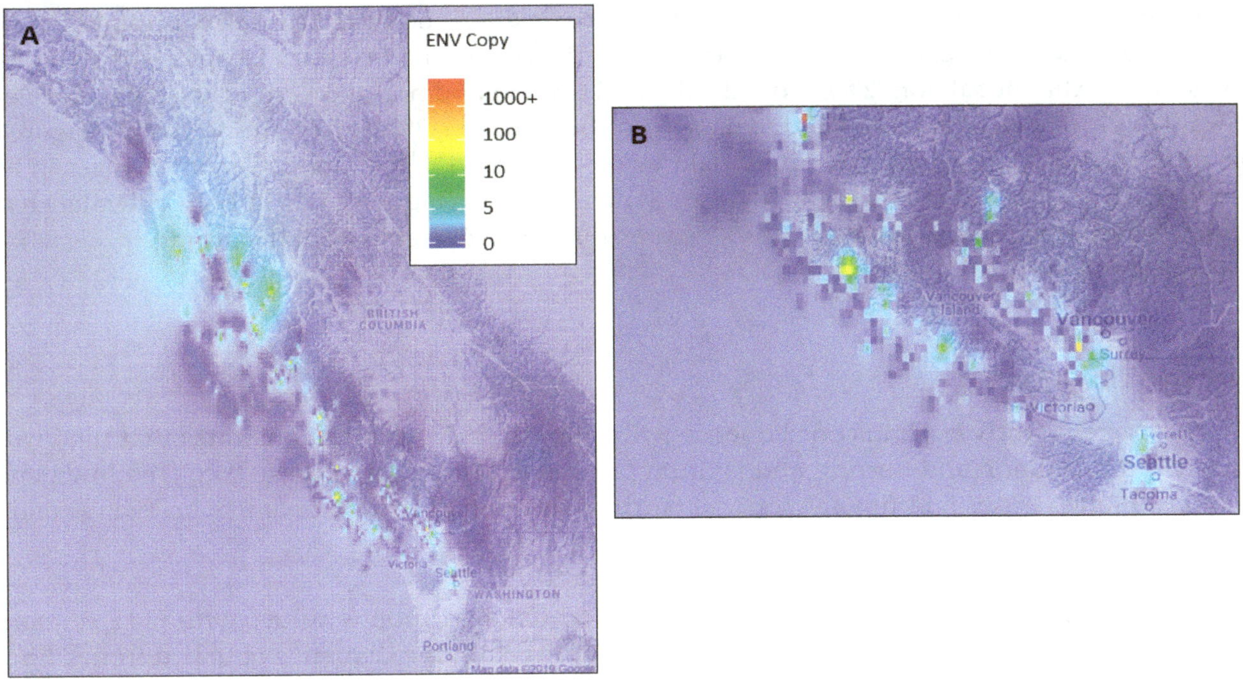

Figure 3. (**A**) Full sampling area heat map of the mean calculated ENV copy number, based on interpolated values. Copy number values are binned into 6 numerical categories (see color legend), (**B**) shows Vancouver Island inset map. Adults and smolts of all species are shown and LOD criteria are not applied.

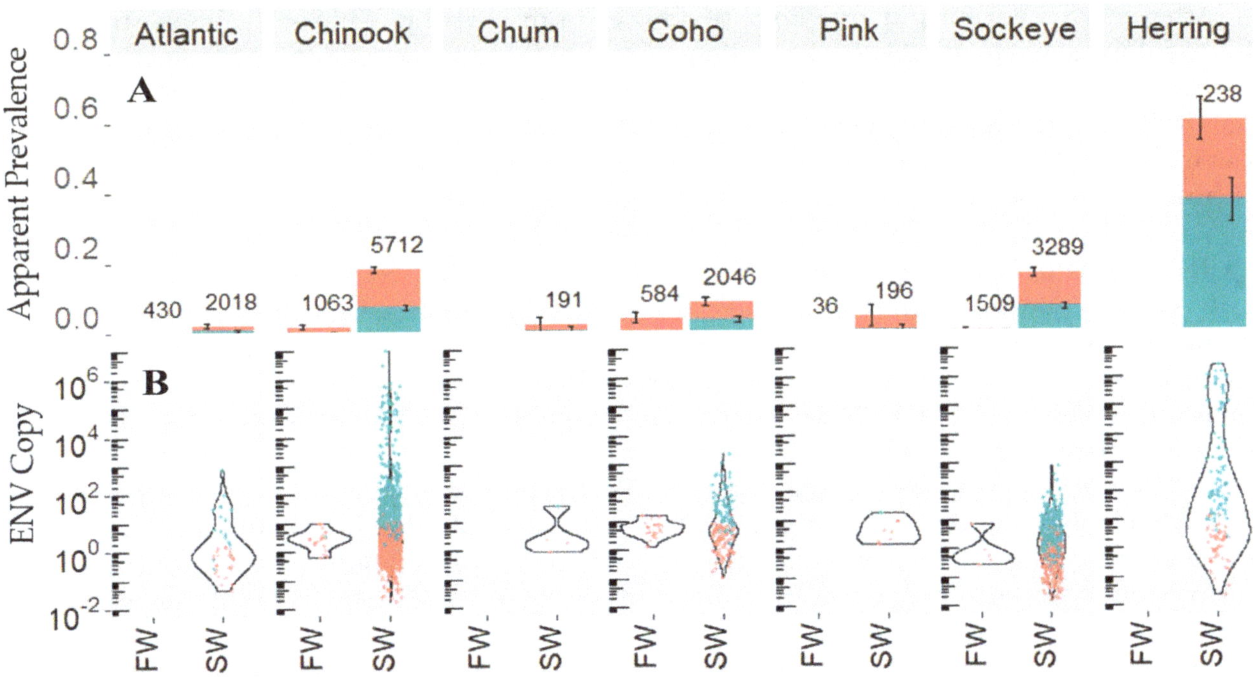

Figure 4. Saline (SW) and freshwater (FW) ENV prevalence (**A**) and load (**B**) among salmon species and herring. Samples with LOD criteria applied are in blue and samples without LOD criteria applied are in coral. Printed values indicate sample sizes and error bars indicate 95% confidence intervals. Only smolts are shown.

3.3. Seasonal and Yearly Variation

In Atlantic salmon, ENV prevalence showed seasonal variation, with the lowest prevalence occurring in August, for both smolt and adult fish (Figure S3). A similar seasonal trend was observed in smolt sockeye and Chinook salmon. A Spearman correlation coefficient of 0.87 was found between monthly prevalence of ENV in sockeye and Atlantic salmon ($p = 0.004$). Despite very low ENV prevalence in Atlantic salmon (2%), farmed coho and Chinook salmon showed higher ENV prevalence (29% and 48%, respectively) compared to their wild counterparts (3% and 7%, respectively) (Figure S4). This difference was significant in Chinook salmon ($p = 2.08 \times 10^{-50}$). Among salmon, changes in ENV prevalence from year to year appear to occur synchronously (Figure S5) and ENV prevalence also varies significantly by year ($p < 2.2 \times 10^{-16}$, Figure S6). A 63% decrease from the average prevalence of 5% is observed after 2013.

4. Discussion

The current study expands on sequence available for ENV and demonstrates that the virus is widespread in salmon and herring and is often present at high load in these fish. The low genetic variation among viruses infecting salmon and herring has implications for potential host range and the taxonomic classification of these viruses.

Nucleotide variation of ENV from herring and Chinook salmon was low, with ENV sequences from Pacific herring [2,4] having >99% nucleotide identity with the sequence obtained from Chinook salmon in this study, indicating that both viruses belong within the same genus, and likely the same species. High sequence similarity between herring and Chinook salmon could be suggestive of viral spillover between hosts. Moreover, the phylogenetic analysis clearly places ENV within the *Iridoviridae*, but distinct from other viruses within the family.

After enrichment for putative ENV sequences in the herring, Chinook salmon, and Atlantic salmon samples, we found 35 of 117 predicted protein-encoding sequences, suggesting that not all ENV transcripts are present in every infected fish. This may depend on expression levels and infection stage. Indeed, variable gene expression during different stages of infection occurs commonly among iridoviruses [49–52]. Moreover, our estimates of prevalence are conservative, as the assay is sensitive to sequence variation, so related strains of ENV could be missed.

ENV was widely distributed geographically and among fish species in marine waters of western North America. Despite high ENV prevalence and load in coastal marine environments, viral load and prevalence were consistently low in freshwater environments. Of the 3622 fish analyzed from freshwater, ENV was only detected in one Chinook salmon specimen after LOD criteria were applied. Interestingly, there were few ENV detections in fish from the open waters north of Vancouver Island (Figure 3); whereas ENV was common in coastal areas, i.e., channels, inlets, and straits along coastal British Columbia and southern Alaska (Figure 3). We hypothesize that interspecies transmission may be more likely in these areas, where there are higher densities of salmon and other fish. Greater ENV prevalence in aquaculture Pacific salmon than in wild counterparts (Figure S4) supports the idea that transmission of the virus may be more common in coastal and high-density environments. There is some support for this hypothesis in the literature as previous studies have shown that chum salmon may contract VEN via waterborne exposure [8] and that increased stock density is a predictor of disease progression in fish infected with other iridoviruses [53,54]. We also report significantly lower ENV prevalence among salmon smolts and in adult herring when compared to respective adult and smolt age classes for these species, suggesting that susceptibility to ENV varies by host species and age class. ENV prevalence may not, however, directly correlate with disease manifestation mediated by the virus, as species with low ENV load and prevalence (<0.5%) in this study, including chum and pink salmon, are more susceptible to VEN than other salmon species [55,56]. Thus, salmon species in which ENV prevalence and load were the greatest (Chinook and coho salmon) may be able to sustain higher viral loads and display fewer clinical symptoms.

We hypothesize that salmon likely contract the virus from marine reservoirs, given the low detection of ENV in freshwater salmon, high sequence similarity between ENV in salmon and herring, and high viral prevalence in several species of marine fish including herring, anchovy, pollock, and sand lance. Furthermore, the prevalence of ENV in Pacific salmon is much higher than in Atlantic salmon, suggesting that farmed Atlantic salmon are not a significant reservoir for ENV transmission, despite high-density rearing in this species. Previous research has shown that overall infectious agent diversity and burden increases when sockeye salmon enter the ocean [16], suggesting that ENV could contribute to increased infection stress experienced by out-migrating salmon smolts.

Salmon migration, which varies among species and populations, may help explain yearly and seasonal variation in ENV prevalence. Peaks in ENV prevalence occur during spring and late fall for Chinook and sockeye salmon, with significant drops seen during peak salmon river runs in July and August. A similar pattern is observed in Atlantic salmon, which are stationary throughout the year and therefore may provide a useful sentinel to study seasonal ENV dynamics in wild salmon. Indeed, there is a significant correlation in monthly ENV prevalence between Atlantic and sockeye salmon. However, variation in migration patterns among species and stocks complicates the analysis of seasonal ENV prevalence changes. Migration routes and timing among Chinook salmon stocks, for example, is highly variable [57,58]. Furthermore, several sockeye salmon stocks were sampled during different parts of the year. Increases in ENV prevalence in winter may also arise as a result of herring migrations to coastal regions during this time [59]. Overall, monthly ENV prevalence dynamics in salmon were similar to those observed previously in herring [1], further substantiating the hypothesis that interactions with herring promote infection dynamics in wild salmon.

Previously, little was known about the epidemiology of ENV in salmon and marine fish, as most studies focused on ENV in herring. Hershberger et al. detected ENV in up to 67% of herring, with similar seasonal variation to that which we observed in salmon, with the greatest proportion of fish testing positive for ENV in summer months [1]. They also reported that ENV epizootics can arise and dissipate spontaneously in geographically isolated regions along the North Pacific coastline. Additionally, Teffer et al. investigated ENV prevalence in returning Chinook salmon and detected ENV in 16% of tagged males and 25% of females in the Chilliwack River [20]. Together, our research and these studies indicate that the virus is widely distributed on the west coast of British Columbia and Alaska and that salmon are likely infected once they enter the ocean, with herring or other marine fish likely acting as a reservoir for ENV.

Implications

Detection of ENV has not been conclusively linked to disease onset and further studies are required to characterize this relationship. VEN is a poorly characterized disease in Chinook and sockeye salmon, yet it is relatively common in wild at-risk populations of these species. In contrast, ENV was relatively rare in pink and chum salmon, even though these species are more susceptible to VEN than Chinook and sockeye salmon in challenge studies [5]. However, relatively few pink ($n = 222$) and chum ($n = 191$) salmon were sampled in our study. There were significant differences in viral loads among species, with lowest mean ENV loads occurring in Atlantic salmon. Low viral loads and prevalence could indicate higher virulence, which may lower the chance of transmission and detection compared to persistently infected fish [60]. Alternatively, species with large ranges in ENV copy number, such as herring and Chinook salmon, could carry a persistent infection which becomes virulent at higher loads. These species may transmit the virus to susceptible species such as pink and chum salmon. Future challenge studies which further characterize Chinook salmon infection may elucidate whether infection dynamics appear similar to those of herring, which have the most similar distribution of viral load.

ENV prevalence was lower in salmon smolts and adult herring. Similarly, Hershberger et al. reported more frequent VEN epizootics in juvenile herring, compared to adults [1]. Presumably, lower viral prevalence in smolt salmon arises because fewer smolt have been exposed to marine waters, where we propose the virus originates. Previous studies reported that osmoregulatory stress, such as transitions between saline and freshwater environments, could be implicated in herring mortality in fish infected with VEN [15,61]. If viral infection does, indeed, impact osmoregulatory capacity and adaptation, the relatively high prevalence and load of ENV detected in salmon soon after ocean entry could diminish their ability to properly acclimate to changes in salinity in their environment.

Numerous studies [10–14] have reported greater disease severity caused by iridoviruses that are closely related to ENV when the temperature increases. It has been suggested that below 20 °C, iridoviruses may remain dormant in teleost hosts [14]. Other members of *Iridoviridae* that infect fish and show high infection mortality typically occur in warmer climates, such as Southeast Asia and Australia. Similarly, VEN progression is most severe during the summer in Pacific salmon [5]. Changes in temperature were not investigated in this study, but seasonal and yearly variation in ENV prevalence suggests that environmental variables, such as temperature, may be important. A significant drop in ENV prevalence following 2013 coincides with a shift to a positive Pacific Decadal Oscillation Index and warming temperatures in the study region [62].

It is possible that disease progression intensifies at warmer temperatures, such that fewer fish harboring the virus survive. This interpretation is consistent with a decrease in the overall prevalence of salmon infectious agents in the region from 2012 to 2013 reported by Nekouei et al. [16]. Alternatively, fish infected with the virus may be weakened or more susceptible to other diseases at suboptimal temperatures. In the context of climate change, this is an interesting avenue of future research, and directly relevant to salmon populations, as evidence suggests that increasing coastal and oceanic temperatures can have significant and detrimental effects on salmon migration and spawning [63,64]. If VEN has a temperature-dependent onset similar to other diseases caused by iridoviruses, ENV-mediated mortality could further stress at-risk populations of salmon and herring in the NE Pacific Ocean.

5. Conclusions

This research demonstrates that ENV is highly prevalent in the NE Pacific Ocean. Low ENV prevalence in freshwater, high prevalence in marine fish, and seasonal variability corresponding to marine migrations of salmon and herring suggest that ENV originates from the marine environment. High prevalence in several marine fish species suggests that the virus is endemic and that these species are reservoirs of the virus. Moreover, the similarity between ENV sequences from Chinook salmon and those from Pacific herring indicates that transmission between these species is possible. Finally, we present 89 new protein-encoding sequences attributed to ENV in this study.

Author Contributions: Conceptualization, G.J.M., K.M.M., and C.A.S.; Methodology, V.A.P., G.J.M., A.D.S., K.H.K., T.J.M., S.L., and A.K.T.; Formal analysis, V.A.P.; Investigation, V.A.P., G.J.M., A.D.S.; Resources, K.M.M. and C.A.S.; Data curation, A.T.; Writing—original draft preparation, V.A.P.; Writing—review and editing, G.J.M., C.A.S., and K.M.M.; Visualization, V.A.P.; Supervision, G.J.M., K.M.M., and C.A.S.; Project administration, G.J.M., C.A.S., and K.M.M.

Acknowledgments: We would like to thank Christoph Deeg for feedback on phylogenetic analyses and insightful suggestions.

References

1. Hershberger, P.K.; Elder, N.E.; Grady, C.A.; Gregg, J.L.; Pacheco, C.A.; Greene, C.; Rice, C.; Meyers, T.R. Prevalence of Viral Erythrocytic Necrosis in Pacific Herring and Epizootics in Skagit Bay, Puget Sound, Washington. *J. Aquat. Anim. Health* **2009**, *21*, 1–7. [CrossRef] [PubMed]
2. Emmenegger, E.J.; Glenn, J.A.; Winton, J.R.; Batts, W.N.; Gregg, J.L.; Hershberger, P.K. Molecular Identification of Erythrocytic Necrosis Virus (ENV) from the Blood of Pacific Herring *(Clupea pallasii)*. *Vet. Microbiol.* **2014**, *174*, 16–26. [CrossRef] [PubMed]
3. Winton, J.R.; Hershberger, P.K. Viral Erythrocytic Necrosis: Chapter 2.2.7. In *FHS Blue Book: Suggested Procedures for the Detection and Identification of Certain Finfish and Shellfish Pathogens*; American Fisheries Society: Bethesda, MA, USA, 2014.
4. Purcell, M.K.; Pearman-Gillman, S.; Thompson, R.L.; Gregg, J.L.; Hart, L.M.; Winton, J.R.; Emmenegger, E.J.; Hershberger, P.K. Identification of the Major Capsid Protein of Erythrocytic Necrosis Virus (ENV) and Development of Quantitative Real-time PCR Assays for Quantification of ENV DNA. *J. Vet. Diagn. Investig.* **2016**, *28*, 382–391. [CrossRef] [PubMed]
5. Evelyn, T.P.T.; Traxler, G.S. Viral Erythrocytic Necrosis: Natural Occurrence in Pacific Salmon and Experimental Transmission. *J. Fish. Res. Board Can.* **1978**, *35*, 903–907. [CrossRef]
6. Rohovec, J.S.; Amandi, A. Incidence of Viral Erythrocytic Necrosis among Hatchery Reared Salmonids of Oregon. *Fish Pathol.* **1981**, *15*, 135–141. [CrossRef]
7. Haney, D.C.; Hursh, D.A.; Mix, M.C.; Winton, J.R. Physiological and Hematological Changes in Chum Salmon Artificially Infected with Erythrocytic Necrosis Virus. *J. Aquat. Anim. Health* **1992**, *4*, 48–57. [CrossRef]
8. MacMillian, J.R.; Mulcahy, D. Artificial Transmission to and Susceptibility of Puget Sound Fish to Viral Erythrocytic Necrosis (VEN). *J. Fish. Res. Board Can.* **1979**, *36*, 1097–1101. [CrossRef]
9. Eaton, W.D. Artificial Transmission of Erythrocytic Necrosis Virus (ENV) from Pacific Herring in Alaska to Chum, Sockeye, and Pink Salmon. *J. Appl. Ichthyol.* **1990**, *6*, 136–141. [CrossRef]
10. International Committee on Taxonomy of Viruses; King, A.M.Q.; International Union of Microbiological Societies; Virology Division. *Virus Taxonomy: Classification and Nomenclature of Viruses: Ninth Report of the International Committee on Taxonomy of Viruses*; Academic Press: London, UK, 2012; ISBN 978-0-12-384684-6.
11. Huang, S.M.; Tu, C.; Tseng, C.H.; Huang, C.C.; Chou, C.-C.; Kuo, H.-C.; Chang, S.-K. Genetic Analysis of Fish Iridoviruses Isolated in Taiwan during 2001–2009. *Arch. Virol.* **2011**, *156*, 1505–1515. [CrossRef]
12. Wang, C.S.; Shih, H.H.; Ku, C.C.; Chen, S.N. Studies on Epizootic Iridovirus Infection Among Red Sea Bream, *Pagrus major* (Temminck & Schlegel), Cultured in Taiwan. *J. Fish Dis.* **2003**, *26*, 127–133. [PubMed]
13. Liu, H.I.; Lin, Y.C.; Chiou, P.P.; Chou, H.-Y. Temperature-Dependent Pathogenicity of Grouper Iridovirus of Taiwan (TGIV). *J. Mar. Sci. Technol.* **2016**, *24*, 637–644.
14. Wang, Y.Q.; Lü, L.; Weng, S.P.; Huang, J.N.; Chan, S.-M.; He, J.G. Molecular Epidemiology and Phylogenetic Analysis of a Marine Fish Infectious Spleen and Kidney Necrosis Virus-like (ISKNV-like) Virus. *Arch. Virol.* **2007**, *152*, 763–773. [CrossRef] [PubMed]
15. Meyers, T.R.; Hauck, A.K.; Blankenbeckler, W.D.; Minicucci, T. First report of Viral Erythrocytic Necrosis in Alaska, USA, Associated with Epizootic Mortality in Pacific Herring, *Clupea harengus pallasi* (Valenciennes). *J. Fish Dis.* **1986**, *9*, 479–491. [CrossRef]
16. Nekouei, O.; Vanderstichel, R.; Ming, T.; Kaukinen, K.H.; Thakur, K.; Tabata, A.; Laurin, E.; Tucker, S.; Beacham, T.D.; Miller, K.M. Detection and Assessment of the Distribution of Infectious Agents in Juvenile Fraser River Sockeye Salmon, Canada, in 2012 and 2013. *Front. Microbiol.* **2018**, *9*. [CrossRef] [PubMed]
17. Tucker, S.; Li, S.; Kaukinen, K.H.; Patterson, D.A.; Miller, K.M. Distinct Seasonal Infectious Agent Profiles in Life-history Variants of Juvenile Fraser River Chinook Salmon: An Application of High-Throughput Genomic Screening. *PLoS ONE* **2018**, *13*, e0195472. [CrossRef] [PubMed]
18. Thakur, K.K.; Vanderstichel, R.; Li, S.; Laurin, E.; Tucker, S.; Neville, C.; Tabata, A.; Miller, K.M. A Comparison of Infectious Agents Between Hatchery-Enhanced and Wild Out-Migrating Juvenile Chinook Salmon (Oncorhynchus tshawytscha) from Cowichan River, British Columbia. *FACETS* **2018**. [CrossRef]
19. Laurin, E.; Jaramillo, D.; Vanderstichel, R.; Ferguson, H.; Kaukinen, K.H.; Schulze, A.D.; Keith, I.R.; Gardner, I.A.; Miller, K.M. Histopathological and Novel High-Throughput Molecular Monitoring Data from Farmed Salmon (*Salmo salar* and *Oncorhynchus* spp.) in British Columbia, Canada, from 2011–2013. *Aquaculture* **2019**, *499*, 220–234. [CrossRef]

20. Teffer, A.K.; Bass, A.L.; Miller, K.M.; Patterson, D.A.; Juanes, F.; Hinch, S.G. Infections, Fisheries Capture, Temperature, and Host Responses: Multistressor Influences on Survival and Behaviour of Adult Chinook Salmon. *Can. J. Fish. Aquat. Sci.* **2018**, *75*, 2069–2083. [CrossRef]
21. DFO. Stock Assessment Report on Pacific Herring in British Columbia. Canadian Science Advisory Secretariat Science Advisory Report 2009/059. Available online: http://www.dfo-mpo.gc.ca/Library/338390.pdf (accessed on 17 April 2019).
22. Varpe, Ø.; Fiksen, Ø.; Slotte, A. Meta-Ecosystems and Biological Energy Transport from Ocean to Coast: The Ecological Importance of Herring Migration. *Oecologia* **2005**, *146*, 443. [CrossRef] [PubMed]
23. Pacific Salmon Commission. *2010 Annual Report of Catches and Escapements*; Report TCChinook (11)-2; Pacific Salmon Commission: Vancouver, BC, Canada, 2011.
24. Government of Canada. Summary of COSEWIC Wildlife Species Assessments. November 2018. Available online: https://www.canada.ca/en/environment-climate-change/services/committee-status-endangered-wildlife/assessments/wildlife-species-assessment-summary-nov-2018.html (accessed on 28 January 2019).
25. Miller, K.M.; Teffer, A.; Tucker, S.; Li, S.; Schulze, A.D.; Trudel, M.; Juanes, F.; Tabata, A.; Kaukinen, K.H.; Ginther, N.G.; et al. Infectious Disease, Shifting Climates, and Opportunistic Predators: Cumulative Factors Potentially Impacting Wild Salmon Declines. *Evol. Appl.* **2014**, *7*, 812–855. [CrossRef] [PubMed]
26. Miller, K.M.; Gardner, I.A.; Vanderstichel, R.; Burnley, T.; Schulze, A.D.; Li, S.; Tabata, A.; Kaukinen, K.H.; Ming, T.J.; Ginther, N.G. *Report on the Performance Evaluation of the Fluidigm BioMark Platform for High Throughput Microbe Monitoring in Salmon*; Fisheries and Oceans Canada: Ottawa, ON, Canada, 2016.
27. Di Cicco, E.; Ferguson, H.W.; Kaukinen, K.H.; Schulze, A.D.; Li, S.; Tabata, A.; Günther, O.P.; Mordecai, G.; Suttle, C.A.; Miller, K.M. The Same Strain of Piscine Orthoreovirus (PRV-1) is Involved in the Development of Different, but Related, Diseases in Atlantic and Pacific Salmon in British Columbia. *FACETS* **2018**, *3*, 599–641. [CrossRef]
28. Bass, A.L.; Hinch, S.G.; Teffer, A.K.; Patterson, D.A.; Miller, K.M. A Survey of Microparasites Present in Adult Migrating Chinook Salmon (*Oncorhynchus tshawytscha*) in South-Western British Columbia Determined by High-Throughput Quantitative Polymerase Chain Reaction. *J. Fish Dis.* **2017**, *40*, 453–477. [CrossRef]
29. Bolger, A.M.; Lohse, M.; Usadel, B. Trimmomatic: A Flexible Trimmer for Illumina Sequence Data. *Bioinforma. Oxf. Engl.* **2014**, *30*, 2114–2120. [CrossRef] [PubMed]
30. Davidson, W.S.; Koop, B.F.; Jones, S.J.; Iturra, P.; Vidal, R.; Maass, A.; Jonassen, I.; Lien, S.; Omholt, S.W. Sequencing the Genome of the Atlantic Salmon (*Salmo salar*). *Genome Biol.* **2010**, *11*, 403. [PubMed]
31. Li, H.; Durbin, R. Fast and Accurate Short Read Alignment with Burrows-Wheeler Transform. *Bioinform. Oxf. Engl.* **2009**, *25*, 1754–1760. [CrossRef] [PubMed]
32. Li, H.; Handsaker, B.; Wysoker, A.; Fennell, T.; Ruan, J.; Homer, N.; Marth, G.; Abecasis, G.; Durbin, R. 1000 Genome Project Data Processing Subgroup the Sequence Alignment/Map Format and SAMtools. *Bioinforma. Oxf. Engl.* **2009**, *25*, 2078–2079. [CrossRef] [PubMed]
33. Bankevich, A.; Nurk, S.; Antipov, D.; Gurevich, A.A.; Dvorkin, M.; Kulikov, A.S.; Lesin, V.M.; Nikolenko, S.I.; Pham, S.; Prjibelski, A.D.; et al. SPAdes: A New Genome Assembly Algorithm and its Applications to Single-Cell Sequencing. *J. Comput. Biol. J. Comput. Mol. Cell Biol.* **2012**, *19*, 455–477. [CrossRef] [PubMed]
34. Buchfink, B.; Xie, C.; Huson, D.H. Fast and Sensitive Protein Alignment using DIAMOND. *Nat. Methods* **2015**, *12*, 59–60. [CrossRef] [PubMed]
35. Caporaso, J.G.; Kuczynski, J.; Stombaugh, J.; Bittinger, K.; Bushman, F.D.; Costello, E.K.; Fierer, N.; Peña, A.G.; Goodrich, J.K.; Gordon, J.I.; et al. QIIME Allows Analysis of High-Throughput Community Sequencing Data. *Nat. Methods* **2010**, *7*, 335–336. [CrossRef]
36. Borodovsky, M.; Mills, R.; Besemer, J.; Lomsadze, A. Prokaryotic Gene Prediction Using GeneMark and GeneMark.hmm. *Curr. Protoc. Bioinform.* **2003**, *4*, 4.5.1–4.5.16. [CrossRef] [PubMed]
37. Altschul, S.F.; Gish, W.; Miller, W.; Myers, E.W.; Lipman, D.J. Basic Local Alignment Search Tool. *J. Mol. Biol.* **1990**, *215*, 403–410. [CrossRef]
38. Yutin, N.; Wolf, Y.I.; Raoult, D.; Koonin, E.V. Eukaryotic Large Nucleo-Cytoplasmic DNA Viruses: Clusters of Orthologous Genes and Reconstruction of Viral Genome Evolution. *Virol. J.* **2009**, *6*, 223. [CrossRef] [PubMed]
39. Kearse, M.; Moir, R.; Wilson, A.; Stones-Havas, S.; Cheung, M.; Sturrock, S.; Buxton, S.; Cooper, A.; Markowitz, S.; Duran, C.; et al. Geneious Basic: An Integrated and Extendable Desktop Software Platform for the Organization and Analysis of Sequence Data. *Bioinformatics* **2012**, *28*, 1647–1649. [CrossRef] [PubMed]

40. Guindon, S.; Dufayard, J.-F.; Lefort, V.; Anisimova, M.; Hordijk, W.; Gascuel, O. New Algorithms and Methods to Estimate Maximum-Likelihood Phylogenies: Assessing the Performance of PhyML 3.0. *Syst. Biol.* **2010**, *59*, 307–321. [CrossRef] [PubMed]
41. Thompson, J.D.; Higgins, D.G.; Gibson, T.J. CLUSTAL W: Improving the Sensitivity of Progressive Multiple Sequence Alignment Through Sequence Weighting, Position-Specific Gap Penalties and Weight Matrix Choice. *Nucleic Acids Res.* **1994**, *22*, 4673–4680. [CrossRef] [PubMed]
42. R Core Team R: A Language and Environment for Statistical Computing. *Ed. Found. Stat. Comput. Vienna Austria* **2013**.
43. Pebesma, E.J. Multivariable Geostatistics in S: The gstat Package. *Comput. Geosci.* **2004**, *30*, 683–691. [CrossRef]
44. Heuvelink, G.B.M. Spatio-Temporal Interpolation Using gstat by Benedikt Gräler, Edzer Pebesma and Gerard Heuvelink. *R J.* **2016**, *8*, 204–218.
45. Kahle, D.; Wickham, H. ggmap: Spatial Visualization with ggplot2. *R J.* **2013**, *5*, 144. [CrossRef]
46. Mangiafico, S. rcompanion: Functions to Support Extension Education Program Evaluation [R Statistical Package]. 2016. Available online: https://cran.r-project.org/web/packages/rcompanion/ (accessed on 17 April 2018).
47. Bates, D.; Mächler, M.; Bolker, B.; Walker, S. Fitting Linear Mixed-Effects Models Using lme4. *J. Stat. Softw.* **2015**, *1*. [CrossRef]
48. Hothorn, T.; Bretz, F.; Westfall, P. Simultaneous Inference in General Parametric Models. *Biom. J.* **2008**, *50*, 346–363. [CrossRef]
49. Teng, Y.; Hou, Z.; Gong, J.; Liu, H.; Xie, X.; Zhang, L.; Chen, X.; Qin, Q.W. Whole-Genome Transcriptional Profiles of a Novel Marine Fish Iridovirus, Singapore Grouper Iridovirus (SGIV) in Virus-Infected Grouper Spleen Cell Cultures and in Orange-Spotted Grouper, *Epinephulus coioides*. *Virology* **2008**, *377*, 39–48. [CrossRef] [PubMed]
50. Chen, L.M.; Wang, F.; Song, W.; Hew, C.L. Temporal and Differential Gene Expression of Singapore Grouper Iridovirus. *J. Gen. Virol.* **2006**, *87*, 2907–2915. [CrossRef] [PubMed]
51. İnce, İ.A.; Özcan, O.; Ilter-Akulke, A.Z.; Scully, E.D.; Özgen, A. Invertebrate Iridoviruses: A Glance Over the Last Decade. *Viruses* **2018**, *10*, 161. [CrossRef]
52. Sample, R.; Bryan, L.; Long, S.; Majji, S.; Hoskins, G.; Sinning, A.; Olivier, J.; Chinchar, V.G. Inhibition of Iridovirus Protein Synthesis and Virus Replication by Antisense Morpholino Oligonucleotides Targeted to the Major Capsid Protein, the 18 kDa Immediate-Early Protein, and a Viral Homolog of RNA Polymerase II. *Virology* **2007**, *358*, 311–320. [CrossRef] [PubMed]
53. Kurobe, T.; MacConnell, E.; Hudson, C.; McDowell, T.S.; Mardones, F.O.; Hedrick, R.P. Iridovirus Infections among Missouri River Sturgeon: Initial Characterization, Transmission, and Evidence for Establishment of a Carrier State. *J. Aquat. Anim. Health* **2011**, *23*, 9–18. [CrossRef]
54. Georgiadis, M.P.; Hedrick, R.P.; Johnson, W.O.; Yun, S.; Gardner, I.A. Risk Factors for Outbreaks of Disease Attributable to White Sturgeon Iridovirus and White Sturgeon Herpesvirus-2 at a Commercial Sturgeon Farm. *Am. J. Vet. Res.* **2000**, *61*, 1232–1240. [CrossRef]
55. Traxler, G.; Bell, G. Pathogens Associated with Impounded Pacific Herring *Clupea harengus pallasi*, with Emphasis on Viral Erythrocytic Necrosis (VEN) and Atypical *Aeromonas salmonicida*. *Dis. Aquat. Organ.* **1988**, *5*, 93–100. [CrossRef]
56. Hershberger, P.; Hart, A.; Gregg, J.; Elder, N.; Winton, J. Dynamics of Viral Hemorrhagic Septicemia, Viral Erythrocytic Necrosis and Ichthyophoniasis in Confined Juvenile Pacific Herring *Clupea pallasii*. *Dis. Aquat. Organ.* **2006**, *70*, 201–208. [CrossRef]
57. Weitkamp, L.A. Marine Distributions of Chinook Salmon from the West Coast of North America Determined by Coded Wire Tag Recoveries. *Trans. Am. Fish. Soc.* **2010**, *139*, 147–170. [CrossRef]
58. Myers, K.W.; Klovach, N.V.; Gritsenko, O.F.; Urawa, S.; Royer, T.C. *Stock-Specific Distributions of Asian and North American Salmon in the Open Ocean, Interannual Changes, and Oceanographic Conditions*; NPAFC: Vancouver, BC, Canada, 2007.
59. Hourston, A.S.; Haegele, C.W. *Herring on Canada's Pacific Coast*; Department of Fisheries and Oceans: Longueuil, QC, USA, 1980.
60. Schroeder, D.C.; Martin, S.J. Deformed wing virus. *Virulence* **2012**, *3*, 589–591. [CrossRef] [PubMed]

61. Bell, Q.R.; Traxler, G.S. First Record of Viral Erythrocytic Necrosis and *Ceratomyxa Shasta* Noble, 1950 (Myxozoa: Myxosporea) in Feral Pink Salmon (*Oncorhynchus gorbuscha* Walbaum). *J. Wildl. Dis.* **1985**, *21*, 169–171. [CrossRef] [PubMed]
62. Peterson, W.; Morgan, C.; Casillas, E.; Peterson, J.; Fisher, J.; W Ferguson, J. Ocean Ecosystem Indicators of Salmon Marine Survival in the Northern California Current. 2019. Available online: https://www.fwspubs.org/doi/suppl/10.3996/042010-JFWM-009/suppl_file/10.3996_042010-jfwm-009.s6.pdf (accessed on 17 April 2019).
63. Goniea, T.; Bjornn, T.C.; Peery, C.A.; Bennett, D.H.; Stuehrenberg, L.C. Behavioral Thermoregulation and Slowed Migration by Adult Fall Chinook Salmon in Response to High Columbia River Water Temperatures. *Trans. Am. Fish. Soc.* **2006**, *135*, 408–419. [CrossRef]
64. Beacham, T.D.; Murray, C.B. Effect of Female Size, Egg Size, and Water Temperature on Developmental Biology of Chum Salmon (*Oncorhynchus keta*) from the Nitinat River, British Columbia. *Can. J. Fish. Aquat. Sci.* **1985**, *42*, 1755–1765. [CrossRef]

5

Interaction between Two Iridovirus Core Proteins and their Effects on Ranavirus (RGV) Replication in Cells from Different Species

Xiao-Tao Zeng [1] **and Qi-Ya Zhang** [1,2,*]

1 State Key Laboratory of Freshwater Ecology and Biotechnology, Institute of Hydrobiology, Wuhan 430072, China; zengxt@ihb.ac.cn
2 The Innovation Academy of Seed Design, Chinese Academy of Sciences, Beijing 100101, China
* Correspondence: zhangqy@ihb.ac.cn

Abstract: The two putative proteins RGV-63R and RGV-91R encoded by *Rana grylio* virus (RGV) are DNA polymerase and proliferating cell nuclear antigen (PCNA) respectively, and are core proteins of iridoviruses. Here, the interaction between RGV-63R and RGV-91R was detected by a yeast two-hybrid (Y2H) assay and further confirmed by co-immunoprecipitation (co-IP) assays. Subsequently, RGV-63R or RGV-91R were expressed alone or co-expressed in two kinds of aquatic animal cells including amphibian Chinese giant salamander thymus cells (GSTCs) and fish *Epithelioma papulosum cyprinid* cells (EPCs) to investigate their localizations and effects on RGV genome replication. The results showed that their localizations in the two kinds of cells are consistent. RGV-63R localized in the cytoplasm, while RGV-91R localized in the nucleus. However, when co-expressed, RGV-63R localized in both the cytoplasm and the nucleus, and colocalized with RGV-91R in the nucleus. 91R△NLS represents the RGV-91R deleting nuclear localization signal, which is localized in the cytoplasm and colocalized with RGV-63R in the cytoplasm. qPCR analysis revealed that sole expression and co-expression of the two proteins in the cells of two species significantly promoted RGV genome replication, while varying degrees of viral genome replication levels may be linked to the cell types. This study provides novel molecular evidence for ranavirus cross-species infection and replication.

Keywords: *Rana grylio* virus (RGV); iridovirus core proteins; protein interaction; aquatic animals; cross-species transmission; yeast two-hybrid (Y2H); co-immunoprecipitation (Co-IP)

---

1. Introduction

A large number of aquatic viruses regulate population dynamics and community interactions in aquatic ecosystems [1]. They are also involved in aquatic animal diseases. As pathogens, aquatic animal viruses often infect shellfishes, fishes, amphibians, reptiles and aquatic mammals [2–5]. Ranaviruses (*Rana grylio* virus, RGV) are members of the family *Iridoviridae* that are large, double-stranded DNA viruses and infect ectothermic vertebrates [6]. Importantly, ranaviruses are capable of crossing species barriers of numerous ectothermic vertebrates and can spread between different species [7–9]. Recent reports have revealed more about the molecular mechanism of aquatic viral disease, including ranaviral disease [10]. Science and technology have been applied to a wide range of studies on aquatic viruses [11]. Different approaches, such as co-immunoprecipitation (co-IP) assays [12], fluorescence microscopy, and yeast two hybrid (Y2H), have been widely used in the investigation of protein–protein relationships or interactions [13]. Recently, the Y2H assay was used to analyze protein–protein interactions among the structural proteins of *Chilo* iridescent virus, a member of genus *Iridovirus* [14].

DNA polymerases play multiple roles [15]. Its main function is DNA replication and is capable of catalyzing DNA synthesis [16]. DNA polymerases of some large DNA viruses, such as herpesvirus and vaccinia virus, play a crucial role in virus genome replication [17,18]. Iridoviral DNA polymerases (RGV-63R and its homologous proteins) are believed to be essential components for virus DNA replication [19,20]. RGV-91R and its homologous proteins are considered proliferating cell nuclear antigens (PCNAs). This kind of protein is also found in humans as a cofactor of DNA polymerase, which can increase the processivity of DNA strand synthesis during replication [21]. In eukaryotes, PCNA as a sliding clamp protein forms a ring-shaped homo-trimer that encircles double-stranded DNA. It can confer high processivity with respect to replicative DNA polymerase. Moreover, it forms a mobile platform on DNA that recruits many of the proteins involved in DNA replication, repair, and recombination. For example, PCNA interacts with DNA polymerase δ, a member of family B [22,23]. There are 26 core proteins shared in *Iridoviridae* [24], including RGV-63R, RGV-91R, and some other proteins related to transcription, replication, and nucleotide metabolism. Several RGV encoded proteins involved in gene transcription, viral infection, and assembly have been identified previously [25–28].

However, the ranavirus protein–protein interactions and their effects on virus replication in cross-species transmission remain unaddressed. The goal of this study was to investigate the interactions between ranavirus core proteins, and their effects on virus replication in both fish cells (*Epithelioma papulosum cyprinid* cells, EPCs) and amphibian cells (Chinese giant salamander thymus cells, GSTCs) through multiple technical approaches, such as Y2H, co-IP assays, fluorescence microscopy, Western blotting and qPCR. Y2H was used to test that the gene products can physically bind to each other, while a fluorescence experiment was used to test whether the gene products had similar co-localization in different host cells. Co-IP was performed to confirm the results from Y2H.

2. Materials and Methods

2.1. Cell Lines and Virus

Epithelioma papulosum cyprinid cells (EPCs) and Chinese giant salamander thymus cells (GSTCs) [29] were cultured in Medium 199 supplemented with 10% fetal bovine serum (FBS) at 25 °C. They were used for all work needing cells, except the co-immunoprecipitation (co-IP) assay. *Rana grylio* virus (RGV) was used in the study and the virus was repeatedly cultured on EPCs for long-term preservation. Human embryonic kidney (HEK293T) cells were grown at 37 °C in 5% CO2 in Dulbecco's modified Eagle's medium (DMEM, gibco, Thermo Fisher Scientific, Waltham, MA, USA) supplemented with 10% FBS, and were only used for the co-immunoprecipitation (co-IP) assay.

2.2. Plasmid Construction

RGV propagation in EPC cell cultures, purification, and its genomic DNA preparation were performed as described previously [27]. Different kinds of nucleic-acid fragments were amplified with the corresponding primers (Table 1). The plasmids used for yeast two-hybrid (Y2H), co-immunoprecipitation (co-IP), fluorescence microscopy and quantitative polymerase chain reaction (qPCR) were constructed. All constructs used in this study were confirmed by DNA sequencing. The operations are as follows:

Table 1. Primers used in this study.

Name	Sequence (5′ to 3′) [a]	Usage
63R-F	GGGAATTCCATATGGATCTCTTTGTGTACCAGTG (*NdeI*)	pGBKT7-*63R*/Y2H
63R-R	CCGGAATTCTTACTTTTTCTTGAACGACA (*EcoRI*)	
1R-F	CCGGAATTCATGGCATTCTCGGCAGAAGA (*EcoRI*)	pGADT7-*1R*/Y2H
1R-R	CCGCTCGAGTCATAGGGGGGTAAACTTCC (*XhoI*)	
2L-F	CCGGAATTCATGTCCATCATCGGAGCGAC (*EcoRI*)	pGADT7-*2L*/Y2H
2L-R	CGCGGATCCTTACCATCTCACTGTAGAGA (*BamHI*)	
9R-F	GGGAATTCCATATGGAAATGTTTGCATCTAAATC (*NdeI*)	pGADT7-*9R*/Y2H
9R-R	CCGGAATTCTCATCGCCACTCAAAGGATT (*EcoRI*)	
10L-F	CCGGAATTCATGGACACATCACCCTACGA (*EcoRI*)	pGADT7-*10L*/Y2H
10L-R	CCGCTCGAGTCAGGCAAACTTGCCCCTCC (*XhoI*)	
13L-F	GGGAATTCCATATGTGCTCCAAACTCGTAGAGAT (*NdeI*)	pGADT7-*13L*/Y2H
13L-R	CCGGAATTCTTAGAAACCCATGGTCTCGA (*EcoRI*)	
16R-F	CCGGAATTCATGGAACAAGTACCCATAAA (*EcoRI*)	pGADT7-*16R*/Y2H
16R-R	CCGCTCGAGCTAATCGTCCAAGTCCGACT (*XhoI*)	
21R-F	GGGAATTCCATATGGCTACAAATTACTGTGACGA (*NdeI*)	pGADT7-*21R*/Y2H
21R-R	CCGGAATTCTTACATCGTGAAGCTCTCAA (*EcoRI*)	
23L-F	CCGGAATTCATGGAAACCATAGTGCTGGT (*EcoRI*)	pGADT7-*23L*/Y2H
23L-R	CCGCTCGAGTTACGACGAGGACCCAAATG (*XhoI*)	
24R-F	CCGGAATTCATGGCTAACGCTACCATAAA (*EcoRI*)	pGADT7-*24R*/Y2H
24R-R	CGCGGATCCCTACTCTTGCTGCTCGGCTC (*BamHI*)	
29R-F	GGGAATTCCATATGGCCAATTTTCTACAAGATGT (*NdeI*)	pGADT7-*29R*/Y2H
29R-R	CCGGAATTCTCAATGACGCTCCTTGGCCC (*EcoRI*)	
40R-F	CCGGAATTCATGCAAGTTTTTCTAGATTT (*EcoRI*)	pGADT7-*40R*/Y2H
40R-R	CGCGGATCCTCACCTCCTCCTGCTCCTGC (*BamHI*)	
44R-F	GGGAATTCCATATGAGAGTCGTGGTAAACGCAAA (*NdeI*)	pGADT7-*44R*/Y2H
44R-R	CCGGAATTCTCACATCAGAAGAGACACGT (*EcoRI*)	
53R-F	CCGGAATTCATGGGAGCAGCGGAATCTAT (*EcoRI*)	pGADT7-*53R*/Y2H
53R-R	CCGCTCGAGTTAACCCCTGTGGGCCGGAA (*XhoI*)	
60R-F	GGGAATTCCATATGGCAATGGTTTCCAACGTAAA (*NdeI*)	pGADT7-*60R*/Y2H
60R-R	CCGGAATTCCTACAGGCTCTTTAGGATAA (*EcoRI*)	
63R-F	GGGAATTCCATATGGATCTCTTTGTGTACCAGTG (*NdeI*)	pGADT7-*63R*/Y2H
63R-*R2*	CCGGAATTCTTACTTTTTCTTGAACGACA (*EcoRI*)	
65L-F	CGCGGATCCTGTCCAGGGGCATGACTACC (*BamHI*)	pGADT7-*65L*/Y2H
65L-R	CCGCTCGAGTCACTTGAAGGCTATGGAAA (*XhoI*)	
73L-F	CCGGAATTCATGTTTCCTCACGTCACCAT (*EcoRI*)	pGADT7-*73L*/Y2H
73L-R	CGCGGATCCTTAGATGTCCAGGGGTTCGT (*BamHI*)	

Table 1. *Cont.*

Name	Sequence (5′ to 3′) [a]	Usage
87L-F	CCGGAATTCATGGAAGGTTGGTTGGGAAA (*EcoRI*)	pGADT7-*87L*/Y2H
87L-R	CCGCTCGAGCTAGACTCCCTTGGCATGAA (*XhoI*)	
88R-F	CCGGAATTCATGTCTTTTCAGAGAGATTA (*EcoRI*)	pGADT7-*88R*/Y2H
88R-R	CCGCTCGAGCTACCTGGTCCACCTCTTGC (*XhoI*)	
91R-F	GGGAATTCCATATGCTGTGGGAAGCCGTAACAGA (*NdeI*)	pGADT7-*91R*/Y2H
91R-R	CCGGAATTCTTAGCCCTCAAAGAGAGTCA (*EcoRI*)	
92R-F	GGGAATTCCATATGAGCATCCCTACAGTCATAGC (*NdeI*)	pGADT7-*92R*/Y2H
92R-R	CCGGAATTCTTACCGCACATTTCTAGACA (*EcoRI*)	
95R-F	CCGGAATTCATGCACGGTTGCAATTGTAA (*EcoRI*)	pGADT7-*95R*/Y2H
95R-R	CCGCTCGAGTCAGTTAAAAGTGCTCGTAT (*XhoI*)	
97R-F	CCGGAATTCATGTCTTCTGTAACTGGTTC (*EcoRI*)	pGADT7-*97R*/Y2H
97R-R	CCGCTCGAGGACCCATGACGGAAAAGACT (*XhoI*)	
98R-F	CCGGAATTCATGGCAAACTTTGTGACAGA (*EcoRI*)	pGADT7-*98R*/Y2H
98R-R	CCGCTCGAGTTAGGCTCTGACCACAAACA (*XhoI*)	
101L-F	CCGGAATTCATGGATCCAGAAGGAATGAT (*EcoRI*)	pGADT7-*101L*/Y2H
101L-R	CCGCTCGAGTCACAGCACCTTTCTCAGGT (*XhoI*)	
102R-F	CCGGAATTCATGGGCATAAAAGGACTGAA (*EcoRI*)	pGADT7-*102R*/Y2H
102R-R	CCGCTCGAGTCACTTGCGCTTGCACTTCT (*XhoI*)	
63R-F2	CCCAAGCTTATGGATCTCTTTGTGTACCA (*HindIII*)	pcDNA3.1-*63R-3Flag*
63R-3Flag-R	CCGGAATTCTTACTTATCGTCGTCATCCTTGTAATCGATCTT ATCGTCGTCATCCTTGTAATCTCCCTTATCGTCGTCATCCTT GTAATCCTTTTTCTTGAACGACACAA (*EcoRI*)	
91R-F2	CCCAAGCTTATGCTGTGGGAAGCCGTAAC (*HindIII*)	pcDNA3.1-*91R-HA*
91R-HA-R	CCGGAATTCTTAAGCGTAATCTGGAACATCGTATGGGTAC ATGCCCTCAAAGAGAGTCACGG (*EcoRI*)	
63R-R3	CCGGAATTCGACTTTTTCTTGAACGACAC (*EcoRI*)	pEGFP-*63R*/colocalization
91R-F3	CCCAAGCTTCAATGCTGTGGGAAGCCGTA (*HindIII*)	pDsRed2-*91R*/colocalization
91R-R2	ACGCGTCGACTTAGCCCTCAAAGAGAGTCA (*SalI*)	
91R∆NLS1-R	CCCATGAGCCTCAGCGTCACGTAGCTGGTAAAGACCGATG	pDsRed2-*91R∆NLS1*
91R∆NLS1-F	CATCGGTCTTTACCAGCTACGTGACGCTGAGGCTCATGGG	
91R∆NLS2-R	GAAAAGGGTCCCATCTTGACCACTCCTCCGGACGCCACCA	pDsRed2-*91R∆NLS2*
91R∆NLS2-F	TGGTGGCGTCCGGAGGAGTGGTCAAGATGGGACCCTTTTC	
MCP-F	ATGGTTGTGGAGCAGGTG	qPCR
MCP-R	TGACGCAGGTGTAATTGGAG	

[a] Sequences of restriction sites are underlined.

To construct the bait plasmid for screening the interacting protein of RGV-63R, the gene that encodes RGV-63R was amplified with primers *63R*-F/R and cloned into pGBKT7 to produce bait plasmid pGBKT7-*63R*. To construct the prey plasmids containing 26 RGV core genes (x represents the genes, such as *91R*, *97R*, *98R*, and *102R*), fragments were amplified from RGV genomic DNA and cloned into pGADT7 to obtain prey plasmids pGADT7-x.

To generate the plasmids for coimmunoprecipitation (co-IP) assays, *RGV-63R* was amplified with primers *63R*-F2/*63R-3Flag*-R and cloned into pcDNA3.1(+) to construct pcDNA3.1-*63R-3Flag*. *RGV-91R* was amplified with primers *91R*-F2/*91R-HA*-R and cloned into pcDNA3.1(+) to construct pcDNA3.1-*91R-HA*.

To analyze the colocalization between RGV-63R and RGV-91R, the *RGV-63R* was amplified with primers *63R*-F/*63R*-R3 and cloned into pEGFP-N3 to construct pEGFP-*63R*. *RGV-91R* was amplified with primers *91R*-F3/*91R*-R2 and cloned into pDsRed2-C1 to produce pDsRed2-*91R*.

The RGV-91R (245aa) sequence and structure domains are as follows. Grey shades indicate nuclear localization signals (NLS1 and NLS2):

MLWEAVTDKPVKLKGLLELLLNNMDSARLVVTSQSVSVVDYQSNMAVTASMPSSVFTSYVYKSD
AECLYAGLPHAALPDLKSFKAKCNVTLRLMGDPECGQYTMKIIIANASHMSTSINMVVDHGKKEA
DRGHPEGAGKSFTLTQQEFNTLCKTFKQGPVNLGVFGGVLVASGGVDGIKVKEVAFGAPDCVTP
HVKLCVHAEKMSRLVKMGPFSAGSLTVCVAQGSVTVSTRGHLGSLTVTLFEG.

To analyze the effect of the nuclear localization signal (NLS) on the subcellular location of RGV-91R, two DNA fragments *91R-N1* and *91R-C1* were amplified from RGV genomic DNA with primers *91R-F3/91R△NLS1-R* and *91R△NLS1*-F/*91R*-R2, respectively. The DNA fragment *91R△NLS1* was amplified via overlap PCR using *91R-N1* and *91R-C1* as templates and cloned into plasmid pDsRed2-C1 to produce pDsRed2-*91R△NLS1*. As described above, primers *91R*-F3/*91R△NLS2*-R and *91R△NLS2*-F/*91R*-R2 were used to construct the plasmid pDsRed2-*91R△NLS2*.

The previously constructed plasmid pMD-*MCP* [30] was used for a standard curve of qPCR.

2.3. Yeast Two Hybrid (Y2H) Assays

The RGV-63R gene was cloned into bait vector pGBKT7 and expressed in the yeast Y2HGold strain and cultured on SD/-Trp plates. Twenty-six RGV core genes (x represents) including *91R*, *97R*, *98R*, and *102R*, which were considered as iridovirus core genes, were cloned into prey vectors pGADT7 and expressed in the yeast Y187 strain, and cultured on SD/-Leu plates. However, no colony formed when yeast Y187 was separately transformed with pGADT7-*73L*/*95R*/*101L*. The interactions between bait protein pGBKT7-63R and 23 prey proteins pGADT7-x were then identified using the yeast two-hybrid (Y2H) system and were performed according to the Matchmaker Gold Yeast Two-Hybrid System User Manual (Clontech, Clontech Laboratories, Inc., Mountain View, CA, USA). Three parallel experiments were performed. Y2HGold/pGBKT7-*63R* and Y187/pGADT7-x were placed in 2×YPDA media and incubated at 30 °C for 24 h. The mating yeast cells were collected by centrifugation, and resuspended with 0.5×YPDA media, and they were then cultured on 2Q (SD/-Trp/-Leu) and 3Q/X (SD/-Trp/-Leu/-His/X-α-Gal) plates for 6 days at 30 °C. As a positive control, yeast mating between Y187/pGADT7-*T* and Y2HGold/pGBKT7-*P53* was carried out in the same way, since T interacts with P53. As a negative control, yeast mating between Y187/pGADT7-*T* and Y2HGold/pGBKT7-*Lam* was carried out in the same way, since T does not interact with Lam. Interactions take place if colonies form on both 2Q and 3Q/X plates. As a positive control, many colonies formed on 2Q and 3Q/X plates, and as a negative control, no colonies formed on 3Q/X.

2.4. Co-IP Assays

To further confirm interaction between RGV-63R and RGV-91R detected by Y2H, co-IP assays were performed. Briefly, HEK293T cells seeded in 10 cm dishes were cotransfected with 5 μg of pcDNA3.1-*91R-HA* and 5 μg of pcDNA3.1-*63R-3Flag*. As a control, 5 μg of pcDNA3.1-*91R-HA* and

5 μg of empty vector pcDNA3.1 were cotransfected in parallel. At 24 h post-transfection (hpt), cells were lysed with a radio immunoprecipitation assay (RIPA) buffer containing protease inhibitor cocktail (Roche), and cell lysates were incubated with Red Anti-HA Affinity Gel (Sigma, Sigma-Aldrich LLC., Santa Clara, CA, USA) overnight at 4 °C. The precipitates were collected by centrifugation, washed with ice-cold phosphate-buffered saline (PBS) three times, eluted with 40 μL of PBS, and finally subjected to Western blot analysis.

2.5. Western Blot Analysis

Western blot analysis was performed as described previously [30]. Protein samples were resolved by SDS-PAGE, followed by electroblotting to polyvinylidene difluoride (PVDF) membrane. The blots were probed with anti-HA mouse monoclonal antibody (1: 1000, Sigma, Sigma-Aldrich LLC., Santa Clara, CA, USA) or anti-Flag rabbit monoclonal antibody (1: 1000, Abbkine, Abbkine Scientific Co., Ltd. Wuhan, China). Peroxidase-conjugated goat anti-mouse or anti-rabbit IgG (H+L) antibody was used as the secondary antibody. The signals were detected with a chemiluminescent horseradish peroxidase (HRP) substrate (Millipore, Millipore Corporation, Billerica, MA, USA).

2.6. Fluorescence Microscopy

Colocalization assays were performed to confirm the interaction between RGV-63R and RGV-91R. EPCs or GSTCs transfected with plasmid pEGFP-*63R* or pDsRed2-*91R* alone or cotransfected with both plasmids were fixed, permeabilized, stained with Hoechst 33342, and observed under a Leica DM IRB fluorescence microscope (objective 100×), as described previously [31].

2.7. Quantitative Analysis of RGV Genomic DNA

To investigate the effects of RGV-63R and RGV-91R on RGV replication in cells from different species, the DNA level of RGV in EPCs or GSTCs expressing one protein alone or coexpressing both proteins were measured by qPCR, respectively. A total of 2×10^5 EPCs or GSTCs grown in 24-well plates were transfected with 0.5 μg of empty vector pcDNA3.1 or cotransfected with 0.25 μg of pcDNA3.1-*63R-3Flag* and 0.25 μg of pcDNA3.1, 0.25 μg of pcDNA3.1-*91R-HA* and 0.25 μg of pcDNA3.1, or 0.25 μg of pcDNA3.1-*63R-3Flag* and 0.25 μg of pcDNA3.1-*91R-HA*. The transfected cells at 24 hpt were harvested and subjected to Western blot analysis. For qPCR, at 24 hpt, the transfected cells were infected with RGV with an MOI of 0.5. The infected cells at 48 h post-infection (hpi) were harvested and washed with PBS three times. The viral genomic DNA was then extracted according to the instructions of TaKaRa MiniBEST Universal Genomic DNA Extraction Kit. Each DNA sample was dissolved in 100 μL of elution buffer, and 1 μL of each DNA sample was used for each qPCR reaction. Specific primers *MCP*-F/*MCP*-R for RGV MCP gene was used for qPCR. qPCR was carried out using Fast SYBR Green Master Mix with the StepOneTM Real-Time PCR System (Applied Biosystems, Foster City, CA, USA), as described previously [32].

3. Results

3.1. Interaction between RGV-63R and RGV-91R was First Screened by Y2H

The interactions between RGV-63R and other 23 proteins from RGV were analyzed by yeast two hybrid. The results of detecting interactions between RGV-63R and 91R, 97R, 98R, or 102R are shown in Figure 1. Only RGV-91R showed interaction with RGV-63R. When the interaction between RGV-63R and RGV-91R was detected, colonies formed on SD/-Trp/-Leu (2Q) and SD/-Trp/-Leu/-His/X-α-Gal (3Q/X) plates. When detecting the interactions between RGV-63R and 97R, 98R, or 102R, colonies formed on 2Q plates, but no colony formed on 3Q/X plates.

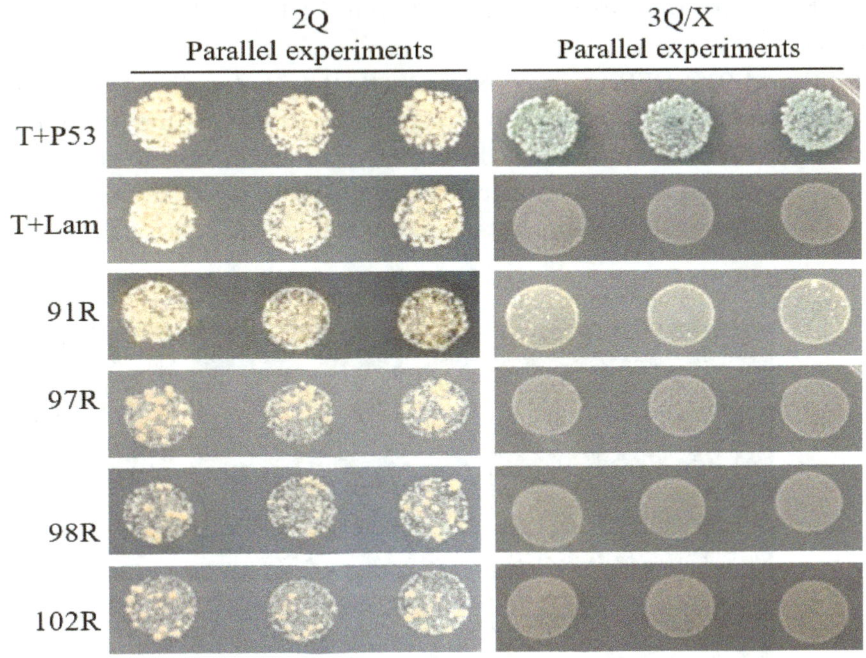

Figure 1. The formed yeast colonies during testing interactions between RGV-63R and four proteins by Y2H. Three parallel experiments were performed. 2Q: SD/-Trp/-Leu. 3Q/X: SD/-Trp/-Leu/-His/X-α-Gal. 91R, 97R, 98R, and 102R: RGV proteins, which are encoded by iridovirus core genes, respectively. T+P53: the positive control; T+Lam: the negative control.

3.2. Confirmation of Interaction between RGV-63R and RGV-91R by Co-IP Followed by Western Blotting

Co-IP assays were used to confirm the interaction between RGV-63R and RGV-91R. The proteins 63R-3Flag and 91R-HA were co-expressed and anti-HA antibody affinity gel was used to precipitate the protein complexes. The result of co-IP followed by western blotting is shown in Figure 2; 63R-3Flag and 91R-HA were detected in cell lysates and immunoprecipitated (IP) protein complexes from HEK293T cells cotransfected with pcDNA3.1-*63R-3Flag* and pcDNA3.1-*91R-HA*. However, only 91R-HA was detected in cell lysates and immunoprecipitated protein complexes from pcDNA3.1 and pcDNA3.1-*91R-HA* cotransfected cells. This further confirmed that RGV-63R interacts with RGV-91R.

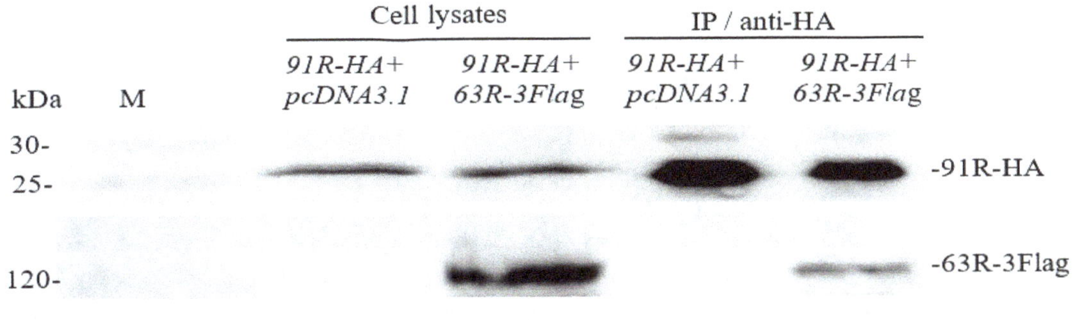

Figure 2. Western blot analysis for sample of testing interaction between RGV-63R and RGV-91R by co-IP. Cell lysates from HEK293T cells cotransfected with indicated plasmids (*91R-HA+ pcDNA3.1, 91R-HA+ 63R-3Flag*) and IP (immunoprecipitated) protein complexes with indicated plasmids (*91R-HA+ pcDNA3.1, 91R-HA+ 63R-3Flag*) are subjected to Western blot analysis using anti-HA and anti-Flag. Cells lysates and IP showed the bands of 91R-HA and 63R-3Flag. M: protein molecular mass marker.

3.3. Localization of RGV-63R and RGV-91R in Cells of Different Species

Fluorescence microscopy was carried out. The protein 63R-EGFP expressed alone localized in the cytoplasm of GSTCs or EPCs. The protein 91R-RFP expressed alone localized in the nucleus of GSTCs or EPCs. Obviously, the localizations of the same protein in GSTCs or EPCs from different species were consistent (Figure 3A).

When 63R-EGFP and 91R-RFP were co-expressed, they colocalized in the nucleus of GSTCs. They consistently expressed in EPCs, as shown in Figure 3B. These results not only further confirmed the interaction between RGV-63R and RGV-91R, which alters RGV-63R localization, but also exactly showed the same phenomenon that was observed in cells from two different species. These results indicate that the two proteins of RGV can play the same role in the cells of different species.

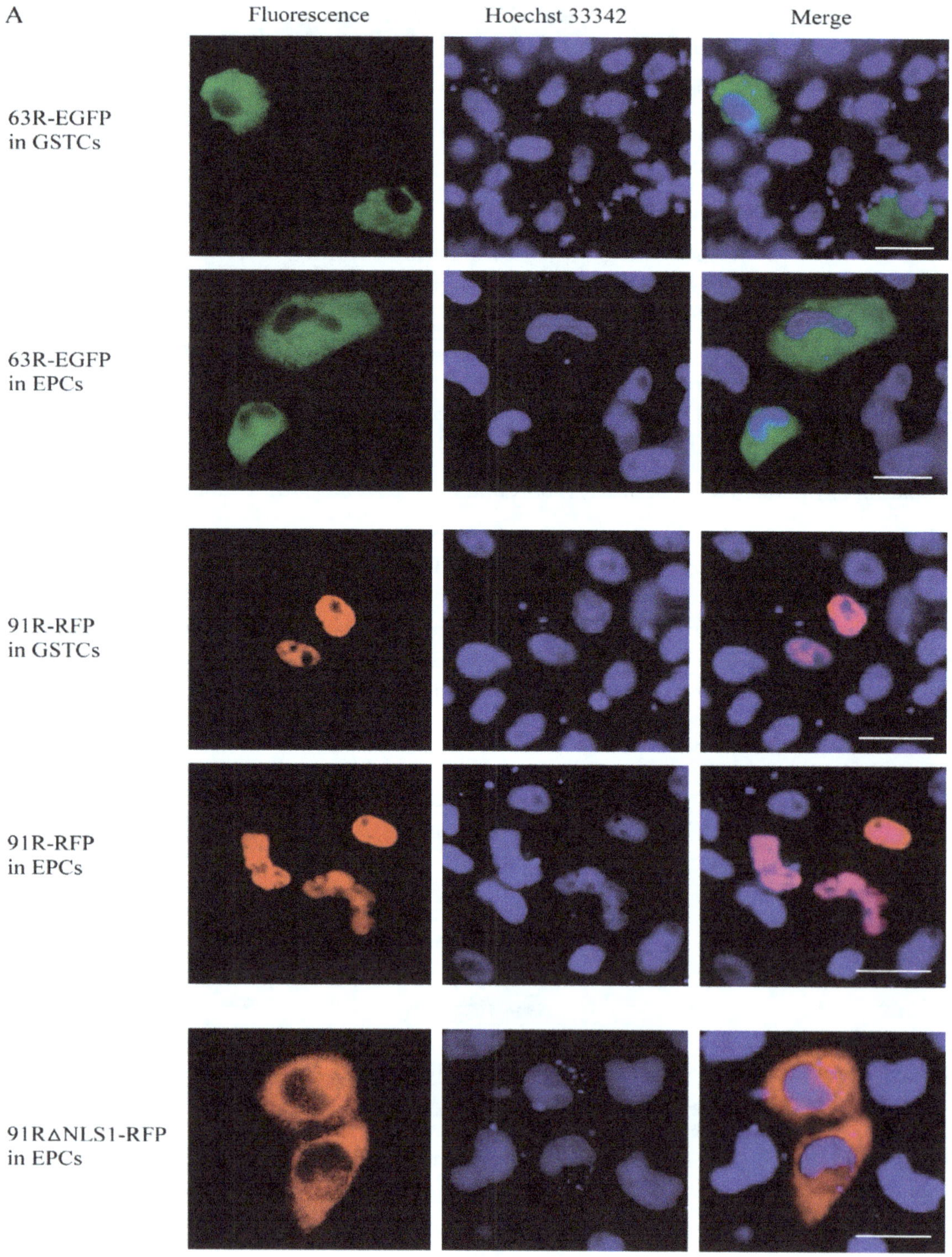

Figure 3. *Cont.*

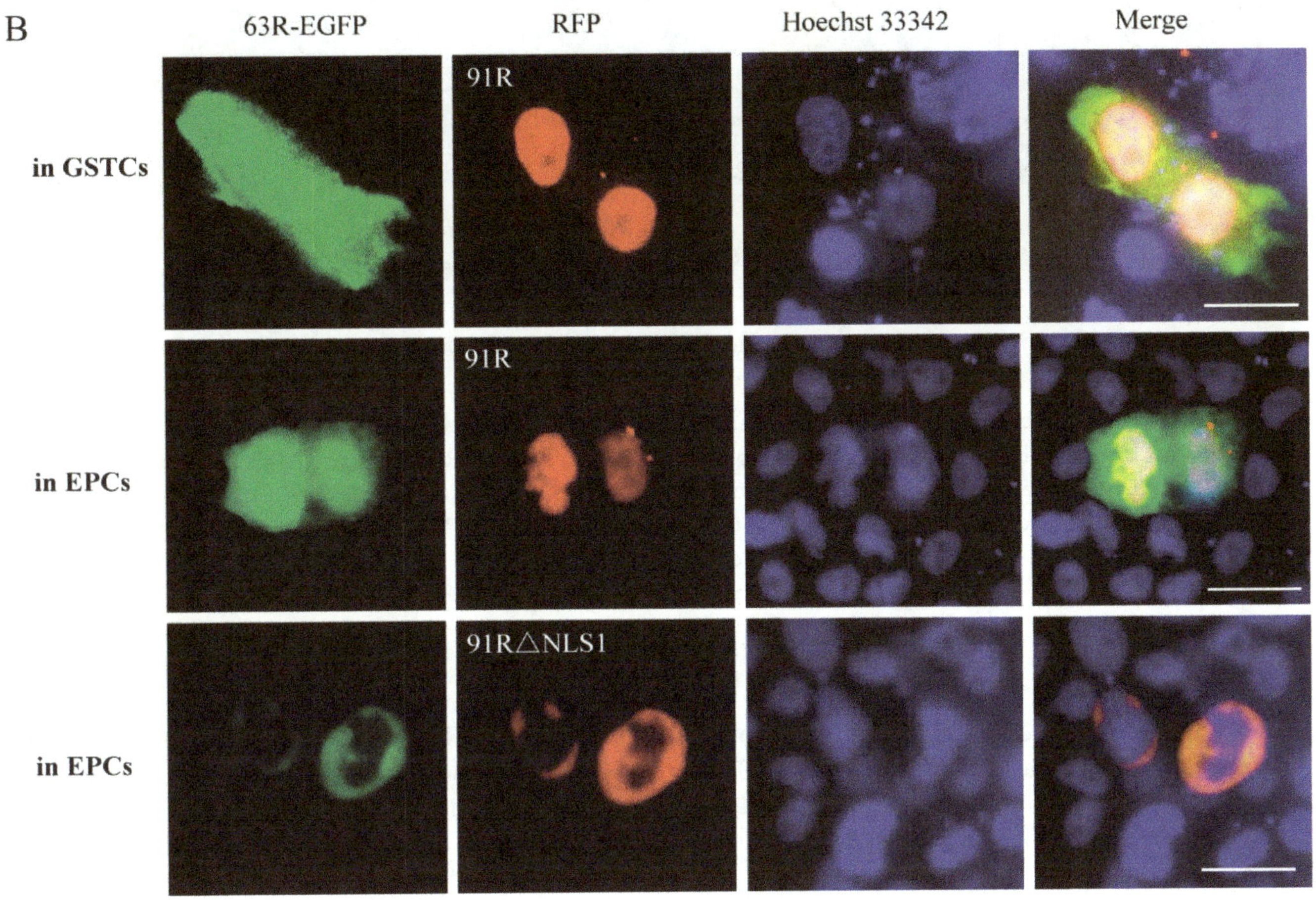

Figure 3. Fluorescence micrographs of cells expressing a single protein or co-expressing two proteins. (**A**) Expressing proteins 63R-EGFP, 91R-RFP, and 91R△NLS1-RFP alone in giant salamander thymus cells (GSTCs) or *Epithelioma papulosum cyprinid* cells (EPCs), respectively. (**B**) Co-expressing two proteins, 63R-EGFP + 91R-RFP and 63R-EGFP + 91R△NLS1-RFP, in GSTCs or EPCs, respectively. 63R-EGFP (green), 91R-RFP (red), 91R△NLS1-RFP (red), nucleus (blue), and colocalization (yellow). Scale bar: 10 μm.

RGV-91R was predicted to contain two NLSs, namely, NLS1 (60~88 aa) and NLS2 (176~207 aa). When the protein 63R-EGFP was co-expressed with NLS deletion proteins (91R△NLS1-RFP or 91R△NLS2-RFP), they colocalized in the cytoplasm rather than in the nucleus (Figure 3B). These revealed that the NLS domain of RGV-91R is essential to transporting RGV-63R into the nucleus.

3.4. *Detection of RGV-63R and RGV-91R Expressions by Western Blotting*

The plasmids pcDNA3.1-*63R-3Flag* and pcDNA3.1-*91R-HA* were mixed when they were cotransfected into cells. The transfection efficiency was about 10%. There is about 7.6% cells cotransfected with plasmids pcDNA3.1-*63R-3Flag* and pcDNA3.1-*91R-HA*, and about 2.4% cells transfected with alone plasmid by counting the number of transfected cells using immunofluorescence.

The expressions of RGV-63R and RGV-91R in GSTCs or EPCs at 24 hpt were detected by Western blot analysis. The results were shown in Figure 4. RGV-63R and RGV-91R were detected in GSTCs or EPCs expressing one protein alone or co-expressing two proteins, indicating that RGV-63R and RGV-91R were both successfully expressed.

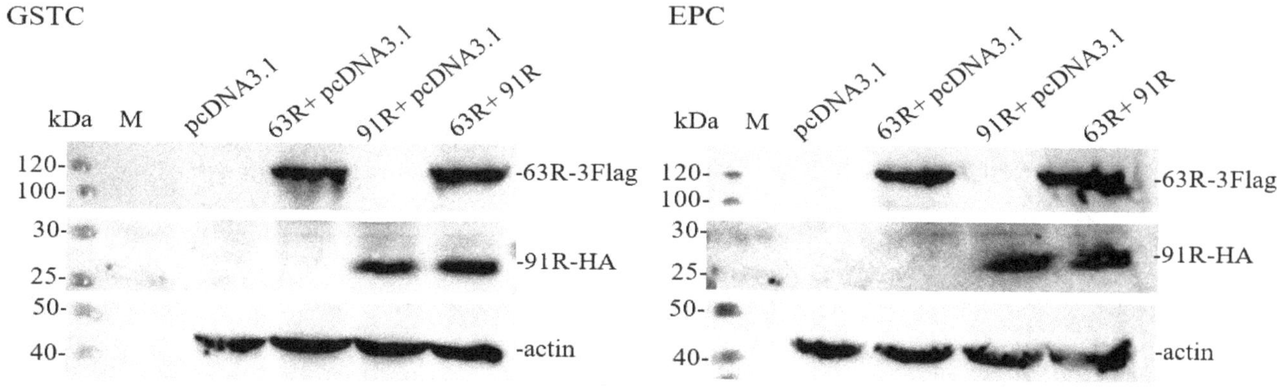

Figure 4. Western blot analysis of protein samples from cells transfected with plasmids for 24 h using anti-Flag and anti-HA. GSTC: in GSTCs expressing protein RGV-63R (63R + pcDNA3.1) or RGV-91R (91R + pcDNA3.1) alone or co-expressing two proteins RGV-63R and RGV-91R (63R + 91R), respectively. EPC: in EPCs expressing protein RGV-63R (63R + pcDNA3.1) or RGV-91R (91R + pcDNA3.1) alone or co-expressing two proteins RGV-63R and RGV-91R (63R + 91R), respectively. The empty vector pcDNA3.1 was used as a control. M: protein molecular mass marker.

3.5. RGV-63R and RGV-91R Promote RGV Genome Replication

The effect of RGV-63R or RGV-91R on RGV genome replication was analyzed by expressing one protein alone or co-expressing two proteins in GSTCs and EPCs, respectively. qPCR analysis showed that RGV genomic copies in both GSTCs and EPCs expressing RGV-63R or RGV-91R alone or co-expressing the two proteins in EPCs were significantly higher than those in the control (Figure 5). A similar promoting activity is indicated when the two proteins are expressed alone in cells from different species.

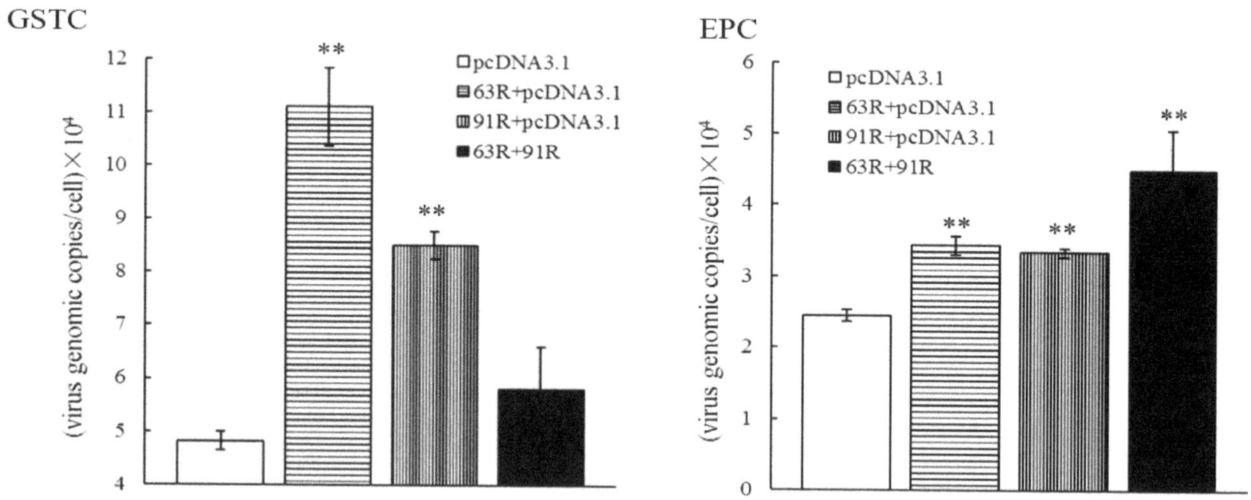

Figure 5. qPCR analysis of RGV genomic copies in GSTCs or EPCs. GSTC: in GSTCs expressing protein RGV-63R (63R + pcDNA3.1) or RGV-91R (91R + pcDNA3.1) alone or co-expressing two proteins RGV-63R and RGV-91R (63R + 91R), respectively. EPC: in EPCs expressing protein RGV-63R (63R + pcDNA3.1) or RGV-91R (91R + pcDNA3.1) alone or co-expressing two proteins RGV-63R and RGV-91R (63R + 91R), respectively. The empty vector pcDNA3.1 was used as a control. The transfected cells were infected with RGV, and RGV genome DNA was extracted from the cells at 48 hpi and quantified by qPCR. Each data point represents the average value of three independent infections. Error bars indicate standard deviations. ** represents $p < 0.01$.

The RGV genomic copies in EPCs co-expressing the two proteins were higher than those expressed alone. However, the genomic copies were lower when the two proteins were co-expressed rather than expressed alone in GSTCs (Figure 5), which was in line with expectations because of the long-term preservation of RGV in EPCs, not in GSTCs. This suggests that ranaviruses cross-species transmission were closely related with their adaptability to the host, but how it is affected is still an important unresolved area that needs to be studied.

4. Discussion

In this study, the interactions of RGV-63R with other iridoviral core proteins from RGV were tested by a Y2H assay, showing that RGV-63R interacts with RGV-91R. However, only white colonies formed on both SD/-Trp/-Leu and SD/-Trp/-Leu/-His/X-α-Gal plates while detecting the interaction between RGV-63R and RGV-91R by Y2H. It is possible that the interaction between RGV-63R and RGV-91R in yeast is weak and only activates the reporter gene *HIS3*, but does not activate the reporter gene *MEL-1* which encodes α-galactosidase hydrolyzing chromogenic substrate X-α-Gal. Subsequently, the interaction between the two proteins was further confirmed by co-IP.

In this study, 63R-EGFP protein only localized in cytoplasm of GSTCs or EPCs transfected with pEGFP-*63R*. When 63R-EGFP and 91R-RFP were co-expressed in GSTCs or EPCs, 63R-EGFP localized in both cytoplasm and the nucleus and colocalized with 91R-RFP in the nucleus. This not only further confirmed the interaction between the two proteins, but also implied that the nuclear import of RGV-63R mediated by RGV-91R might be very important for the virus genome synthesis within the nucleus. Nucleus import of macromolecules larger than 60 kDa is an active, energy-dependent process mediated by sequence-specific motif and NLSs [33]. RGV-63R was predicted to encode DNA polymerase with a predicted molecular mass of 114 kDa. This implied that the nucleus import of RGV-63R might be mediated by NLSs. Fluorescent microscopy showed that the 91R△NLS-RFP protein localized in cytoplasm and did not mediate the nuclear import of RGV-63R, suggesting that the nuclear import of RGV-63R was dependent on the NLSs of RGV-91R.

Iridoviral DNA polymerase has been considered to be involved in virus genome replication [19]. Iridoviral PCNA was also considered to be involved in nucleic acid synthesis [34]. RGV-91R was identified as a late protein, and its homologue in a member of genus *Ranavirus* was identified as a structural protein [35,36]. In this study, the effects of RGV-63R, RGV-91R, and the interaction between the two proteins on RGV replication in GSTCs or EPCs were further analyzed. The results showed that overexpression of RGV-63R or RGV-91R in both GSTCs and EPCs significantly increased RGV genome copies. This indicated that RGV-63R and RGV-91R both promote RGV replication.

An interesting phenomenon is that the number of RGV genomic copies was lower in GSTCs when the two proteins were coexpressed compared to proteins expressed alone, but it was higher in EPCs when the two proteins were coexpressed. It has been reported that some host proteins inhibit ranavirus replication [37]. It has also been reported that antiviral activities of some cellular proteins are species-specific [38,39]. Whether co-expression of the two proteins in GSTCs or EPCs is associated with these similar host proteins needs further research. It also may result from the preparation of virus stocks. Compared with virus replication in EPCs, the virus was always multiplied in GSTCs, implying that the RGV genome replication in GSTCs might be more efficient. However, more experiments are required to confirm and explain this mechanism.

In conclusion, this report is the first confirmation and description of ranaviral protein–protein interactions that promote virus genome replication in cells from different species, which provides novel molecular evidence and insights for ranavirus cross-species infection.

Author Contributions: Q.-Y.Z. conceived and designed the study. X.-T.Z. performed experiments. Q.-Y.Z. and X.-T.Z. analyzed the data and wrote the manuscript. Both authors have read and approved the final manuscript.

References

1. Zhang, Q.Y.; Gui, J.F. Diversity, evolutionary contribution and ecological roles of aquatic viruses. *Sci. China Life Sci.* **2018**, *61*, 1486–1502. [CrossRef]
2. Chinchar, V.G.; Hick, P.; Ince, I.A.; Jancovich, J.K.; Marschang, R.; Qin, Q.W.; Subramaniam, K.; Waltzek, T.B.; Whittington, R.; Williams, T.; Zhang, Q.Y. ICTV Report Consortium. ICTV Virus Taxonomy Profile: *Iridoviridae. J. Gen. Virol.* **2017**, *98*, 890–891. [PubMed]
3. Zhang, Q.Y.; Gui, J.F. Virus genomes and virus-host interactions in aquaculture animals. *Sci. China Life Sci.* **2015**, *58*, 156–169. [CrossRef] [PubMed]
4. Zhang, Q.Y.; Gui, J.F. *Atlas of Aquatic Viruses and Viral Diseases*; Science Press: Beijing, China, 2012.
5. Gui, L.; Zhang, Q.Y. A brief review on aquatic animal virology researches in China. *J. Fisheries China* **2019**, *43*, 168–187.
6. Stohr, A.C.; Lopez-Bueno, A.; Blahak, S.; Caeiro, M.F.; Rosa, G.M.; Alves de Matos, A.P.; Martel, A.; Alejo, A.; Marschang, R.E. Phylogeny and differentiation of reptilian and amphibian ranaviruses detected in Europe. *PLoS ONE* **2015**, *10*, e0118633. [CrossRef] [PubMed]
7. Ke, F.; Gui, J.F.; Chen, Z.Y.; Li, T.; Lei, C.K.; Wang, Z.H.; Zhang, Q.Y. Divergent transcriptomic responses underlying the ranaviruses-amphibian interaction processes on interspecies infection of Chinese giant salamander. *BMC Genom.* **2018**, *19*, 211. [CrossRef]
8. Robert, J.; Jancovich, J.K. Recombinant ranaviruses for studying evolution of host-pathogen interactions in ectothermic vertebrates. *Viruses* **2016**, *8*, 187. [CrossRef]
9. Ke, F.; Zhang, Q.Y. Aquatic animal viruses mediated immune evasion in their host. *Fish. Shellfish Immunol.* **2019**, *86*, 1096–1105. [CrossRef]
10. Gui, L.; Chinchar, V.G.; Zhang, Q.Y. Molecular basis of pathogenesis of emerging viruses infecting aquatic animals. *Aquaculture Fisheries* **2018**, *3*, 1–5. [CrossRef]
11. Liu, J.; Yu, C.; Gui, J.F.; Pang, D.W.; Zhang, Q.Y. Real-time dissecting the entry and intracellular dynamics of single reovirus particle. *Front. Microbiol.* **2018**, *9*, 2797. [CrossRef] [PubMed]
12. Corthell, J.T. *Basic Molecular Protocols in Neuroscience: Tips, Tricks, and Pitfalls*; Elsevier Science: Amsterdam, The Netherland, 2014.
13. Striebinger, H.; Koegl, M.; Bailer, S.M. A high-throughput yeast two-hybrid protocol to determine virus-host protein interactions. In *Virus-Host Interactions. Methods in Molecular Biology (Methods and Protocols)*; Bailer, S., Lieber, D., Eds.; Humana Press: Totowa, NJ, USA, 2013; vol 1064, pp. 1–15.
14. Ozsahin, E.; van Oers, M.M.; Nalcacioglu, R.; Demirbag, Z. Protein–protein interactions among the structural proteins of *Chilo* iridescent virus. *J. Gen. Virol.* **2018**, *99*, 851–859. [CrossRef] [PubMed]
15. Halperin, S.O.; Tou, C.J.; Wong, E.B.; Modavi, C.; Schaffer, D.V.; Dueber, J.E. CRISPR-guided DNA polymerases enable diversification of all nucleotides in a tunable window. *Nature* **2018**, *560*, 248–252. [CrossRef]
16. Gardner, A.F.; Kelman, Z. DNA polymerases in biotechnology. *Front. Microbiol.* **2014**, *5*, 659. [CrossRef]
17. Czarnecki, M.W.; Traktman, P. The vaccinia virus DNA polymerase and its processivity factor. *Virus Res.* **2017**, *234*, 193–206. [CrossRef] [PubMed]
18. Zarrouk, K.; Piret, J.; Boivin, G. Herpesvirus DNA polymerases: structures, functions and inhibitors. *Virus Res.* **2017**, *234*, 177–192. [CrossRef]
19. Chinchar, V.G.; Waltzek, T.B. Ranaviruses: not just for frogs. *PLoS Pathog.* **2014**, *10*, e1003850. [CrossRef]
20. Chinchar, V.G.; Waltzek, T.B.; Subramaniam, K. Ranaviruses and other members of the family *Iridoviridae*: Their place in the virosphere. *Virology* **2017**, *511*, 259–271. [CrossRef] [PubMed]
21. Maga, G.; Villani, G.; Crespan, E.; Wimmer, U.; Ferrari, E.; Bertocci, B.; Hubscher, U. 8-oxo-guanine bypass by human DNA polymerases in the presence of auxiliary proteins. *Nature* **2007**, *447*, 606–608. [CrossRef] [PubMed]
22. Dieckman, L.M.; Freudenthal, B.D.; Washington, M.T. PCNA structure and function: insights from structures of PCNA complexes and post-translationally modified PCNA. In *The Eukaryotic Replisome: A Guide to Protein Structure and Function*; MacNeill, S., Ed.; Springer: Dordrecht, The Netherlands, 2012; pp. 281–299.
23. Baple, E.L.; Chambers, H.; Cross, H.E.; Fawcett, H.; Nakazawa, Y.; Chioza, B.A.; Harlalka, G.V.; Mansour, S.; Sreekantan-Nair, A.; Patton, M.A.; et al. Hypomorphic PCNA mutation underlies a human DNA repair disorder. *J. Clin. Invest.* **2014**, *124*, 3137–3146. [CrossRef] [PubMed]

24. Eaton, H.E.; Metcalf, J.; Penny, E.; Tcherepanov, V.; Upton, C.; Brunetti, C.R. Comparative genomic analysis of the family *Iridoviridae*: re-annotating and defining the core set of iridovirus genes. *Virol. J.* **2007**, *4*, 11. [CrossRef]
25. Zhao, Z.; Ke, F.; Gui, J.F.; Zhang, Q.Y. Characterization of an early gene encoding for dUTPase in *Rana grylio* virus. *Virus Res.* **2007**, *123*, 128–137. [CrossRef] [PubMed]
26. Ke, F.; Zhao, Z.; Zhang, Q.Y. Cloning, expression and subcellular distribution of a *Rana grylio* virus late gene encoding ERV1 homologue. *Mol. Biol. Rep.* **2009**, *36*, 1651–1659. [CrossRef]
27. Lei, X.Y.; Ou, T.; Zhu, R.L.; Zhang, Q.Y. Sequencing and analysis of the complete genome of *Rana grylio* virus (RGV). *Arch. Virol.* **2012a**, *157*, 1559–1564. [CrossRef] [PubMed]
28. Lei, X.Y.; Ou, T.; Zhang, Q.Y. *Rana. grylio* virus (RGV) 50L is associated with viral matrix and exhibited two distribution patterns. *PLoS ONE* **2012b**, *7*, e43033. [CrossRef] [PubMed]
29. Yuan, J.D.; Chen, Z.Y.; Huang, X.; Gao, X.C.; Zhang, Q.Y. Establishment of three cell lines from Chinese giant salamander and their sensitivities to the wild-type and recombinant ranavirus. *Vet. Res.* **2015**, *46*, 58. [CrossRef]
30. Zeng, X.T.; Gao, X.C.; Zhang, Q.Y. *Rana grylio* virus *43R* encodes an envelope protein involved in virus entry. *Virus Genes* **2018**, *54*, 779–791. [CrossRef]
31. Wang, J.; Gui, L.; Chen, Z.Y.; Zhang, Q.Y. Mutations in the C-terminal region affect subcellular localization of crucian carp herpesvirus (CaHV) GPCR. *Virus Genes* **2016**, *52*, 484–494. [CrossRef]
32. Chen, Z.Y.; Li, T.; Gao, X.C.; Wang, C.F.; Zhang, Q.Y. Protective immunity induced by DNA vaccination against ranavirus infection in chinese giant salamander *Andrias davidianus*. *Viruses* **2018**, *10*, 52. [CrossRef]
33. Peters, R. Translocation through the nuclear pore: Kaps pave the way. *Bioessays* **2009**, *31*, 466–477. [CrossRef]
34. Chinchar, V.G.; Yu, K.H.; Jancovich, J.K. The molecular biology of frog virus 3 and other iridoviruses infecting cold-blooded vertebrates. *Viruses* **2011**, *3*, 1959–1985. [CrossRef]
35. Sun, W. Molecular cloning and characterization of two novel viral genes, *3β-HSD* and *PCNA* from *Rana grylio* virus. Ph.D. Thesis, University of Chinese Academy of Sciences, Wuhan, China, 2007.
36. Li, W.; Zhang, X.; Weng, S.P.; Zhao, G.X.; He, J.G.; Dong, C.F. Virion-associated viral proteins of a Chinese giant salamander (*Andrias davidianus*) iridovirus (genus *Ranavirus*) and functional study of the major capsid protein (MCP). *Vet. Microbiol.* **2014**, *172*, 129–139. [CrossRef] [PubMed]
37. Wei, J.G.; Guo, M.L.; Ji, H.S.; Yan, Y.; Ouyang, Z.L.; Huang, X.H.; Hang, Y.H.; Qin, Q.W. Grouper translationally controlled tumor protein prevents cell death and inhibits the replication of Singapore grouper iridovirus (SGIV). *Fish. Shellfish Immunol.* **2012**, *33*, 916–925. [CrossRef] [PubMed]
38. Lodermeyer, V.; Ssebyatika, G.; Passos, V.; Ponnurangam, A.; Malassa, A.; Ewald, E.; Stürzel, C.M.; Kirchhoff, F.; Rotger, M.; Falk, C.S.; et al. The antiviral activity of the cellular glycoprotein LGALS3BP/90K is species specific. *J. Virol.* **2018**, *92*. [CrossRef] [PubMed]
39. Eunhye, K.; Hyunjhung, J.; Joohee, K.; Joohee, K.; Unjoo, P.; Seunghyun, J.; Areum, K.; Sinae, K.; Tan, T.N.; Yongsun, K.; et al. Species specific antiviral activity of porcine interferon-α8 (IFNα8). *Immune Netw.* **2017**, *17*, 424–436.

6

Artemia spp., a Susceptible Host and Vector for Lymphocystis Disease Virus

Estefania J. Valverde, Alejandro M. Labella, Juan J. Borrego and Dolores Castro *

Departamento de Microbiología, Facultad de Ciencias, Universidad de Málaga, 29017 Málaga, Spain; ejv@uma.es (E.J.V.); amlabella@uma.es (A.M.L.); jjborrego@uma.es (J.J.B.)
* Correspondence: dcastro@uma.es

Abstract: Different developmental stages of *Artemia* spp. (metanauplii, juveniles and adults) were bath-challenged with two isolates of the Lymphocystis disease virus (LCDV), namely, LCDV SA25 (belonging to the species *Lymphocystis disease virus 3*) and ATCC VR-342 (an unclassified member of the genus *Lymphocystivirus*). Viral quantification and gene expression were analyzed by qPCR at different times post-inoculation (pi). In addition, infectious titres were determined at 8 dpi by integrated cell culture (ICC)-RT-PCR, an assay that detects viral mRNA in inoculated cell cultures. In LCDV-challenged *Artemia,* the viral load increased by 2–3 orders of magnitude (depending on developmental stage and viral isolate) during the first 8–12 dpi, with viral titres up to 2.3×10^2 Most Probable Number of Infectious Units (MPNIU)/mg. Viral transcripts were detected in the infected *Artemia*, relative expression values showed a similar temporal evolution in the different experimental groups. Moreover, gilthead seabream (*Sparus aurata*) fingerlings were challenged by feeding on LCDV-infected metanauplii. Although no Lymphocystis symptoms were observed in the fish, the number of viral DNA copies was significantly higher at the end of the experimental trial and major capsid protein (*mcp*) gene expression was consistently detected. The results obtained support that LCDV infects *Artemia* spp., establishing an asymptomatic productive infection at least under the experimental conditions tested, and that the infected metanauplii are a vector for LCDV transmission to gilthead seabream.

Keywords: Lymphocystis disease virus; *Artemia* spp.; viral infection; *Sparus aurata*; viral transmission

1. Introduction

The family *Iridoviridae* comprises two subfamilies and six genera [1]. Three of them, *Lymphocystivirus*, *Megalocytivirus*, and *Ranavirus* (subfamily *Alphairidovirinae*), infect ectothermic vertebrates (amphibians, reptiles and bony fish), whereas the hosts for the other three genera, *Iridovirus*, *Chloriridovirus*, and *Decapodiridovirus* (subfamily *Betairidovirinae*), are invertebrates (primarily insects and crustaceans) [1–3].

Members of the genus *Lymphocystivirus*, collectively named as Lymphocystis disease virus (LCDV), are the causative agents of the Lymphocystis disease (LCD) affecting a wide variety of freshwater, brackish, and marine fish species [4]. The characteristic lesions of LCD are small pearl-like nodules on fish skin and fins, grouped in raspberry-like clusters of tumorous appearance [5,6]. Although this disease is rarely fatal, affected fish cannot be commercialized, provoking important economic losses [7]. LCD is the main viral infection reported in gilthead seabream (*Sparus aurata*) aquaculture [8] and it is caused by *Lymphocystis disease virus 3* (LCDV-Sa). Two more species have been recognized in the genus *Lymphocystivirus*—*Lymphocystis disease virus 1* (LCDV1) and *Lymphocystis disease virus 2* (LCDV-C), that infect European flounder (*Platichthys flesus*) and Japanese flounder (*Paralichthys olivaceus*), respectively—and a number of isolates have been obtained from other LC-diseased fish species, but their taxonomic position is unclear [1].

It is assumed that LCDV transmission occurs through the skin and gills of fish by direct contact or by waterborne exposure [9,10]. However, viral transmission via the alimentary canal has been demonstrated for gilthead seabream larvae feeding on LCDV-positive rotifers [11].

The brine shrimp *Artemia* (Crustacea, Branchiopoda, Anostraca) is an aquatic crustacean frequently used for the feeding of postlarvae fish in aquaculture practice [12,13]. Several authors have considered *Artemia* spp. nauplii as a possible source for the introduction of microorganisms into the rearing systems, including bacteria, viruses, and protozoa [14–19], and in most studies, a mechanical carrier stage, in which the pathogen does not multiplicate into the brine shrimp but accumulates in its alimentary canal, has been proposed [16,19,20].

LCDV has been detected by PCR-based methods in *Artemia* cysts and nauplii/metanauplii collected in gilthead seabream hatcheries [21,22]. In a previous study, we demonstrated that the nauplii become easily contaminated with LCDV by immersion challenge. Furthermore, infectious LCDV persists along the *Artemia* life cycle, with viral genome and antigens detected not only in the gut of adult specimens, which could be related to a viral bioaccumulation process, but also in the ovisac in females [21]. These findings suggest that *Artemia* might act as a reservoir of LCDV and could support viral replication.

The aim of the present study was to investigate the susceptibility of different developmental stages of *Artemia* spp. to LCDV and to elucidate its role as a vector for LCDV transmission to gilthead seabream.

2. Materials and Methods

2.1. Brine Shrimp Culture

Artemia spp. cysts (Artemia AF, INVE Aquaculture Inc., Salt Lake City, UT, USA) were decapsulated using a mixture of sodium hypochlorite (0.5 g active chlorine per gram of cysts) and sodium hydroxide (0.15 g/g cysts), following a standard procedure [23]. Residual hypochlorite was neutralized with sodium thiosulfate (0.1%, *w/v*, 5 min). Decapsulated cysts were hatched in sterile seawater (33 g/L salinity) at 26 °C [12]. After a 48 h incubation, hatched instar II nauplii were separated from the unhatched and empty cysts and transferred to aquaria with fresh sterile seawater. Nauplii were reared to the adult stage at 26 °C, with continuous aeration and a 24 h photoperiod, and fed with a commercial phytoplankton-based food (Mikrozell-Hobby, Dohse Aquaristik GmbH, Grafschaft-Gelsdorf, Germany). Different developmental stages (nauplii, metanauplii, juveniles and adults) were taken from this stock at 4, 8, 14, and 21 d post-hatching (dph), respectively, and used in the experimental infections described below.

2.2. LCDV Infection in Brine Shrimp

The infectivity of LCDV to different developmental stages of *Artemia* (metanauplii, juveniles and adults) was tested by immersion challenge, using an inoculum of 10^2 $TCID_{50}$/mL during 24 h as specified by Cano et al. [21]. After the challenge, animals were filtered through a synthetic net, washed three times for 5 min each in sterile seawater, transferred to aquaria with fresh sterile seawater, and maintained as specified above.

Two LCDV isolates were used for the challenges—LCDV SA25 from gilthead seabream (belonging to genotype VII and identified as LCDV-Sa) [24,25], and LCDV strain Leetown NFH (ATCC VR-342; genotype VIII). Brine shrimps at the same developmental stages inoculated with Leibovitz's L-15 medium (Gibco, Life Technologies, Carlsbad, CA, USA) were used as control groups.

Pooled samples of brine shrimp, approximately 100 mg in weight, were collected from each experimental group at several times post-inoculation (pi) (1, 3, 5, 8, 12, 15, and 23 dpi). The animals were washed with sterile seawater as specified above, gently dried on sterile filter paper, and frozen in liquid nitrogen. Samples were ground in liquid nitrogen using a Mixer Mill MM400 (Retsch GmbH, Haan, Germany), and subsequently used for both nucleic acid extraction and virological analysis.

2.3. *Gilthead Seabream Challenge with LCDV-Infected Artemia*

Gilthead seabream fingerlings (0.5–1 g) were obtained from a research marine aquaculture facility with no record of LCD. Prior to the experiment, 10 fish were randomly collected and analyzed by real-time PCR (qPCR) [22] to ensure that they were LCVD-free. The fish were divided into two groups (50 individuals per group) and stocked at a density of 2 g/L in aquaria with filtered seawater. Fish were maintained at 20–22 °C and a 12-h photoperiod, and fed with commercial pellets (Gemma PG 0.8, Skreeting, Burgos, Spain) at a feeding rate of approximately 5% fish body weight per day.

Artemia nauplii were inoculated by immersion with LCDV isolate SA25 or Leibovitz's L-15 medium as previously specified. At 8 dpi, metanauplii were washed with sterile seawater and used to feed gilthead seabream fingerlings. The presence of LCDV in brine shrimp (two samples of 100 mg per experimental group) was determined by qPCR following the procedure described in Section 2.6.

For the oral challenge, fingerlings were fed once with metanauplii that had been inoculated with LCDV or L-15 medium (challenged and control groups, respectively) at a concentration of 0.2 g/L. The following day, the commercial diet was resumed, and fish were maintained at the conditions indicated above for 30 d. Fingerlings were euthanized by anaesthetic overdose (150 mg/mL MS-222, Sigma-Aldrich, St. Louis, MO, USA). All procedures were carried out following the European Union guidelines for the protection of animals used for scientific purposes (Directive 2010/63/UE).

Seven gilthead seabream fingerlings from both the oral challenged and control groups were randomly sampled at 7, 12, and 24 dpi. Samples, consisting of the caudal part of the body (approximately the posterior one third of the fish body), were homogenized in L-15 medium (10%, *w*/*v*) [26] and used for nucleic acid extraction.

2.4. *Virological Analysis*

A total of 50 mg of brine shrimp tissue powder was suspended in 1 mL of Leibovitz's L-15 medium supplemented with 2% L-glutamine, 1% penicillin-streptomycin and 2% foetal bovine serum, and clarified by centrifugation (10,000× *g* for 5 min at 4 °C) (Rotina 38R centrifuge, Hettich, Kirchlengern, Germany). These homogenates were used to inoculate SAF-1 cells [27], BF-2 cells for homogenates from LCDV ATCC VR-342 infected animals, or were kept at −20 °C until used for virus titration. Cell cultures were maintained at 20 °C until the appearance of cytopathic effects (CPE) (up to 14 dpi). Infectious titres were determined in SAF-1 or BF-2 cells using the ICC-RT-PCR assay described by Valverde et al. [25]. Briefly, cells were inoculated in triplicate with ten-fold serial dilutions of the homogenates and harvested at 5 dpi for total RNA extraction using a commercial kit. After DNase I treatment, one step RT-PCR was performed using primers targeting the major capsid protein (*mcp*) gene. Amplification products were denatured and detected by blot-hybridization using a specific DNA probe. Viral titre, expressed as MPNIU/mL, was estimated using an MPN table with a confidence level of 95%.

2.5. *DNA and RNA Extraction and cDNA Synthesis*

Total DNA and RNA were extracted from 20 mg of tissue powder (brine shrimp samples) or 200 μL of homogenate (fish samples) using the Illustra triplePrep Kit (GE Healthcare, Chicago, IL, USA), following the manufacturer's instructions. Total RNA was treated with RNase-free DNase I (Sigma-Aldrich) for 30 min at 37 °C. RNA purity and quantity were determined using a NanoDrop 1000 (Thermo Scientific, West Palm Beach, FL, USA). After DNase treatment, total RNA was used in the qPCR reaction in order to control for the absence of viral genomic DNA. First-strand DNA synthesis was carried out with 1 μg of total RNA and random hexamer primers using the Transcriptor First Strand cDNA Synthesis Kit (Roche Life Science, Indianapolis, IN, USA). DNA and cDNA were stored at −20 °C until used as template for qPCR.

2.6. LCDV DNA Quantification and Gene Expression

Viral DNA quantification was carried out by qPCR using the methodology described by Valverde et al. [22]. Viral loads were expressed as copies of viral DNA per milligram of tissue.

The *mcp* gene expression was analyzed as an indicator of viral productive infection. Relative quantification of *mcp* gene expression was carried out by RT-qPCR, following the protocol mentioned above but using 20 μL final volume reactions and cDNA generated from 200 ng of the original RNA template.

For LCDV-challenged *Artemia*, relative viral gene expression values were calculated using the comparative delta-Ct method with *Artemia actin* expression used for normalization. Primers for *Artemia actin* gene detection by qPCR (Art-actin-F: 5′-GGTCGTGACTTGACGGACTATCT-3′, and Art-actin-R: 5′- AGCGGTTGCATTTCTTGTT-3′) were designed using Primer Express Software v3.0 (Applied Biosystems, Life Technologies, Carlsbad, CA, USA) based on the sequence obtained from GenBank (accession no. X52602.1). No significant differences in Ct values were observed for this housekeeping gene between different experimental groups during the course of the infection (CV = 1.02%, Kruskal-Wallis test H = 2.27, $p = 0.13$).

In the case of gilthead seabream fingerlings challenged by feeding, normalized relative *mcp* expression levels were calculated by applying the formula $F = \log_{10} [(E + 1)^{40-Ct}/N]$ [28], where E is the amplification efficiency of the qPCR, Ct (threshold cycle) corresponds to the PCR cycle number, N is the maximal number of viral DNA copies/mg of tissue detected minus the number of viral DNA copies/mg of tissue determined by absolute qPCR for the sample, and Ct of 40 arbitrarily corresponds to "no Ct" by qPCR. In this challenge, results obtained for viral DNA quantification and relative gene expression were analyzed using a Mann–Whitney U test followed by a Holm–Bonferroni correction for multiple comparisons.

3. Results

3.1. Infection of Artemia spp. by LCDV

To establish if LCDV replicates in *Artemia* spp. cells, experimental infections were carried out using LCDV SA25 and three developmental stages of *Artemia*. The time course of the experimental infection was studied by analyzing viral load and *mcp* gene expression in parallel. In challenged metanauplii, the viral load increased by more than two orders of magnitude from the first to the 8th dpi (from 7.6×10^0 to 1.7×10^3 copies of viral DNA/mg of tissue). The viral load remained above 10^2 copies of viral DNA/mg during the entire sampling period (Figure 1A). In juveniles and adults, the time course of the infection was similar to that obtained for metanauplii, reaching the maximal value at 8 dpi (6.7×10^2 and 7×10^2 copies of viral DNA/mg of tissue, respectively) (Figure 1A). Relative expression of viral *mcp* transcripts showed a similar temporal evolution for the three experimental groups analyzed, reaching the highest value at 8 dpi (Figure 1B). Neither LCDV genomes nor mRNA were detected in brine shrimp inoculated with L-15 medium (control groups).

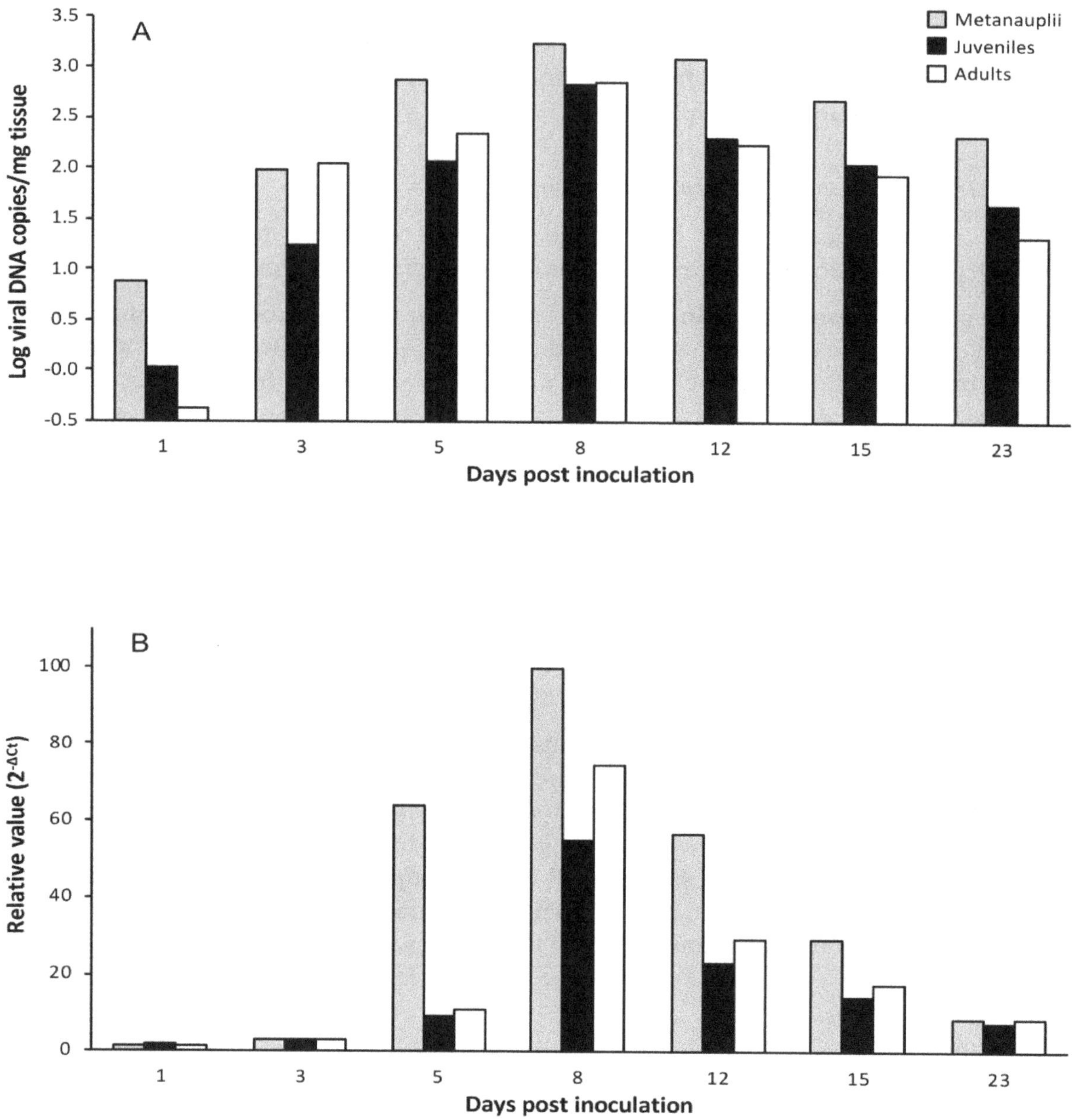

Figure 1. Temporal evolution of viral loads (**A**) and relative major capsid protein (*mcp*) gene expression values (**B**) in different developmental stages of *Artemia* inoculated with Lymphocystis disease virus (LCDV) SA25.

No CPE could be observed in cell cultures inoculated with LCDV-infected *Artemia* homogenates and maintained up to 14 dpi. Nevertheless, by using the ICC-RT-PCR assay, viral infectious titre determination was carried out at 8 dpi. The estimated viral titres were 9.3×10^1 MPNIU/mg for metanauplii and juveniles, and 2.3×10^2 MPNIU/mg for infected adults.

Viral load and *mcp* gene expression were also investigated in *Artemia* metanauplii challenged with LCDV ATCC VR-342. Viral load reached the maximal value at 12 dpi (1.7×10^2 copies of viral DNA/mg of tissue), and the same was observed for relative viral gene expression (Figure 2). In this case, the viral titre at 8 dpi was 7.5 MPNIU/mg, one order of magnitude lower than obtained in metanauplii infected by LCDV SA25.

In any of the experimental groups, clinical signs or mortality were not observed in the *Artemia* cultures.

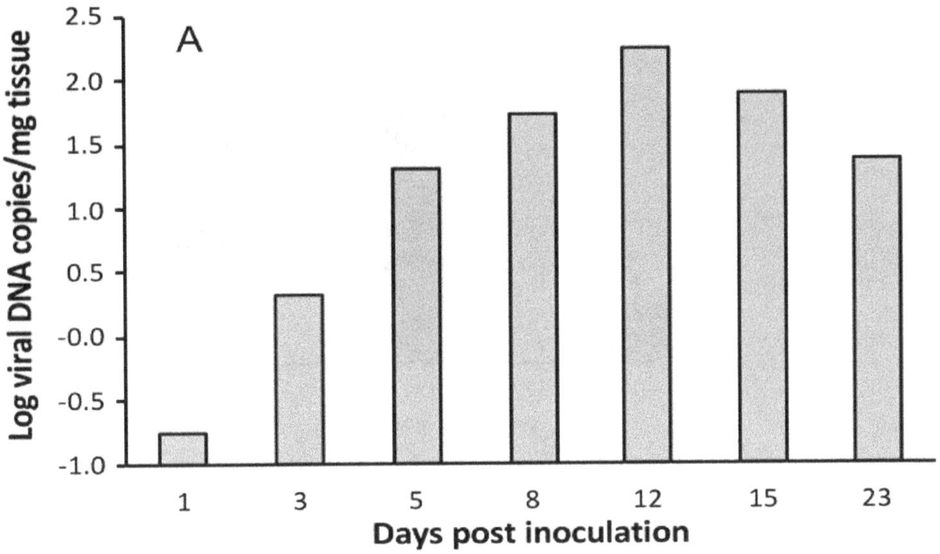

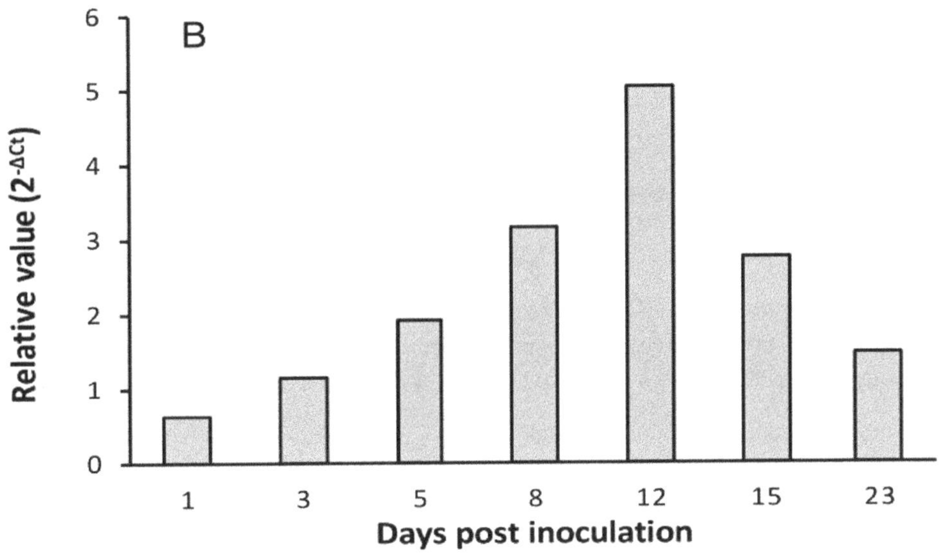

Figure 2. Temporal evolution of viral loads (**A**) and relative *mcp* gene expression values (**B**) in *Artemia* metanauplii inoculated with LCDV ATCC VR-342.

3.2. LCDV Transmission to Gilthead Seabream Fingerlings

Artemia metanauplii used for fingerlings feeding were infected by LCDV, as demonstrated by qPCR, with an estimated viral load of 1.2 ± (0.1) × 10^3 copies of viral DNA/mg of tissue, whereas metanauplii in the control group remained LCDV-negative.

In fingerlings fed on the LCDV-positive metanauplii (challenged group), LCDV was detected by qPCR in all fish and at all time points analyzed. At 7 dpi, the estimated viral load ranged between 10.6 and 26.8 copies of viral DNA per mg of tissue. Five days later, viral loads were significantly higher ($p < 0.01$), and they remained at similar values at 24 dpi (Figure 2A). No LCD symptoms were observed in these fish at the end of the experiment (30 dpi). The *mcp* gene expression was also detected in all fish analyzed, with the highest F-value observed at 12 dpi (Figure 3B). Neither LCDV genomes nor mRNA were detected in fish from the control group (i.e., fed on metanauplii inoculated with L-15 medium).

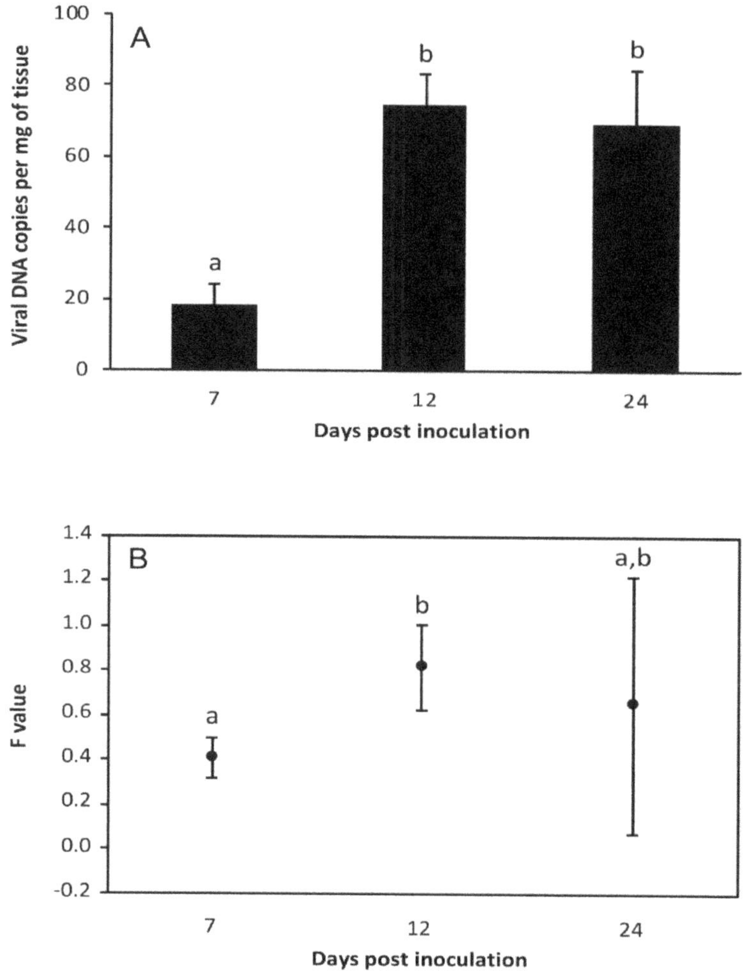

Figure 3. Viral loads (**A**) and relative *mcp* gene expression values (**B**) in gilthead seabream fingerlings orally challenged with LCDV-positive *Artemia* metanauplii (mean ± standard deviation; $n = 7$). Different letters indicate significant differences ($p < 0.01$) (Mann–Whitney U-test, Holm–Bonferroni correction).

4. Discussion

A number of studies have confirmed the role of *Artemia* nauplii as vectors for several crustacean viruses, such as *Macrobrachium rosenbergii* nodavirus (MrNV), hepatopancreatic parvo-like virus (HPV), white spot syndrome virus (WSSV), and infectious myonecrosis virus (IMNV) [29–32]. In addition, *Artemia* appears to be susceptible to some of these viruses, including WSSV and MrNV, with the infection being asymptomatic [33,34]. Regarding fish pathogens, *Artemia* nauplii have proven to be a mechanical vector only in the case of microsporidia and *Vibrio anguillarum* [35,36], although some studies have shown that they could also accumulate viral pathogens and protozoa [16,19,37].

Previous studies demonstrated that infectious virus could be detected in *Artemia* nauplii inoculated with LCDV-Sa by immersion and the virus persisted to the adult stage, and from adults to reproductive cysts [21]. These results led us to consider the hypothesis that *Artemia* spp. could be susceptible to LCDV infection, acting as reservoir and biological vector for LCDV.

The results obtained in the experimental infections carried out demonstrated that *Artemia* spp. could be infected by LCDV at different developmental stages, since viral loads increased during the course of the experiments. In addition, viral transcripts were also detected, showing a similar temporal evolution. Thus, *Artemia* spp. seems to be a susceptible host for LCDV, at least in experimental conditions, with the resulting infection being asymptomatic. This is the first description of a fish virus that also infects invertebrates. Viral loads and infectious titres estimated in LCDV-infected *Artemia* were higher than those previously obtained for subclinically infected gilthead seabream fingerlings or juveniles [25,28,38].

During the course of the experimental infections, particularly in those performed with metanauplii and juveniles, brine shrimps kept growing, doubling or tripling in size, and completed their life cycle, becoming adults. Thus, viral loads expressed per mg of tissue are difficult to interpret, and do not reflect actual viral loads per individual. Taking this into account, the number of genome copies probably increased in each brine shrimp specimen during the experimental trial.

Viral replication kinetics were similar in the experimental infections carried out with different developmental stages of *Artemia*. Nevertheless, relative viral gene expression values were higher in metanauplii compared to juveniles or adults infected by LCDV SA25, although this difference was not reflected in infectious titres. Differences in relative viral expression values were observed in metanauplii infected by both viral isolates, which might indicate that viral infectivity is variable among LCDV genotypes or that *Artemia* susceptibility to these isolates differs.

On the other hand, gilthead seabream fingerlings fed on LCDV-infected metanauplii became infected, as demonstrated by the increase in the number of LCDV DNA copies during the experimental period, and the detection of viral transcripts in these animals. Estimated viral loads were consistent with those previously reported for asymptomatic infections in gilthead seabream [25]. These results indicate that *Artemia* metanauplii can be a vector for LCDV, participating in viral transmission to gilthead seabream via the alimentary route under laboratory conditions that mimics those used in aquaculture farms. Whether *Artemia* is also a reservoir host for LCDV remains to be investigated.

In conclusion, the results demonstrate that LCDV establishes a productive infection in *Artemia* spp., at least under the experimental conditions tested, which extend the host range of the genus *Lymphocystivirus* to crustaceans. Furthermore, our study confirms that *Artemia* metanauplii act as a vector for LCDV transmission to gilthead seabream fingerlings.

Author Contributions: Conceptualization and experimental design, J.J.B. and D.C.; investigation, E.J.V.; formal analysis and visualization, E.J.V. and A.M.L.; writing—original draft, J.J.B. and D.C.; writing—review and editing, J.J.B., D.C. and A.M.L.; all authors read and approved the final manuscript.

References

1. Chinchar, V.G.; Hick, P.; Ince, I.A.; Jancovich, J.K.; Marschang, R.E.; Qin, Q.; Subramaniam, K.; Waltzek, T.B.; Whittington, R.J.; Williams, T.; et al. ICTV virus taxonomy profile: *Iridoviridae*. *J. Gen. Virol.* **2017**, *98*, 890–891. [CrossRef] [PubMed]
2. Chinchar, V.G.; Kwang, H.Y.; Jancovich, J.K. The molecular biology of Frog virus 3 and other iridoviruses infecting cold-blooded vertebrates. *Viruses* **2011**, *3*, 1959–1985. [CrossRef] [PubMed]
3. Jancovich, J.K.; Qin, Q.; Zhang, Q.-Y.; Chinchar, V.G. Ranavirus replication: Molecular, cellular, and immunological events. In *Ranaviruses: Lethal Pathogens of Ectotermic Vertebrates*; Gray, M.J., Chinchar, V.G., Eds.; Springer OPEN: Heidelberg, Germany, 2015; pp. 71–104.
4. Borrego, J.J.; Valverde, E.J.; Labella, A.M.; Castro, D. Lymphocystis disease virus: Its importance in aquaculture. *Rev. Aquac.* **2017**, *9*, 179–193. [CrossRef]
5. Sarasquete, C.; Gonzalez de Canales, M.L.; Arellano, J.; Perez-Prieto, S.I.; Garcia-Rosado, E.; Borrego, J.J. Histochemical study of lymphocystis disease in skin of gilthead seabream, *Sparus aurata*, from the South Atlantic coast of Spain. *Histol. Histopathol.* **1998**, *13*, 37–45. [PubMed]
6. Smail, D.A.; Munro, A.L.S. The virology of teleosts. In *Fish Pathology*, 3rd ed.; Roberts, R.J., Ed.; W.B. Saunders: Edinburgh, UK, 2011; pp. 169–253.
7. Masoero, L.; Ercolini, C.; Caggiano, M.; Rossa, A. Osservazioni preliminary sulla linfocisti in una maricoltura intensive italiana. *Riv. Ital. Piscic. Ittiopatol.* **1986**, *21*, 70–74.

8. Colorni, A.; Padrós, F. Diseases and health management. In *Sparidae: Biology and Aquaculture of Gilthead Sea Bream and Other Species*; Pavlidis, M., Mylonas, C., Eds.; Wiley-Blackwell: Oxford, UK, 2011; pp. 321–357.
9. Bowser, P.R.; Wooster, G.A.; Getchell, R.G. Transmission of walleye dermal sarcoma and lymphocystis via the waterborne exposure. *J. Aquat. Anim. Health* **1999**, *11*, 158–161. [CrossRef]
10. Wolf, K. *Fish Viruses and Fish Viral Diseases*; Cornell University Press: Ithaca, NY, USA, 1988.
11. Cano, I.; Valverde, E.J.; Garcia-Rosado, E.; Alonso, M.C.; Lopez-Jimena, B.; Ortiz-Delgado, J.B.; Borrego, J.J.; Sarasquete, C.; Castro, D. Transmission of lymphocystis disease virus to cultured gilthead seabream, *Sparus aurata* L., larvae. *J. Fish Dis.* **2013**, *36*, 569–576. [CrossRef] [PubMed]
12. Lavens, P.; Sorgeloos, P. *Manual on the Production and Use of Live Food for Aquaculture*; FAO Fisheries Technical Paper No. 361; FAO: Rome, Italy, 1996; 295p.
13. Sorgeloos, P.; Dhert, P.; Candreva, P. Use of the brine shrimp, *Artemia* spp., in marine fish larviculture. *Aquaculture* **2001**, *200*, 147–159. [CrossRef]
14. Austin, B.; Allen, D.A. Microbiology of laboratory hatched brine shrimp (*Artemia*). *Aquaculture* **1982**, *26*, 369–383. [CrossRef]
15. Mendez-Hermida, F.; Gomez-Couso, H.; Ares-Mazas, E. *Artemia* is capable of spreading oocysts of *Cryptosporidium* and the cysts of *Giardia*. *J. Eukaryot. Microbiol.* **2006**, *53*, 432–434. [CrossRef]
16. Mortensen, S.; Evensen, O.; Rodseth, O.; Hjeltnes, B. The relevance of infectious pancreatic necrosis virus IPNV in farmed Norwegian turbot *Scophthalmus maximus*. *Aquaculture* **1993**, *115*, 245–252. [CrossRef]
17. Muroga, K.; Higashi, M.; Keitiku, H. The isolation of intestinal microflora of farmed red seabream *Pagrus major* and black seabream *Acanthopagrus schlegeli* larval juvenile stages. *Aquaculture* **1987**, *65*, 79–88. [CrossRef]
18. Nicolas, J.L.; Robic, E.; Ansquer, D. Bacterial flora associated with a trophic chain consisting of microalgae, rotifers and turbot larvae: Influence of bacteria on larval survival. *Aquaculture* **1989**, *83*, 237–248. [CrossRef]
19. Skliris, G.P.; Richards, R.H. Assessment of the susceptibility of the brine shrimp *Artemia salina* and rotifer *Brachionus plicatilis* to experimental nodavirus infections. *Aquaculture* **1998**, *169*, 133–141. [CrossRef]
20. Overstreet, R.M.; Stuck, K.C.; Krol, R.A.; Hawkins, W.E. Experimental infection with *Baculovirus penaei* in white shrimp *Penaeus vannamei* (Crustacea: Decapoda) as a bioassay. *J. World Aquac. Soc.* **1988**, *19*, 175–187. [CrossRef]
21. Cano, I.; Lopez-Jimena, B.; Garcia-Rosado, E.; Ortiz-Delgado, J.B.; Alonso, M.C.; Borrego, J.J.; Sarasquete, M.C.; Castro, D. Detection and persistence of lymphocystis disease virus in *Artemia* sp. *Aquaculture* **2009**, *291*, 230–236. [CrossRef]
22. Valverde, E.J.; Cano, I.; Labella, A.; Borrego, J.J.; Castro, D. Application of a new real-time polymerase chain reaction assay for surveillance studies of lymphocystis disease virus in farmed gilthead seabream. *BMC Vet. Res.* **2016**, *12*, 71. [CrossRef]
23. Moretti, A.; Fernandez-Criado, M.P.; Cittolin, G.; Guidastri, R. *Manual on Hatchery Production of Seabass and Gilthead Seabream*; Food and Agriculture Organization of the United Nations (FAO): Rome, Italy, 1999; Volume 1, pp. 75–82.
24. Cano, I.; Valverde, E.J.; Lopez-Jimena, B.; Alonso, M.C.; Garcia-Rosado, E.; Sarasquete, M.C.; Borrego, J.J.; Castro, D. A new genotype of *Lymphocystivirus* isolated from cultured gilthead seabream, *Sparus aurata* L., and Senegalese sole, *Solea senegalensis* (Kaup). *J. Fish Dis.* **2010**, *33*, 695–700. [CrossRef]
25. Valverde, E.J.; Borrego, J.J.; Castro, D. Evaluation of an integrated cell culture RT-PCR assay to detect and quantify infectious lymphocystis disease virus. *J. Virol. Methods* **2016**, *238*, 62–65. [CrossRef]
26. Alonso, M.C.; Cano, I.; Garcia-Rosado, E.; Castro, D.; Lamas, J.; Barja, J.L.; Borrego, J.J. Isolation of lymphocystis disease virus from sole, *Solea senegalensis* Kaup, and blackspot sea bream, *Pagellus bogaraveo* (Brünnich). *J. Fish Dis.* **2005**, *28*, 221–228. [CrossRef]
27. Bejar, J.; Borrego, J.J.; Alvarez, M.C. A continous cell line from the cultured marine fish gilt-head sea bream (*Sparus aurata* L.). *Aquaculture* **1997**, *150*, 143–153. [CrossRef]
28. Valverde, E.J.; Borrego, J.J.; Sarasquete, M.C.; Ortiz-Delgado, J.B.; Castro, D. Target organs for lymphocystis disease virus replication in gilthead seabream (*Sparus aurata*). *Vet. Res.* **2017**, *48*, 21. [CrossRef] [PubMed]
29. Da Silva, S.M.B.C.; Lavander, H.D.; De Santana Luna, M.M.; Da Silva, A.O.M.E.; Galvez, A.O.; Coimbra, M.R.M. *Artemia franciscana* as a vector for infectious myonecrosis virus (IMNV) to *Litopenaeus vannamei* juvenile. *J. Invertebr. Pathol.* **2015**, *126*, 1–5. [CrossRef] [PubMed]

30. Sivakumar, V.K.; Sarathi, M.; Venkatesan, C.; Sivaraj, A.; Sahul Hameed, A.S. Experimental exposure of *Artemia* to hepatopancreatic parvo-like virus and subsequent transmission to post-larvae of *Penaeus monodon*. *J. Invertebr. Pathol.* **2009**, *102*, 191–195. [CrossRef] [PubMed]
31. Sudhakaran, R.; Yoganadhan, K.; Ishaq Ahmed, V.P.; Sahul Hameed, A.S. *Artemia* as a possible vector for *Macrobrachium rosenbergii* nodavirus (MrNV) and extra small virus transmission (XSV) to *Macrobrachium rosenbergii* post-larvae. *Dis. Aquat. Org.* **2006**, *70*, 161–166. [CrossRef] [PubMed]
32. Zhang, J.S.; Dong, S.L.; Dong, Y.W.; Tian, X.L.; Cao, Y.C.; Li, Z.J.; Yan, D.C. Assessment of the role of brine shrimp *Artemia* in white spot syndrome virus (WSSV) transmission. *Vet. Res. Commun.* **2010**, *34*, 25–32. [CrossRef]
33. Li, Q.; Zhang, J.; Chen, Y.; Yang, F. White spot syndrome virus (WSSV) infectivity for *Artemia* at different developmental stages. *Dis. Aquat. Org.* **2003**, *57*, 261–264. [CrossRef]
34. Sudhakaran, R.; Ishaq Ahmed, V.P.; Haribabu, P.; Mukherjee, S.C.; Sri Widada, J.; Bonami, J.R.; Sahul-Hameed, A.S. Experimental vertical transmission of *Macrobrachium rosenbergii* nodavirus (MrNV) and extra small virus (XSV) from brooders to progeny in *Macrobrachium rosenbergii* and *Artemia*. *J. Fish Dis.* **2007**, *30*, 27–35. [CrossRef]
35. Grisez, L.; Chair, M.; Sorgeloos, P.; Ollevier, F. Mode of infection and spread of *Vibrio anguillarum* in turbot *Scophthalmus maximus* larvae after oral challenge through live feed. *Dis. Aquat. Org.* **1996**, *26*, 181–187. [CrossRef]
36. Olson, R.E. Laboratory and field studies on *Glugea stephani* (Hagenmuller), a microsporidan parasite of pleuronectid flatfishes. *J. Protozool.* **1976**, *23*, 158–164. [CrossRef]
37. Mendez-Hermida, F.; Gomez-Couso, H.; Ares-Mazas, E. Possible involvement of *Artemia* as live diet in the transmission of cryptosporidiosis in cultured fish. *Parasitol. Res.* **2007**, *101*, 823–827. [CrossRef] [PubMed]
38. Valverde, E.J.; Cano, I.; Castro, D.; Paley, R.K.; Borrego, J.J. Rapid and sensitive detection of Lymphocystis disease virus genotype VII by loop-mediated isothermal amplification. *Food Environ. Virol.* **2017**, *9*, 114–122. [CrossRef] [PubMed]

7

Susceptibility of *Exopalaemon carinicauda* to the Infection with Shrimp Hemocyte Iridescent Virus (SHIV 20141215), a Strain of Decapod Iridescent Virus 1 (DIV1)

Xing Chen [1,2], Liang Qiu [1], Hailiang Wang [1], Peizhuo Zou [1], Xuan Dong [1], Fuhua Li [3] and Jie Huang [1,2,*]

1 Laboratory for Marine Fisheries Science and Food Production Processes, National Laboratory for Marine Science and Technology (Qingdao); Key Laboratory of Maricultural Organism Disease Control, Ministry of Agriculture and Rural Affairs; Qingdao Key Laboratory of Mariculture Epidemiology and Biosecurity; Yellow Sea Fisheries Research Institute, Chinese Academy of Fishery Sciences, Qingdao 266071, China; chenxing910520@163.com (X.C.); qiuliang@ysfri.ac.cn (L.Q.); whl846130@126.com (H.W.); zoupz1992@163.com (P.Z.); dongxuan@ysfri.ac.cn (X.D.)

2 College of Fisheries and Life Science, Shanghai Ocean University, Shanghai 201306, China

3 Key Laboratory of Experimental Marine Biology, Institute of Oceanology, Chinese Academy of Sciences, Qingdao 266071, China; fhli@ms.qdio.ac.cn

* Correspondence: huangjie@ysfri.ac.cn

Abstract: In this study, ridgetail white prawns—*Exopalaemon carinicauda*—were infected per os (PO) with debris of *Penaeus vannamei* infected with shrimp hemocyte iridescent virus (SHIV 20141215), a strain of decapod iridescent virus 1 (DIV1), and via intramuscular injection (IM with raw extracts of SHIV 20141215. The infected *E. carinicauda* showed obvious clinical symptoms, including weakness, empty gut and stomach, pale hepatopancreas, and partial death with mean cumulative mortalities of 42.5% and 70.8% by nonlinear regression, respectively. Results of TaqMan probe-based real-time quantitative PCR showed that the moribund and surviving individuals with clinical signs of infected *E. carinicauda* were DIV1-positive. Histological examination showed that there were darkly eosinophilic and cytoplasmic inclusions, of which some were surrounded with or contained tiny basophilic staining, and pyknosis in hemocytes in hepatopancreatic sinus, hematopoietic cells, cuticular epithelium, etc. On the slides of in situ DIG-labeling-loop-mediated DNA amplification (ISDL), positive signals were observed in hematopoietic tissue, stomach, cuticular epithelium, and hepatopancreatic sinus of infected prawns from both PO and IM groups. Transmission electron microscopy (TEM) of ultrathin sections showed that icosahedral DIV1 particles existed in hepatopancreatic sinus and gills of the infected *E. carinicauda* from the PO group. The viral particles were also observed in hepatopancreatic sinus, gills, pereiopods, muscles, and uropods of the infected *E. carinicauda* from the IM group. The assembled virions, which mostly distributed along the edge of the cytoplasmic virogenic stromata near cellular membrane of infected cells, were enveloped and approximately 150 nm in diameter. The results of molecular tests, histopathological examination, ISDL, and TEM confirmed that *E. carinicauda* is a susceptible host of DIV1. This study also indicated that *E. carinicauda* showed some degree of tolerance to the infection with DIV1 per os challenge mimicking natural pathway.

Keywords: SHIV; DIV1; *Decapodiridovirus*; *Exopalaemon carinicauda*; susceptibility; host; ISDL

1. Introduction

The iridescent virus family *Iridoviridae* contains large icosahedral double-stranded DNA viruses, which is divided into two subfamilies (i.e., *Alphairidovirinae* and *Betairidovirinae*) and composed of five known genera: *Lymphocystivirus, Megalocytivirus, Ranavirus, Chloriridovirus,* and *Iridovirus*. Of them, the genera *Lymphocystivirus, Megalocytivirus,* and *Ranavirus* belong to subfamily *Alphairidovirinae,* and *Chloriridovirus* and *Iridovirus* belong to subfamily *Betairidovirinae* [1–3]. Iridescent viruses of *Alphairidovirinae* lead to a high mortality rate in significant fish and amphibians [3], while iridescent viruses of *Betairidovirinae* infect insects and crustaceans. The first discovery of a possible iridescent virus in crustaceans was published in 1993 [4]. In the same year, Lightner and Redman reported another suspected iridescent virus in penaeid shrimp *Protrachypene precipua* [5]. Subsequently, Miao et al. [6] reported that a suspected iridescent virus was observed in lymphoid cell cultures from the Chinese white shrimp *Penaeus chinensis* in 1999. Several years later, Tang et al. [7] identified an iridoviru—i.e., Sergestid iridovirus (SIV)—diseased sergestid shrimp *Acetes erythraeus,* which is a likely causative agent evidenced through in situ hybridization and PCR test. Xu et al. [8] reported *Cherax quadricarinatus* iridovirus (CQIV) isolated from diseased red claw crayfish *Cherax quadricarinatus*. Qiu et al. [9] identified an iridescent virus named shrimp hemocyte iridescent virus (SHIV), which was isolated from farmed *Penaeus vannamei* in 2014 and also detected in *P. chinensis* and *Macrobrachium rosenbergii*. In March 2019, the Executive Committee of the International Committee on Taxonomy of Viruses (ICTV) approved the proposal made by Chinchar et al. [10] that proposed a new species of decapod iridescent virus 1 in a new genus *Decapodiridovirus* to include SHIV 20141215 and CQIV CN01 as two isolates.

The range of susceptible hosts is an important part of epidemiology and pathogen ecology. It is also of significant concerned for international trade and control of the disease. In 2014, the World Organization for Animal Health (OIE) has adopted a chapter in *The Aquatic Animal Health Code* to provide the criteria for determining susceptibility of aquatic animal species to infection with a specific pathogen with emphasis on that the route of transmission is consistent with natural pathways for the infection [11]. Ridgetail white prawn—*Exopalaemon carinicauda*—is one of the major economic crustaceans in China, which can propagate all year round, and is naturally distributes around the coast of Yellow Sea and Bohai Sea of China [12,13]. *E. carinicauda* accounts for one-third of the total yields of polyculture ponds in the eastern China [12]. It is also a species of wild prawns and commonly exists in the ponds farming *M. rosenbergii* or penaeid shrimp. To date, there is no report pointing out if *E. carinicauda* can be infected with iridescent viruses. Owing to the common existence of *E. carinicauda* in intensive shrimp farming ponds, verifying its susceptibility to DIV1 is a prerequisite for evaluation of the potential transmission routes and possible reservoirs of DIV1 in the nature. As there is no crustacean cell line available, bioassays with susceptible species are the only way to study the infectivity of crustacean viruses. Establishment of infection model with susceptible animals will provide approaches for crustacean virological studies. In this study, *E. carinicauda* were challenged per os to mimic natural infection according to the OIE standards, while intramuscular injection was used as a positive control. A combination of TaqMan real-time PCR, histopathological observation, in situ DIG-labeling-loop-mediated DNA amplification (ISDL), and ultra-thin transmission electron microscopy were used to confirm the infection with DIV1 in *E. carinicauda*.

2. Materials and Methods

All protocols for animal breeding, handling and sampling were approved by the Animal Care and Ethics Committee, Yellow Sea Fisheries Research Institute, Chinese Academy of Fishery Sciences. Efforts were made to provide a comfortable living environment for the animals and to minimize the suffering of animals during the sampling process, according to recommendations proposed by the European Commission (1997). The study was carried out in accordance with the approved protocol. All the methods were applied in accordance with relevant guidelines.

2.1. Animals

Specific-pathogen-free (SPF) ridgetail white prawns *E. carinicauda*, (3.6 ± 0.4) cm in body length, were reared and bred in the Key Laboratory of Experimental Marine Biology, Institute of Oceanology, Chinese Academy of Sciences. Prawns *E. carinicauda* were cultivated in a 40 L plastic tanks containing 20 L seawater of 30 ppt in salinity with 90% daily exchange rate, at a water temperature of 27 °C, with continuous aeration, and fed three times per day with the formula feed for a week in our wet lab. For multiplication of DIV1, 30 healthy white leg shrimp *P. vannamei* of (10.2 ± 0.8) cm in body length were purchased from a shrimp farm in Weifang of Shandong Province and then held in 40 L seawater of 30 ppt salinity with same management above mentioned.

Before and after the challenge testing, the prawns and the shrimp were sampled and examined for potential pathogens, including white spot syndrome virus (WSSV), taura syndrome virus (TSV), yellow head virus (YHV), infectious hypodermal and hematopoietic necrosis virus (IHHNV), and acute hepatopancreatic necrosis disease-causing *Vibrio parahaemolyticus* (Vp_{AHPND}), using the nested PCR or TaqMan probe based quantitative real-time PCR (TaqMan qPCR) methods [14], covert mortality nodavirus (CMNV) using TaqMan qPCR methods [15], and DIV1 using nested PCR [16].

2.2. Preparation of Viral Inoculum

To prepare material for the challenge test, 30 healthy shrimp *P. vannamei* were infected by feeding with DIV1 (strain SHIV 20141215) infected tissue, which was derived from diseased *P. vannamei* collected at a farm in Zhejiang Province in December 2014 [9]. After 5 days post-infection (dpi), 2.5 g cephalothoraxes (removing tergites) were taken from DIV1-infected shrimp and subsequently homogenized in 40 mL pre-cooled PPB-Tris (376.07 mM NaCl, 6.32 mM K_2SO4, 6.4 mM $MgSO_4$, 14.41 mM $CaCl_2$, and 50 mM Tris-HCl, pH 6.5–8.0) [17]. The suspension was centrifuged at 10,000 rpm for 10 min at 4 °C. The pellet was resuspended in 25 mL PPB-Tris by homogenization and re-centrifuged. The supernatants from two steps were merged and filtered through a 500-mesh sieve and a 0.45 μm filter. The suspension was sampled to detect various pathogens for exclusion of contamination by other viruses and bacteria. DIV1 dose in the suspension was quantified using TaqMan qPCR method [16]. For the challenge test via intramuscular injection, an inoculum containing 10^4 copies/μL DIV1 was prepared by diluting the suspension with PPB-His (replace 50 mM Tris-HCl with 50 mM histidine in PPB-Tris) [17], and then the inoculum was dispensed in 100 μL aliquots before stored at −80 °C.

2.3. Challenge Tests

Challenge tests were performed with healthy *E. carinicauda* in four groups, including intramuscular injection (IM), per os (PO), control of intramuscular infection (CIM), and control of per os (CPO). Each group had three replicates with 10 individual prawns per replicate. Each prawn in the IM group was injected with 10 μL of viral inoculum (~10^5 copies). Prawns in the PO group were fed with total 3 g debris of DIV1-infected *P. vannamei* tissue (with a viral dose of about 10^{10} copies/g). Prawns in the CIM group and the CPO group were injected with 10 μL sterile PPB-His and commercial formula feed, respectively. After the challenge operation, the rearing conditions were kept as same as those prior to the challenge.

Significance analysis of cumulative mortalities between any two groups was carried out using the *t*-test for homoscedasticity hypothesis of two group samples with the add-in tool of data analysis in Microsoft® Excel® 2016 MSO 64-bit.

For evaluation of the pathogenic model in the infection groups, mortality data were nonlinearly regressed following the three-parameter sigmoid equation

$$M_{(t)} = \frac{a}{1 + e^{-k(t-t_0)}}, \tag{1}$$

which formulates the change of $M_{(t)}$ (cumulative mortality) at t (time, dpi) determined by parameters of a (the maximum mortality), t_0 (the time to half of the maximum mortality), and k (the instant incidence rate). The real pathogenic model shall deduct the average mortality of control groups.

2.4. Detection of DIV1 Using TaqMan-Probe-Based Real-Time Quantitative PCR (TaqMan qPCR)

Total DNA were extracted from the hepatopancreas of samples stored at −20 °C using a TIANamp Marine Animals DNA Kit (Tiangen, Beijing, China). TaqMan qPCR, using the total DNA as template, was conducted as described previously [16] to detect and quantify DIV1. The forward and reverse primers were SHIV-F 5′-AGG AGA GGG AAA TAA CGG GAA AAC-3′, and SHIV-R 5′-CGT CAG CAT TTGGTT CAT CCA TG-3′, respectively. The TaqMan probe (5′-CTG CCC ATC TAA CAC CAT CTC CCG CCC-3′) was labeled with 5′-6-carboxyfluorescein (FAM) and 3′-carboxytetramethylrhodamine (TAMRA). Each PCR mixture in 20 μL contained 10 μL 2× FastStart Essential DNA Probes Master (Roche, Indianapolis, IN, USA), 500 nM primer each, 200 nM probe and 100 ng total DNA template. The program started by an initial denaturation at 95 °C for 10 min, followed by 40 cycles at 95 °C for 10 s and 60 °C for 30 s. The amplification and date analysis were carried out in a CFX-96 Quantitative Fluorescence Instrument (BioRad, Hercules, California, USA).

2.5. Histopathological Examination

The cephalothoraxes of *E. carinicauda*, sampled during the challenge tests, were fixed with Davison's AFA fixative (DAFA) for 24 h and processed for sectioning and hematoxylin and eosin (H&E) staining as described by Bell & Lightner [18]. The histological sections were analyzed and photographed under a light microscopy system (Eclipse 80i, Nikon, Tokyo, Japan).

2.6. Loop-Mediated Isothermal Amplification (LAMP)

A set of specific primers, composed of FIP, BIP, LF, LB, F3, and B3, for LAMP detection of DIV1 was designed to target the gene of the second largest subunit of DNA-directed RNA polymerase II in the DIV1 genomic sequence (GenBank accession No. MF599468), using PrimerExplorerV4 (http://primerexplorer.jp/elamp4.0.0/index.html). These primers compared with the database in NCBI (http://blast.ncbi.nlm.nih.gov/Blast.cgi) to analyze sequence similarities. The primers were synthesized by Sangon Biotech (Shanghai, China). Each LAMP mixture contained 1.6 μM each of inner primers FIP and BIP, 0.8 μM each of loop primers LF and LB, 0.2 μM each of outer primers F3 and B3, 1.4 mM of dNTP mix (TaKaRa, Dalian, China), 1.2 M betaine (Solarbio, Beijing, China), 25 μM calcein (Sigma, St. Louis, MO, USA), 500 μM $MnCl_2$, 6 mM $MgCl_2$, 8 U Bst 2.0 DNA polymerase (New England Biolabs Inc., Beverly, USA), 1× supplied buffer and the specified amount of template DNA in a final volume of 25 μL. The procedure was 60 cycles for 60 °C, following 5 min at 85 °C on the CFX-96 Quantitative Fluorescence Instrument (BioRad, Hercules, CA, USA) using calcein fluorescent channel. Detection specificity of the LAMP primers were examined using 100 ng of the total DNA extracted from uninfected prawns and shrimp infected with other pathogens, including WSSV, Vp_{AHPND}, IHHNV, and EHP.

2.7. In Situ DIG-Labeling-Loop-Mediated DNA Amplification (ISDL)

ISDL followed the method published by Jitrakorn et al. [19] with some modification to target DIV1. Paraffin sections were dewaxed and rehydrated according to the normal DIG-labeled in situ hybridization method [18–20]. Rehydrated slides were added with ddH_2O and denatured on a 100 °C heating block for 2 min, then subsequently placed in a wet box. Total of 150 μL LAMP mixture, which has the same constituents as described in paragraph 2.6, except that the dNTP mix was supplemented with 0.1 mM digoxigenin-11-dUTP (DIG-labeled dUTP) and—without template, calcein, and $MnCl_2$—were added dropwise to each slide. The slides were horizontally incubated at 65 °C for 60 min, followed by 85 °C for 5 min. Subsequent steps were performed in accordance with the post-hybridization steps of a normal in situ hybridization [20]. Tissue sections of healthy prawns were used as the control.

2.8. Transmission Electron Microscopy with Ultrathin Sections

For transmission electron microscopy (TEM) with ultrathin sections, samples were prepared as previous described [21,22]. Briefly, tissues of hepatopancreas, muscle, pereiopods, uropods, and gills of infected *E. carinicauda* were placed in TEM fixative (2% paraformaldehyde, 2.5% glutaraldehyde, 160 mM NaCl and 4 mM $CaCl_2$ in 200 mM PBS, pH 7.2) and cut rapidly into ~1 mm^3 pieces with scalpels, fixed for 1 h at room temperature, and then post-fixed with OsO_4. Specimens were embedded in Spurr's plastic and dyed with uranyl acetate and lead citrate. Ultrathin sections were prepared on collodion-coated grids by the Equipment Center of the Medical College, Qingdao University (Qingdao, China). All grids were examined under a JEM-1200 electron microscope (JEOL, Japan) operating at 80–100 kV.

3. Results

3.1. Clinical Signs and Cumulative Mortality

The challenge test lasted for 15 days. After 3 dpi, prawns in both intramuscular (IM) and per os (PO) groups showed clinically symptoms, including empty stomach and gut and pale hepatopancreas. Additionally, the hematopoietic tissue at the base of rostrum showed slight cloudy white (Figure 1, IM and PO); while prawns in control groups (i.e., CPO and CIM) displayed normal gross signs, including feed-filled stomach and gut and light brown hepatopancreas. However, the hematopoietic tissue was not visible (Figure 1, CPO).

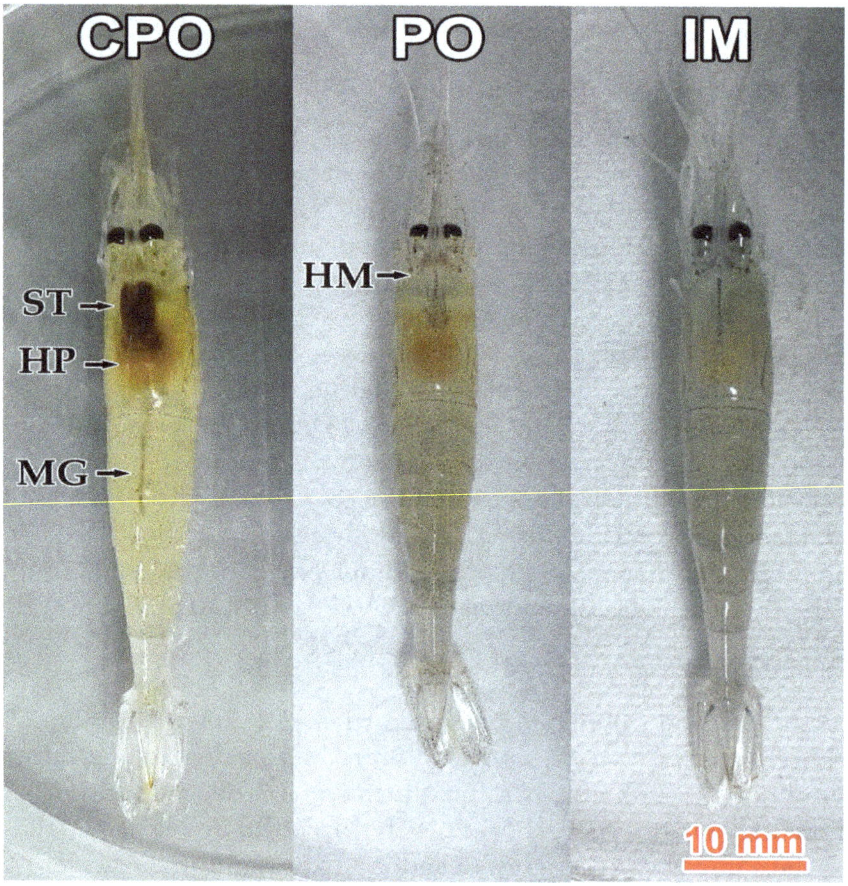

Figure 1. Gross signs of prawns *Exopalaemon carinicauda* in different groups of the challenge test. CPO: Prawn in per os control group; PO: Prawn in per os group; IM: Prawn in intermuscular injection group. Solid arrows indicate stomach (ST), hepatopancreas (HP), midgut (MG), and hematopoietic tissue (HM). Red bar = 10 mm.

Prawns in IM and PO groups suffered from a rapid increase in mortality during the period between 2 dpi and 5 dpi. The average cumulative mortality stabilized at (76.7 ± 18.3)% and (50.0 ± 26.5)% in the IM group after 5 dpi and PO group after 10 dpi, respectively. These rates were both significantly higher ($P < 0.01$) than that of CIM and CPO groups which both had an average mortality of (6.7 ± 5.8)%. Additionally, the cumulative mortality of the IM group was significantly higher ($P < 0.05$) than that of the PO group at 3 dpi and 4 dpi (Figure 2). Significance analysis of the overall data of four groups showed significant differences among IM, PO, and control groups ($P < 0.01$).

Based on the nonlinear regression following the three-parameter sigmoid equation, the pathogenic models of infection with DIV1 on *E. carinicauda* via IM and PO were

$$M_{IM(t)} = \frac{70.8 \pm 0.2}{1 + e^{-(4.76 \pm 2.49)[t-(2.64 \pm 0.21)]}} \quad (2)$$

$$\text{and } M_{PO(t)} = \frac{42.5 \pm 1.2}{1 + e^{-(0.82 \pm 0.51)[t-(3.83 \pm 0.87)]}}, \quad (3)$$

respectively. The functions (2) and (3) indicated the infection of *E. carinicauda* with DIV1 via IM caused (70.8 ± 0.2)% maximum cumulative mortality, while the infection PO caused (42.5 ± 1.2)% maximum cumulative mortality. The former reached the half level of the maximum mortality at (2.64 ± 0.21) dpi with an instant incidence rate at (4.76 ± 2.49)/day; and the latter did at (3.83 ± 0.87) dpi with an instant rate at (0.82 ± 0.51)/day.

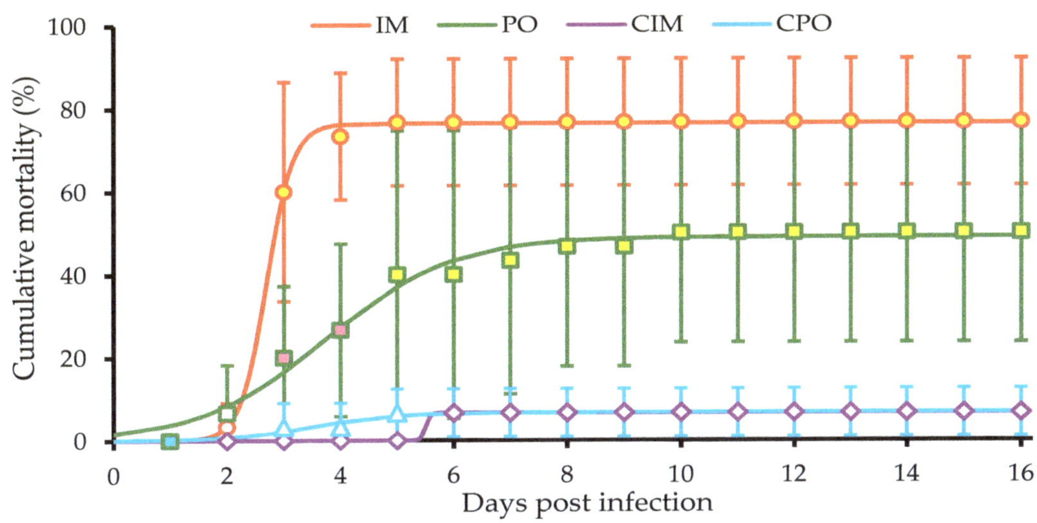

Figure 2. Cumulative mortalities of *Exopalaemon carinicauda* in the challenge test. IM group, prawns were challenged with filtrated viral suspension via intermuscular injection; PO group, prawns were fed with tissues of DIV1-infected *P. vannamei*; CIM group, prawns were injected with sterile PPB-His buffer; CPO group, prawns were fed with commercial feed. Cumulative mortalities are shown as means of data from three replicates for each experimental group (each replicate contained 10 individuals). The mean points with same color indicate no significant difference ($P > 0.05$), and the mean points with different colors indicate a significant difference ($P < 0.05$). Overall analysis indicated there are very significant differences among IM, PO, and the two controls ($P < 0.01$). The curves were drawn based on the nonlinear regression following the three-parameter sigmoid equation.

3.2. TaqMan qPCR Detection

Moribund prawns in IM and PO groups were sampled throughout the experimental duration, and all surviving prawns in each group were collected at the end of the challenge experiment.

All samples were examined for potential pathogens, using the TaqMan qPCR method. The DIV1-positive rates were 80.0% (24/30) and 46.7% (14/30) in the IM group and PO group, respectively. The results showed that all DIV1 negative prawns of the IM and PO groups survived and all prawns of CIM and CPO groups were DIV1 negative. The tests of Vp_{AHPND}, IHHNV and WSSV were negative for all samples. The logarithmic DIV1 loads (in copies/ng-DNA) in positive prawns of the IM and PO groups were 5.65 ± 2.31 and 3.08 ± 0.60, respectively. The viral loads in positive prawns of the IM group were significantly higher ($P < 0.01$) than those of the PO group (Figure 3). The loads of DIV1 in 21 moribund and dead prawns of the IM group were detected to reach $(4.20 \pm 3.88) \times 10^6$ copies/ng-DNA, while that in 3 non-clinical prawns were detected at only $(5.47 \pm 4.16) \times 10^{-1}$ copies/ng-DNA. The loads of DIV1 in 7 dead prawns in the PO group during 5–7 dpi averaged at $(3.63 \pm 1.44) \times 10^4$ copies/ng-DNA, which were much higher than that in 7 surviving prawns in the same group at $(3.57 \pm 3.42) \times 10^2$ copies/ng-DNA (Figure S1). These results showed that prawns in the IM and PO groups were successfully infected by DIV1. In addition, the results suggested that *E. carinicauda* may have some resistance to infection with DIV1 per os mimicking a natural exposure pathway.

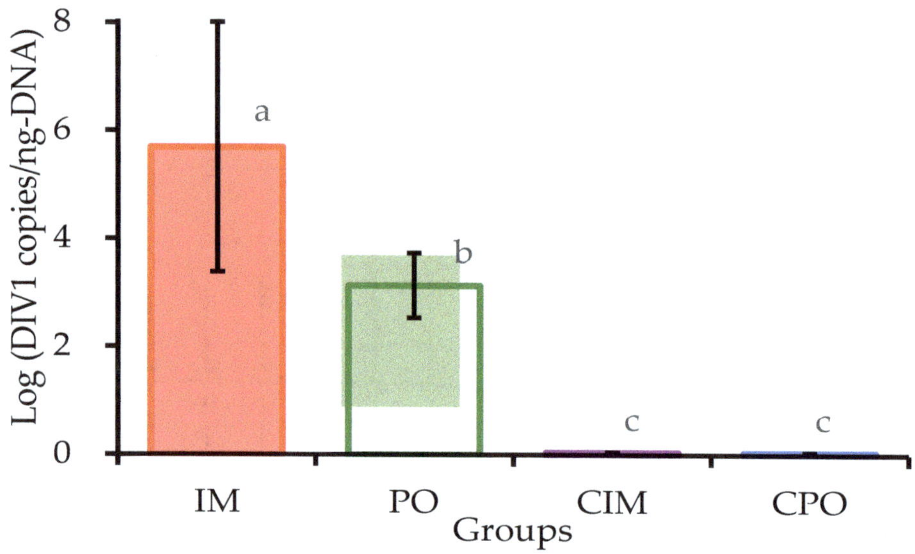

Figure 3. Decapod iridescent virus 1 (DIV1) loads in *Exopalaemon carinicauda* from the challenge test. Different letters above the bars indicate significant difference ($P < 0.01$). IM: prawn group challenged var intramuscular injection, PO: prawn group challenged per os; CIM: control group of injected with sterile buffer; CPO: control group fed with commercial feed. CIM and CPO were negative.

3.3. Histopathology

Histopathological examination of moribund prawns showed nuclear pyknosis and acidophilic inclusions in hemocytes of hepatopancreatic sinus (Figure 4a), hemocytoblasts of hematopoietic tissues (Figure 4c), and cuticular epithelium (Figure 4e); while the control prawns appeared normal in these tissues (Figure 4b,d,f). Typical cytoplasmic and dark eosinophilic inclusions, of which some were surrounded with or contained tiny basophilic staining, appeared beside the nuclei in hematopoietic tissues, hemocytes, and epithelium. The inclusions were more easily found in the infected hematopoietic tissues (Figure 4b).

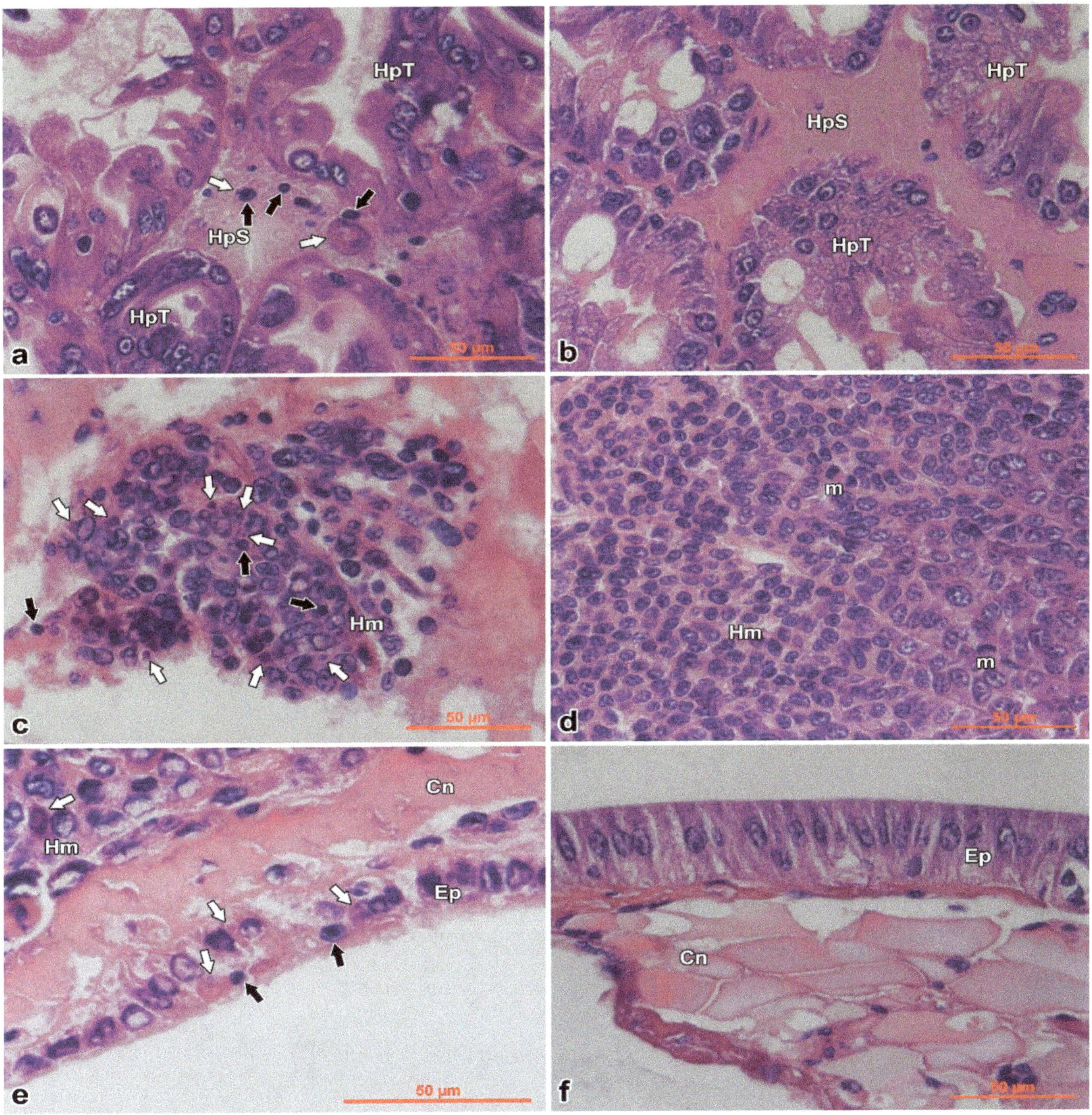

Figure 4. Histopathological examination of *Exopalaemon carinicauda* tissues infected with DIV1 and controls. Black arrows show karyopyknosis and white arrows show eosinophilic inclusions. Pictures (**a**,**b**) are hepatopancreas; (**c**,**d**) are hematopoietic tissues; and (**e**,**f**) are cuticular epithelium on which the cuticles were removed before dehydration. The pictures in the left column (a, c, and e) are the tissues of the infected prawn; the pictures in the right column (b, d, and f) are the tissues of the control prawn. HpT: hepatopancreatic tubule; HpS: hepatopancreatic sinus; Hm: hematopoietic tissue; Ep: epithelium; Cn: connective tissue; m: mitotic phase. Bar = 50 μm.

3.4. Primer Design and LAMP Reaction

Primers for LAMP were designed to target the gene of the second largest subunit DNA-directed RNA polymerase II located between 69047—69263 of SHIV 20141215 genome (GenBank access no. MF599468), of which the sequence was completely complementary to that of CQIV CN01 77416—77200 (Figure 5). In addition, results of LAMP primer specificity assay showed that the reactions were only positive for DIV1-infected shrimp, while the reactions were negative for WSSV, Vp_{AHPND}, IHHNV, or EHP infected shrimps. The reactions indicated that the LAMP primers in this study were specific to DIV1 and should work for ISDL.

Figure 5. Information on primer design and primers used for Loop-mediated isothermal amplification (LAMP), based on the reference sequence of DIV1. Sequences of Shrimp hemocyte iridescent virus (SHIV 20141215 MF599468) and *Cherax quadricarinatus* iridovirus (CQIV CN01 MF197913) were obtained from GenBank.

3.5. ISDL

For ISDL protocol, a LAMP system was applied to in situ amplification to produce DIG-labeled products and subsequently detected with anti-DIG antibody on histological sections. ISDL results showed that blue-violet hybridization signals were detected in the hepatopancreatic sinus, stomach epithelium, cuticular epithelium, and hematopoietic tissues (Figure 6a,c,e,g) of prawns in the IM and PO groups, but no signal was detected in the control groups (Figure 6b,d,f,h). The results of ISDL provided further evidence that prawns in the IM and PO groups were infected by DIV1.

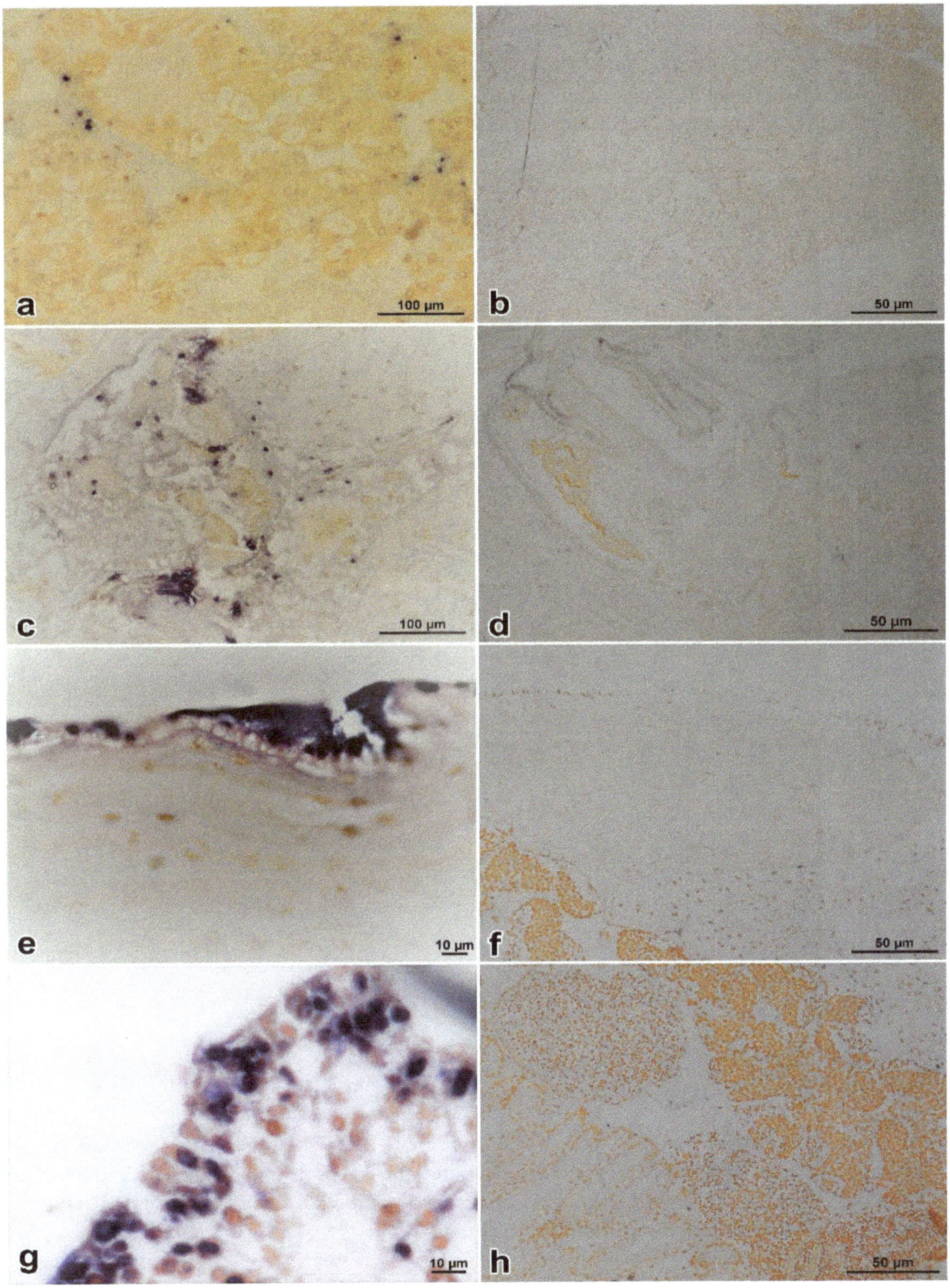

Figure 6. In situ DIG-labeling-loop-mediated DNA amplification (ISDL) micrographs of DIV1-infected and control *Exopalaemon carinicauda*. Hepatopancreas, stomach, cuticular epithelium, and hematopoietic tissue from a challenged prawn of the PO group, respectively (**a**, **c**, **e**, and **g**); hepatopancreas, stomach, cuticular epithelium, and hematopoietic tissue from a healthy prawn of the CPO group, respectively (**b**, **d**, **f**, and **h**). In pictures a, c, e, and g, blue-violet signals were observed in hepatopancreatic sinus, stomach epithelium, cuticular epithelium, and hematopoietic tissues of prawns, respectively. In pictures b, d, f, and h, no hybridization signal was seen in the same tissues of DIV1-negative *E. carinicauda*, except some non-specific signals on gastric sieve of stomach.

3.6. TEM

Icosahedral virions and cytoplasmic inclusions were observed in hepatopancreatic sinus, pereiopods, uropods, muscle, and gills of *E. carinicauda* of the IM group (Figure 7A–E,H). Assembled virions were mostly distributed outside of the assembling area near cellular membrane of the infected cells (Figure 7A,E). Similarly, virions were observed in the hepatopancreatic sinus and gills of prawns of the PO group (Figure 7F,G). Virions were hexagonal (i.e., icosahedral) and about 150 nm in diameter, which matched with characteristics of DIV1.

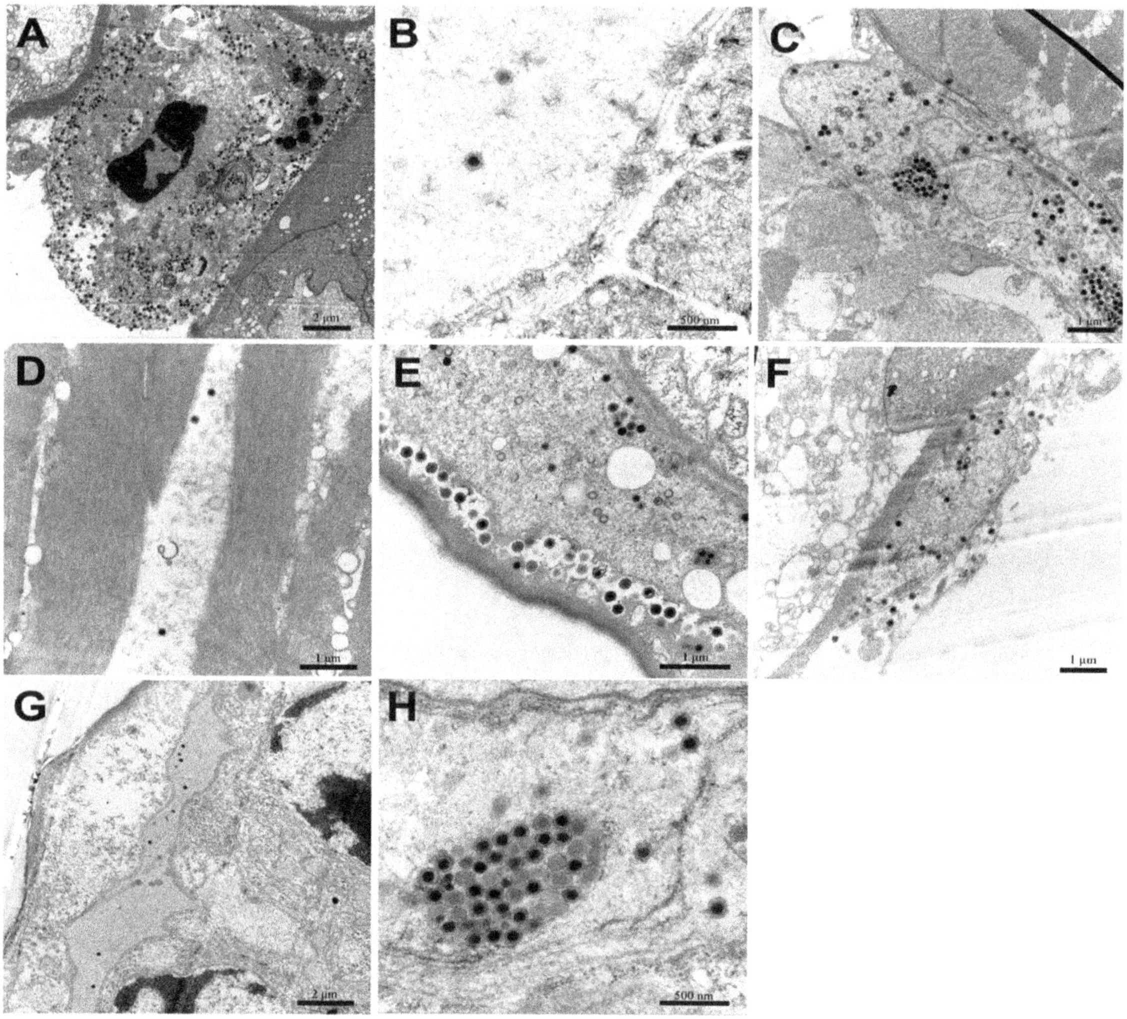

Figure 7. Transmission electron microscopy (TEM) of DIV1-infected *Exopalaemon carinicauda*. (**A–E**): hepatopancreatic sinus, pereiopods, uropods, muscle and gills, respectively, of a prawn in the IM group; (**F** and **G**): hepatopancreatic sinus and gills, respectively, of a prawn in the PO group; (**H**): cytoplasmic inclusions with a cluster of viral particles.

4. Discussion

The susceptibility of infected species and the host range for a newly found virus are important information for studying both host and virus. From the host aspect, knowing an infectious pathogen of a host species will provide important information for prevention of the relevant disease; from the virus aspect, knowledge of a susceptible host range can be used for identification of virus species, discovering a specific viral replication host or cell line, and understanding of viral ecology. Susceptible host range is also important for international trade to prevent the transboundary spreading of a specific disease. For such purpose, the OIE issued criteria in 2014 to determine susceptibility for specific diseases that requires evidence from either natural infection or experimental challenge mimicking natural infection [11]. As a newly found virus, the susceptible host range of DIV1 remains a largely unknown area to be investigated. For the natural infection pathway, positive detections of DIV1 by PCR or qPCR have been reported in different natural samples of crustacean, including *P. vannamei* [9], *P. chinensis* [9], *M. rosenbergii* [9,23], *C. quadricarinatus* [8], *Procambarus clarkii* [23,24], *P. japonicus* [23], *M. nipponense* [23,24], *M. superbum* [24], and Cladocera [24]. Severe disease with high mortality has been reported and infection with DIV1 has been demonstrated in *C. quadricarinatus*, *P. vannamei*, and *M. rosenbergii* [8,9,24]. All of these have been confirmed as being susceptible species for DIV1, fulfilling the natural infection criteria of OIE standards. The susceptibility of *P. chinensis*, *P. japonicus*, and *M. superbum* to DIV1 were not fully confirmed, as the evidence of PCR or qPCR positives [23,24] only supports the existence of DIV1. Confirmation of infection with DIV1 can usually be supported by evidence from histopathological evaluations, TEM, or in situ hybridization. However, they may remain suspicions of susceptibility, as there are confirmed susceptible species in the same genus. Cladocera has been detected as positive by qPCR, but the ISDL result was negative, so that it has been considered as a non-susceptible species [24]. For the experimental infection pathway, experimental challenges of *P. vannamei*, *Pr. clarkii*, Chinese mitten crab *Eriocheir sinensis*, and wild crab *Pachygrapsus crassipes* with CQIV CN01 have been reported [8,25]. Unfortunately, all these experimental infections only used only intramuscular injection, so the results were not enough to confirm of susceptibility for these species because OIE standards require that the challenge pathway to mimics natural infection. This present study is the first report to confirm a susceptible species to the newly found DIV1 by an experimental pathway mimicking the natural infection that fully follows OIE standards.

Due to an absence of a cell line for crustaceans, it is not possible to use normal plaque purification for crustacean viruses. We used a variety of methods to prove that *E. carinicauda* was successfully infected with DIV1, including molecular detection, histopathological and cytopathological analyses, and in situ DIG-labeling-loop-mediated DNA amplification. The original virus strain SHIV 20141215 has been demonstrated by testing for five major shrimp pathogens, including WSSV, YHV, TSV, IHHNV, and Vp_{AHPND} by PCR or RT-PCR methods [9]. In the previous study, the debris of *P. vannamei* and the purified viral inoculum were re-examined using real-time PCR for WSSV, TSV, YHV, IHHNV, Vp_{AHPND}, CMNV, and DIV1 to exclude the possibility of contamination by other viruses and bacteria. In addition, no other viruses, with the exception of DIV1-like, as well as bacteria, were observed with TEM in samples used in this study.

Natively distributing in Yellow Sea and Bohai Sea and widely cultivated broadly in Jiangsu, Zhejiang, Shandong, and Liaoning Provinces of China, *E. carinicauda* is one of important economic crustaceans [12,13]. It has been reported that *E. carinicauda* is a natural host of some viral pathogens, such as WSSV [26] and CMNV [27]. *E. carinicauda* was also considered as a potential model animal for studies of shrimp viral pathogens [28].

In order to investigate the susceptibility of *E. carinicauda* to DIV1, following the OIE standards, the decapod was experimentally challenged with DIV1 per os and intramuscular injection. After 3 dpi, infected prawns of both the PO and IM groups displayed identical clinical symptoms, including debility, empty guts and stomach, pale hepatopancreas, slight whitish hematopoietic tissue, and death. Much higher doses of DIV1 were detected in hepatopancreatic samples of the moribund and deceased prawns in the IM group than those in the PO group. The significant difference in viral dose may be because the injection of DIV1 caused extensive and synchronous infection in cells of different target tissues. The clinical signs observed in *E. carinicauda* were similar to those of *P. vannamei* affected by SHIV 20141215 [9] and also similar to those of *C. quadricarinatus* affected by CQIV CN01 [8]. However, the sign of slightly whitish cloud at base of rostrum was not observed in decapods other than *M. rosenbergii*, which showed a typical white triangle at the hepatopancreatic locus [24]. The invisibility of this symptom in hematopoietic tissue of some DIV1 infected shrimp and crayfish may be due to interspersion and tininess of hematopoietic tissues of *Penaeus* species [18] or the opaque cuticles of crayfish. We may still need to search carefully for this symptom in future observation, as all gross signs in other tissues are not be typical for the infection with DIV1. According to the nonlinear regression of the challenge tests, in the duration of 16 dpi, mean cumulative mortalities of *E. carinicauda* were 70.8% and 42.5% caused by infection with DIV1 via intramuscular and per os, respectively. The pathogenic model (2) and (3) derived from the bioassay revealed that *E. carinicauda* showed some tolerance to the infection with DIV1 per os challenge mimicking natural infection. Compared with the intramuscular challenge, the per os challenge caused only (60.1 ± 6.3)% lower mortality within (145 ± 0.35)% longer time spent at (17.2 ± 14.0)% slower instant speed. The surviving DIV1 positive prawns in per os challenged group possessed viral loads at (3.57 ± 3.42) × 10^2 copies/ng-DNA, which were about 100 folds lower than that detected in dead prawns. Previous studies showed that the mortalities of *P. vannamei*, *C. quadricarinatus*, and *Pr. clarkii* experimentally challenged with DIV1 reached 100% [8,9]. The partial tolerance of *E. carinicauda* to DIV1 may provide an object for future study on mechanisms and breeding of DIV1 resistance. As *E. carinicauda* is a broadly distributed native species, the partial tolerance of the ridgetail white prawn to DIV1 may also provide a possible reservoir of the virus after it transports to species in the area of the Yellow Sea, Bohai Sea, and East China Sea.

Positive results of TaqMan qPCR and ISDL provided further evidences of the infection of *E. carinicauda* by DIV1. According to the results of ISDL, positive signals were detected in the hepatopancreatic sinus, hematopoietic tissue, and cuticular epithelium of infected *E. carinicauda*. Previous studies showed that cuticular epithelium is a major target tissue of WSSV infection [26,29]. However, cuticular epithelium of *P. vannamei* is not a susceptible tissue to DIV1 infection. The results exhibit a difference in tissue tropism between *E. carinicauda* and *P. vannamei* to DIV1 infection. ISDL [19] is a highly sensitive and time-saving method, compared to normal in situ hybridization (ISH). In addition, the histopathology of diseased *E. carinicauda* was consistent or similar to that previously reported in *P. vannamei*, *C. quadricarinatus*, and *Pr. clarkii* [8,9]. Moreover, in this study, massive DIV1 virions with dense cores, characterized by icosahedral shape and measuring ~150 nm in diameter and consistent with the report of Qiu et al. [9], were observed in hemocytes of hepatopancreatic sinus and gills of *E. carinicauda* of the IM and PO groups. The virions primarily existed in cytoplasm of hemocytes and branchial cells. Additionally, in *E. carinicauda* of the IM group, DIV1 virions were also observed in pereiopods, muscles, and uropods. These results expanded the knowledge of tissue tropism of DIV1.

The combined results of clinical observation, molecular biological diagnosis, histological examination, and TEM supported the assertion that *E. carinicauda* is a newly confirmed susceptible species to infection with DIV1, following the criteria for listing species as susceptible to infection with a specific pathogen developed by the OIE [11]. Presently, there are no reports of natural infection of *E. carinicauda* with DIV1 in a farm or in the sea. As *E. carinicauda* has local economic importance, biosecurity strategies [30,31] should be taken into account, considering the risk that DIV1 can cause disease in *E. carinicauda* and the risk that the species may become a possible reservoir of DIV1. It also should be ignored that *E. carinicauda* showed some degree of tolerance to infection with DIV1, especially per os challenge mimicking a natural pathway.

Author Contributions: Conceptualization, J.H. and X.C.; Formal analysis, X.C.; Methodology, X.C., H.W., L.Q., P.Z., and X.D.; Writing—Original Draft Preparation, X.C.; Writing—Review & Editing, J.H., H.W., F.L., and X.C.; Supervision, J.H.; Project administration, J.H. and X.C.; Funding acquisition, J.H.

Acknowledgments: We are grateful to all the laboratory members for their technical advice and helpful discussions.

References

1. Jun, K.; Kazuhiro, N. Megalocytiviruses. *Viruses* **2012**, *4*, 521–538. [CrossRef]
2. Chinchar, V.G.; Yu, K.H.; Jancovich, J.K. The Molecular biology of Frog virus 3 and other Iridoviruses infecting cold-blooded vertebrates. *Viruses* **2011**, *3*, 1959–1985. [CrossRef] [PubMed]
3. Chinchar, V.G.; Hick, P.; Ince, I.A.; Jancovich, J.K.; Marschang, R.; Qin, Q.W.; Subramaniam, K.; Waltzek, T.B.; Whittington, R.; Williams, T.; et al. ICTV virus taxonomy profile: Iridoviridae. *J. Gen. Virol.* **2017**, *98*, 890–891. [CrossRef] [PubMed]

4. Montanie, H.; Bonami, J.R.; Comps, M. Irido-like virus infection in the crab *Macropipus depurator* L. (Crustacea, Decapoda). *J. Invertebr. Pathol.* **1993**, *61*, 320–322. [CrossRef]
5. Lightner, D.V.; Redman, R.M. A putative iridovirus from the penaeid shrimp *Protrachypene precipua* Burkenroad (Crustacea Decapoda). *J. Invertebr. Pathol.* **1993**, *1*, 107–109. [CrossRef]
6. Miao, H.Z.; Tong, S.L.; Xu, B.; Jiang, M.; Liu, X.Y. Viral and Pathological observation in cultured lymphoid tissues of shrimp *Penaeus chinensis*. *J. Fish. China* **1999**, *2*, 169–173. [CrossRef]
7. Tang, K.F.J.; Redman, R.M.; Pantoja, C.R.; Groumellec, M.L.; Duraisamy, P.; Lightner, D.V. Identification of an iridovirus in *Acetes erythraeus* (Sergestidae) and the development of in situ hybridization and PCR method for its detection. *J. Invertebr. Pathol.* **2007**, *96*, 255–260. [CrossRef]
8. Xu, L.M.; Wang, T.; Li, F.; Yang, F. Isolation and preliminary characterization of a new pathogenic iridovirus from redclaw crayfish *Cherax quadricarinatus*. *Dis. Aquat. Organ.* **2016**, *120*, 17–26. [CrossRef]
9. Qiu, L.; Chen, M.M.; Wan, X.Y.; Li, C.; Zhang, Q.L.; Wang, R.Y.; Cheng, D.Y.; Dong, X.; Yang, B.; Wang, X.H.; et al. Characterization of a new member of *Iridoviridae*, Shrimp hemocyte iridescent virus (SHIV), found in white leg shrimp (*Litopenaeus vannamei*). *Sci. Rep.* **2017**, *7*. [CrossRef]
10. Chinchar, V.G.; Yang, F.; Huang, J.; Williams, T.; Whittington, R.; Jancovich, J.; Subramaniam, K.; Waltzek, T.; Hick, P.; Ince, I.A.; et al. One new genus with one new species in the subfamily Betairidovirinae. ICTV taxonomic proposal to the Iridoviridae Study Group of International Committee for Taxonomy of Viruses, 2018.004D. ICTV 2018. Available online: https://talk.ictvonline.org/files/ictv_official_taxonomy_updates_since_the_8th_report/m/animal-dna-viruses-and-retroviruses/8051 (accessed on 10 March 2019).
11. OIE. Chapter 1.5 Criteria for listing species as susceptible to infection with a specific pathogen. In *Aquatic Animal Health Code*, 21nd ed.; World Organisation for Animal Health: Paris, France, 2018; ISBN 978-92-95108-69-1. Available online: http://www.oie.int/index.php?id=171&L=0&htmfile=chapitre_criteria_species.htm (accessed on 10 March 2019).
12. Xu, W.; Xie, J.; Shi, H.; Li, C. Hematodinium infections in cultured ridgetail white prawns, *Exopalaemon carinicauda*, in eastern China. *Aquaculture* **2010**, *300*, 25–31. [CrossRef]
13. Wang, X.Q.; Yan, B.L.; Ma, S.; Dong, S.L. Study on the biology and cultural ecology of *Exopalaemon carinicauda*. *Shandong Fish.* **2005**, *8*, 21–23.
14. OIE. Manual of Diagnostic Tests for Aquatic Animals, 7th Edition. World Organisation for Animal Health. 2016; ISBN 978-92-9044-887-7. Available online: http://www.oie.int/standard-setting/aquatic-manual/access-online/ (accessed on 10 January 2019).
15. Li, X.P.; Wan, X.Y.; Xu, T.T.; Huang, J.; Zhang, Q.L. Development and validation of a TaqMan RT-qPCR for the detection of convert mortality nodavirus (CMNV). *J. Virol. Methods* **2018**, *62*, 65–71. [CrossRef]
16. Qiu, L.; Chen, M.M.; Wan, X.Y.; Zhang, Q.L.; Li, C.; Dong, X.; Yang, B.; Huang, J. Detection and quantification of shrimp hemocyte iridescent virus by TaqMan probe based real-time PCR. *J. Invertebr. Pathol.* **2018**, *154*, 95–101. [CrossRef] [PubMed]
17. Huang, J.; Song, X.L.; Yu, J.; Zhang, L.J. The components of an inorganic physiological buffer for *Penaeus chinensis*. *Methods Cell Sci.* **1999**, *21*, 225–230. [CrossRef]
18. Bell, T.A.; Lightner, D.V. *A Handbook of Normal Penaeid Shrimp Histology*; World Aquaculture Society: Baton Rouge, LA, USA, 1988; ISBN 0-935868-37-2.
19. Jitrakorn, S.; Arunrut, N.; Sanguanrut, P.; Flegel, T.W.; Kiatpathomchai, W.; Saksmerprome, V. In situ DIG-labeling, loop-mediated DNA amplification (ISDL) for highly sensitive detection of infectious hypodermal and hematopoietic necrosis virus (IHHNV). *Aquaculture* **2016**, *456*, 36–43. [CrossRef]
20. Lightner, D.V. *A Handbook of Shrimp Pathology and Diagnostic Procedures for Disease of Cultured Penaeid Shrimp*; World Aquaculture Society: Baton Rouge, LA, USA, 1996; ISBN 0-962452-99-8.
21. Graham, L.; Orenstein, J.M. Processing tissue and cells for transmission electron microscopy in diagnostic pathology and research. *Nat. Protoc.* **2007**, *2*, 2439–2450. [CrossRef] [PubMed]
22. Panphut, W.; Senapin, S.; Sriurairatana, S.; Withyachumnarnkul, B.; Flegel, T.W. A novel integrase-containing element may interact with Laem-Singh virus (LSNV) to cause slow growth in giant tiger shrimp. *BMC Vet. Res.* **2011**, *7*, 18. [CrossRef]
23. Qiu, L.; Dong, X.; Wan, X.Y.; Huang, J. Analysis of iridescent viral disease of shrimp (SHID) in 2017. In *Analysis of Important Diseases of Aquatic Animals in China in 2017*; Fishery and Fishery Administration Bureau under the Ministry of Agriculture and Rural Affairs, National Fishery Technical Extension Center,

Ed.; China Agriculture Press: Beijing, China, 2018; pp. 187–204. ISBN 978-7-109-24522-8. (In Chinese)
24. Qiu, L.; Chen, X.; Zhao, R.H.; Li, C.; Gao, W.; Zhang, Q.L.; Huang, J. Description of a natural infection with decapod iridescent virus 1 in farmed giant freshwater prawn, *Macrobrachium rosenbergii*. *Viruses* **2019**, *11*, 354. [CrossRef]
25. Kun, P.C.; Fang, Y.H.; Tian, W.T.; Feng, Y.; Ming, C.J. Study of *Cherax quadricarinatus* iridescent virus in two crabs. *J. Appl. Oceanogr.* **2017**, *1*, 82–86. [CrossRef]
26. Lei, Z.W.; Huang, J.; Shi, C.Y.; Zhang, L.J.; Yu, K.K. Investigation into the hosts of White spot syndrome virus (WSSV). *Oceanologia et Limnologia Sinica* **2002**, *33*, 250–258.
27. Liu, S.; Li, J.T.; Tian, Y.; Wang, C.; Li, X.P.; Xu, T.T.; Li, J.; Zhang, Q.L. Experimental vertical transmission of covert mortality nodavirus in *Exopalaemon carinicauda*. *J. Gen. Virol.* **2017**, *98*, 652–661. [CrossRef]
28. Zhang, J.; Wang, J.; Gui, T.; Sun, Z.; Xiang, J. A copper-induced metallothionein gene from *Exopalaemon carinicauda* and its response to heavy metal ions. *Int J. Biol. Macromol.* **2014**, *70*, 246–250. [CrossRef]
29. Huang, J.; Song, X.L.; Yu, J.; Yang, C.H. Baculoviral hypodermal and hematopoietic necrosis-study on the pathogen and pathology of the explosive epidemic disease of shrimp. *Mar. Fish. Res.* **1995**, *16*, 1–10. (In Chinese with English abstract)
30. Huang, J.; Zeng, L.B.; Dong, X.; Liang, Y.; Xie, G.S.; Zhang, Q.L. Trend Analysis and Policy Recommendation on Aqutic Biosecurity in China. *Eng. Sci.* **2016**, *18*, 15–21. (In Chinese with English abstract)
31. Huang, J.; Dong, X.; Zhang, Q.L.; Wan, X.Y.; Xie, G.S.; Yang, B.; Wang, H.L.; Yang, Q.; Xu, H.; Wang, X.H. Emerging diseases and biosecurity strategies for shrimp farming. In *#We R Aquaculture, AQUA 2018*; The World Aquaculture Society: Monpellier, France, 25–29 August 2018; p. 356.

8

Localization of *Frog Virus 3* Conserved Viral Proteins 88R, 91R and 94L

Emily Penny and Craig R. Brunetti *

Biology Department, Trent University, 1600 West Bank Dr, Peterborough, ON K9J 7B8, Canada; emily_penny@hotmail.com
* Correspondence: craigbrunetti@trentu.ca

Abstract: The characterization of the function of conserved viral genes is central to developing a greater understanding of important aspects of viral replication or pathogenesis. A comparative genomic analysis of the iridoviral genomes identified 26 core genes conserved across the family *Iridoviridae*. Three of those conserved genes have no defined function; these include the homologs of *frog virus 3* (FV3) open reading frames (ORFs) 88R, 91R, and 94L. Conserved viral genes that have been previously identified are known to participate in a number of viral activities including: transcriptional regulation, DNA replication/repair/modification/processing, protein modification, and viral structural proteins. To begin to characterize the conserved FV3 ORFs 88R, 91R, and 94L, we cloned the genes and determined their intracellular localization. We demonstrated that 88R localizes to the cytoplasm of the cell while 91R localizes to the nucleus and 94L localizes to the endoplasmic reticulum (ER).

Keywords: *Iridoviridae*; frog virus 3; FV3; ranavirus; immunofluorescence; intracellular localization

1. Introduction

The viral family *Iridoviridae* are large (~120–200 nm in diameter) icosahedral dsDNA viruses that replicate in the cytoplasm and nucleus of infected cells [1]. Iridoviral genomes are terminally redundant and circularly permuted [2–4]. Iridoviruses are known to infect invertebrates as well as poikilothermic vertebrates such as: fish, amphibians, and reptiles [5]. Members of the *Iridoviridae*, some of which exhibit very high pathogenicity, have been implicated in significant infection events in wild and captive populations, leading to considerable impacts on wildlife conservation, fish farming, and aquaculture [6]. The family *Iridoviridae* is currently divided into two subfamilies the *Alphairidovirinae* and *Betairidovirinae*. The *Alphairidovirinae* is composed of three genera *Lymphocystivirus*, *Megaloctyivirus*, and *Ranavirus* while the *Betairidovirinae* is composed of the *Chloriridovirus* and *Iridovirus* genera [7,8].

A comparative genomic analysis of members of the family *Iridoviridae* demonstrated that there are 26 open reading frames (ORFs) that are conserved across all members of the family [9]. While novel genes generally define the unique aspects of the life cycle of a particular virus species, conserved genes are usually related to or play a role in essential mechanisms of the viral life cycle [9–16]. There are a number of functions that conserved genes are commonly involved in including: RNA and DNA synthesis and modification, viral protein processing, and viral structure and assembly [9–16]. The characterization and further understanding of the function of conserved viral genes is central to developing a greater understanding of important aspects of viral replication or pathogenesis.

Frog virus 3 (FV3) is the type species of the genus *Ranavirus* and is the most extensively studied iridovirus at the molecular level [6]. The FV3 genome, excluding the terminal redundancy, is 105,903 base pairs long and encodes 98 non-overlapping predicted ORFs, which range in size from 50 to 1293 amino acids in length [15]. Although 26 genes are conserved across the family *Iridoviridae* [9], the products of three of the identified core genes do not appear to have a fully characterized function; these include FV3 ORFs 88R, 91R, and 94L [15]. Though the proteins encoded by these 3 ORFS have yet to be fully characterised within the family *Iridoviridae*, it is likely that their functions are essential based on their highly conserved nature among the iridoviruses.

Sequence homology provides some insight into potential roles for these genes. FV3 88R has similarity to ERV1/ALR (augmenter of liver regeneration) family of proteins [15]. In mammals, ALR is involved in the reduction of renal damage in liver transplant recipients [17]. In viruses, homologues of the ERV1/ALR family of proteins are highly conserved, though poorly characterized [18,19]. Viral ERV1/ALR homologues are thiol oxidoreductases; which are involved in the formation of disulfide bonds and are usually contained in the endoplasmic reticulum (ER) of normal eukaryotic cells [18,20].

In addition to 88R, basic information can be obtained from FV3 91R and 94L sequences. 91R has sequence similarity to immediate-early protein ICP-46 [15]. The FV3 gene 94L has a KKXX-like ER retrieval motif that was identified at the C-terminal of 94L (LRKV). A KKXX motif is responsible for recycling protein from the Golgi complex to the ER [21] suggesting that 94L might be localized to the secretory pathway.

In this paper, we propose to begin to characterize these genes by cloning the putative genes and determining the site of intracellular accumulation of the proteins by ectopic expression and microscopic analysis.

2. Materials and Methods

2.1. Isolation of FV3 DNA

Fathead minnow (FHM) (ATCC CCL-42) cells grown to approximately 80% confluence were infected with FV3 (American Type Culture Collection, Manassas, VA, USA) at a multiplicity of infection (MOI) of 0.1. When cytopathic effects (CPE) were observed, the cells were harvested and resuspended in 400 µL of water. The cells were freeze (−80 °C)-thawed 3 times. An equal volume of phenol/chloroform was added to the cells and after mixing by vortexing, the aqueous phase was transferred to a fresh tube. The phenol/chloroform extraction step was repeated twice more. Following transfer, 10% (*v*/*v*) 5 M sodium acetate and 200% (*v*/*v*) ethanol (100%) was added to the tube and the mixture was incubated on ice for 15 min. The viral DNA was then spun at 10,000× *g* for 10 min. The pellet of viral DNA was air dried and resuspended in 50 µL of water and stored at −20 °C.

2.2. Cloning FV3 88R, 91R, and 94L into pGEM-T Easy Vector

Gene specific oligonucleotide primers were designed for FV3 ORFs 88R, 91R, and 94L. The primers were designed to add a *Hind*III restriction enzyme site at the 5′ end of each gene and a *Xho*I restriction enzyme site at the 3′ end of each gene (Invitrogen, Burlington, Canada) (Table 1). PCR was then performed using these gene specific primers. In a PCR reaction tube, 1× PCR buffer (Invitrogen, Burlington, Canada), 3 mM $MgCl_2$ (Invitrogen, Burlington, Canada), 2.5 units of TAQ DNA polymerase at 5 units/µL (Invitrogen, Burlington, Canada), 0.2 mM dNTPS, 0.1 mM gene specific forward and reverse primers (Table 1), 1 µg FV3 DNA and water to a final volume of 50 µL. The PCR reaction was then placed in a thermocycler for 30 cycles of 94 °C for 30 s, 52 °C for 30 s and 72 °C for 2 min. The PCR product, amplified DNA encoding each target sequence, was ligated into the pGEM-T Easy vector (Promega, Madison, WI, USA) as per the manufacturer's protocol. The reaction was then stored at 4 °C overnight to ensure optimal ligation. Plasmid vectors were then transformed into DH5-α *E. coli* cells (Invitrogen, Burlington, Canada) following the manufacturer's protocol and plated on a LB agar plates prepared with 10 mg/mL tryptone, 5 mg/mL yeast extract, 10 mg/mL NaCl, 1.5% agar and 0.1% ampicillin (50mg/mL). Colonies were isolated and screened for the insert.

Table 1. Oligonucleotide sequences designed to isolate the conserved genes from the *frog virus 3* (FV3) genome. Restriction sites were added 3 base pairs upstream of the start codon (*Hind*III—AAGCTT) and in the place of the stop codon (*Xho*I—CTCGAG) was added at the 3′ end of each gene.

Name	Sequence
88R-F	5′-AAGCTTAAAATGCACGGTTGCAATTG-3′
88R-R	5′-CTCGAGGTTAAAAGTGCTCGTATTTG-3′
91R-F	5′-AAGCTTAACATGGCAAACTTTGTGAC-3′
91R-R	5′-CTCGAGGGCTCTGACCACAAACAG-3′
94L-F	5′-AAGCTTGCAATGGATCCAGAAGGAATG-3′
94L-R	5′-CTCGAGCAGCACCTTTCTCAGGTAC-3′

2.3. Cloning of FV3 88R, 91R, and 94L into a Eukaryotic Expression Vector

Bacteria containing the pGEM T-easy 88R, 91R, or 94L was grown in LB media (10 mg/mL tryptone, 5 mg/mL yeast extract, 10 mg/mL NaCl) containing 100 µg/mL ampicillin. Plasmid DNA from the bacteria was extracted using a PureLink HiPure Plasmid DNA Purification Kit (Invitrogen, Burlington, Canada) as per the manufacturer's protocol. A restriction enzyme digest was performed on the plasmid DNA to extract the DNA encoding each ORF of interest and the expression vector pcDNA3.1. Briefly, 20 µg of pGEM-T Easy plasmid DNA or pCDNA3.1 was digested with 25 units *Xho*I at 10 U/µL (Invitrogen, Burlington, Canada) and 25 units of *Hind*III at 10 U/µL (Invitrogen, Burlington, Canada), 1× React 2 reaction buffer, and water to a final volume of 70 µL. The reaction was incubated at 37 °C overnight. The product of the restriction enzyme digest had 14 µL of gel electrophoresis 20% loading dye added to the reaction tube and the entire reaction was run on a 1% agarose gel. The fragments of DNA corresponding with each ORF of interest and pcDNA3.1 were then excised from the gel and the DNA was extracted from the agarose gel using the QIAEX II gel extraction kit (Qiagen, Mississauga, Canada) as per the manufacturer's protocol. The product of the gel extraction for each gene of interest was then ligated into the pcDNA3.1 expression vector using a high efficiency T4 DNA ligase (Invitrogen, Burlington, Canada) as per the manufacturer's protocol. The ligation of each of the genes of interest into pcDNA3.1 created the expression constructs containing the gene of interest in frame with a myc tag and are called pcDNA3-88R-myc, pcDNA3-91R-myc, and pcDNA3-94L-myc.

2.4. Transfection of Eukaryotic Expression Vector into Eukaryotic Cells

Baby Green Monkey Kidney (BGMK) cells (BGMK-DAF P17, American Type Culture Collection, Manassas, VA, USA) were cultured at 37 °C in Dulbecco's Modified Eagle's Medium (DMEM; Gibco, Burlington, Canada), which is supplemented with 7% (*v*/*v*) heat-inactivated fetal bovine serum, 1% (*v*/*v*) L-glutamine and 1% (*v*/*v*) penicillin-streptomycin, on glass cover slips in a 6-well plate, to 90–95% confluence. Eukaryotic expression vectors, pcDNA3-88R-myc, pcDNA3-91R-myc, or pcDNA3-94L-myc, were transfected into cells using Lipofectamine 2000 (Invitrogen, Burlington, Canada). Specifically, 4 µg of DNA was diluted in 250 µL of serum-free DMEM (1% (*v*/*v*) L-glutamine, 1% (*v*/*v*) penicillin-streptomycin) and mixed gently. 10 µL of Lipofectamine 2000 was diluted in 250 µL of serum-free DMEM and incubated for 5 min. The diluted DNA and diluted Lipofectamine 2000 were combined, for a total volume of 500 µL, mixed gently and incubated at room temperature for 20 min. The 500 µL of complexes was then added to each well, containing cells and 2 mL supplemented DMEM growth medium and mixed by gently rocking the plate back and forth. The transfected cells were incubated at 37 °C for 48 h.

2.5. Indirect Immunofluorescence

The transfected cells were fixed and permeabilized 48 hours post-transfection. Specifically, the wells were each washed twice in 2 mL of phosphate buffered saline solution (Gibco, Burlington, Canada) with a pH of 7.4 for two minutes each wash. 1 mL of a 3.7% paraformaldehyde solution in PBS was applied to each well for 10 min to fix the cells. Upon removal of the paraformaldehyde solution, the wells were washed twice in 2 mL of PBS for 2 min each wash. The cells were then incubated with 1 mL of a 0.1% Triton X-100 in PBS for 4 min. The cells were then washed twice in 2 mL of PBS. The cells were then incubated overnight at 4 °C with 2 mL of block buffer (50 mM Tris HCl (pH 6.8), 150 mM NaCl, 0.5% (*v*/*v*) NP-40, 5.0 mg/mL BSA). The block buffer was removed and the wells were washed twice with 2 mL of Wash Buffer (50 mM Tris HCl (pH6.8), 150 mM NaCl, 0.5% (*v*/*v*) N-40, 1 mg/mL BSA). The cover slips were removed from the well and the primary antibodies were applied to the cells and incubated for 60 min at room temperature. The primary antibodies used were rabbit anti-myc (Sigma, Saint Louse, USA) diluted to 1:400 in wash buffer, mouse anti-c-myc (Roche, Mississauga, Canada) diluted to 1:400 in wash buffer, and rabbit anti-protein disulfide isomerase (PDI) (Sigma, Saint Louis, MO, USA) diluted to 1:500 in wash buffer. The cells were then washed 3 times with Wash Buffer. Secondary antibodies goat anti-mouse conjugated Cy3 (Jackson ImmunoResearch Laboratories, Inc., West Grove, PA, USA) diluted to 1:200 in Wash Buffer and goat anti-rabbit conjugated fluorescein (FITC) (Jackson ImmunoResearch Laboratories, Inc., West Grove, USA) diluted to 1:200 in Wash Buffer. The cells were incubated at room temperature for 60 min and were washed 3 times with Wash Buffer. The cells were then incubated with TO-PRO-3 at a dilution of 1:8000 in water (Molecular Probes, Eugene, OR, USA) for 7 min. The cells were washed 3 times in Wash Buffer and then mounted on slides using Vectashield (Vector Laboratories, Burlingame, CA, USA). The slides were then visualized using a Leica TCS SP2 SE confocal microscope. The localization of each gene was confirmed in at least 3 independent transfections. Multiple examples of each gene were visualized, and results were all consistent. The sample images shown are representative and consistent with the staining pattern observed for each gene.

3. Results

3.1. FV3 ORF 88R Localize to the Cytoplasm of Transfected Cells

To determine the site of 88R localization, BGMK cells were transfected with pcDNA3-88R-myc and the proteins were visualized by indirect immunofluorescence. The expression of 88R was found to be distributed throughout the cytoplasm (Figure 1). In particular, the staining of 88R was punctate, seeming to be evenly distributed throughout the cell (Figure 1). We examined many examples of 88R staining and they all shared similar cytoplasmic staining.

Figure 1. FV3 88R localizes to the cytoplasm. Baby Green Monkey Kidney (BGMK) cells were transfected with pcDNA3-88R. 48 hours post transfection, cells were fixed, and indirect immunofluorescence was performed using rabbit anti-myc antibodies (red) and TO-PRO-3 (blue). Images were captured at 100× magnification using a confocal microscope.

3.2. FV3 ORF 91R Localizes to the Nucleus of Transfected Cells

FV3 91R is another protein known to be conserved across the family *Iridoviridae*. BGMK cells were transfected with pcDNA3-91R-myc and proteins visualized by indirect immunofluorescence. Figure 2 demonstrates that 91R colocalizes with the TO-PRO-3 staining demonstrating that 91R localizes to the nucleus. Interestingly, there are pockets within the nucleus that lack both 91R and TO-PRO-3 expression (white arrows in Figure 2). Since TO-PRO-3 binds to DNA, it is interesting that 91R mimics the expression pattern suggesting that it localizes to the areas of the nucleus where the DNA resides.

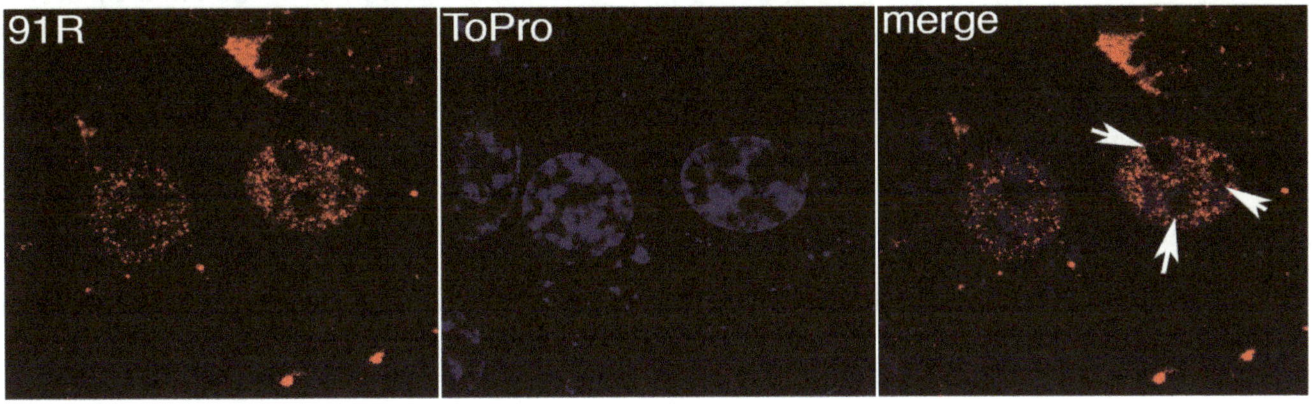

Figure 2. 91R localizes to the nucleus. BGMK cells were transfected with pcDNA3-91R-myc. 48 hours post-transfection, the cells were fixed and indirect immunofluorescence was performed using rabbit anti-myc antibodies (red) and TO-PRO-3 (blue). Images were captured at 100× magnification using a confocal microscope. White arrows highlight nuclear areas that lack 91R and TO-PRO-3.

3.3. FV3 ORF 94L Localizes to the ER in Transfected Cells

To determine the site of intracellular accumulation of 94L, BGMK cells were transfected with pcDNA3-94L-myc. Forty-eight hours post-transfection, the cells were fixed and indirect immunofluorescence was used to visualize 94L-myc protein expression. FV3 94L exhibited strong expression around the nucleus (Figure 3). This perinuclear localization correlated with PDI expression, a marker of the ER [22] (Figure 3). Although the correlation between PDI and 94L expression was not complete, the data does show overlap between the marker and 94L (Figure 3).

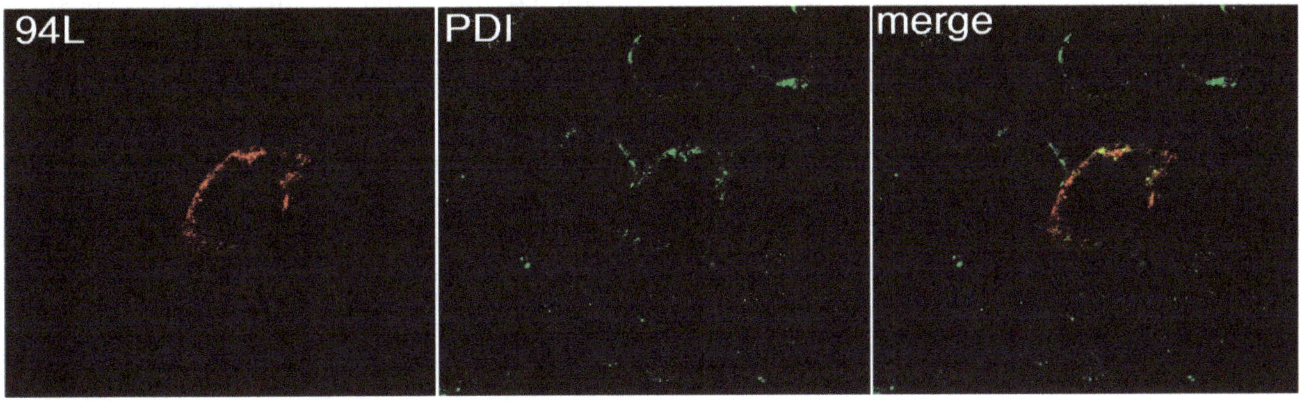

Figure 3. 94L localizes to the Endoplasmic Reticulum. BGMK cells were transfected with pcDNA3-94L-myc. Forty-eight hours post-transfection, the cells were fixed and indirect immunofluorescence was performed using mouse anti-myc antibodies (red) and rabbit anti-PDI antibodies (green). Images were captured at 100× magnification using a confocal microscope.

4. Discussion

The characterization of conserved viral proteins is necessary to gain further insight into the mechanisms of viral infection and replication. Conserved proteins are essential to the viral infection cycle, as their presence in the genomes of multiple members of the same viral family suggest. The purpose of this study was to begin the characterization of the protein products of three FV3 ORFs: 88R, 91R, and 94L.

Cytoplasmic localization was observed for FV3 88R. FV3 88R has similarity to ERV1/ALR family of proteins [15]. Interestingly, the family *Poxviridae* contains a cytoplasmic disulfide bond formation pathway that involves vaccinia virus E10, A2.5, and G4 [18,20]. The vaccinia virus E10 protein is an ERV1/ALR homologue that functions in the cytoplasm as part of the poxvirally encoded cytoplasmic pathway of disulfide bond formation [18,20,23,24]. The localization pattern exhibited by 88R is consistent with the cytoplasmic localization of viral ERV1/ALR homologues [18,20]. Also, it is known that ERV1/ALR homologues are highly conserved in both the *Poxviridae* and *Afarviridae* families, which are some of the most closely related viral families to the family *Iridoviridae* [20,25]. It is possible that FV3 88R performs a similar function to vaccinia virus E10 and there may also be the potential for the identification of a cytoplasmic pathway of disulfide bond formation in iridoviruses.

The second core gene examined, FV3 91R, localized to the nucleus of transfected cells. 91R has sequence similarity to immediate-early protein ICP-46 [15]. Interestingly, transcriptome analysis of FV3 suggested that 91R is an immediate-early gene [26]. FV3 is known to initiate genomic replication early in the viral replication cycle in the nucleus of infected cells [3]. There is a possibility that 91R is involved in an early event in FV3 replication which occurs in the nucleus of infected cells immediately following viral entry. Although the function of 91R may not be limited to being involved in the early events of viral replication. For example, the nuclear localization of 91R might also suggest that 91R regulates viral gene transcription. Therefore there are a number of areas that 91R could function in the nucleus and more research is required to understand the nuclear function of the gene.

The FV3 94L gene colocalized with the ER marker PDI. Many viruses use the ER and secretory pathway [27–29]. For example, both vaccinia virus and African swine fever virus (ASFV) encode viral proteins that are involved in the modulation of the ER and secretory pathway to accommodate viral replication [27–29]. As previously noted, a KKXX-like ER retrieval motif was identified at the C-terminal of 94L (LRKV). A KKXX motif is responsible for recycling protein from the Golgi complex to ER [21]. Retrieval motifs have been found to modulate secretory pathways and it has been experimentally determined that deletion mutant proteins lacking the double lysine motif are transported downstream to the vesicle [21]. The presence of a KKXX-like motif at the C-terminal of 94L further supports the role for 94L in the viral modulation of the secretory pathways of the host cell and adds further support to a role for 94L in the ER. Future research could delete the ER retrieval motif from 94L to see if the localization of the gene is altered.

One caveat of the experiments described is that the genes are ectopically expressed in non-host cells. We used the mammalian cell line, BGMK for a number of reasons. For localization studies, it is much easier to use mammalian cell lines, where a variety of intracellular organelle markers exist. In addition, we have used BGMK cells in the past to localize a variety of FV3 genes, and have found that the localization in BGMK cells mimics the localization in amphibian cell lines [30,31]. It is noteworthy that BGMK cells are infectible by FV3 when the cells are incubated at 30 °C [32]. Although this work was done in a mammalian cell line, it is always possible that the localization of the gene could be altered if the target gene interacts specifically with amphibian proteins. In addition, ectopic expression often results in over-expression of the target protein which may overwhelm intracellular targeting mechanisms resulting in aberrant localization. Finally, it is important to remember that these genes are being expressed outside of an FV3 infection. If 88R, 91R, or 94L interact with other viral proteins, this could also result in altered localization of the gene in and FV3-infected cell.

The results of cellular localization experiments have identified the subcellular expression patterns of the three highly conserved, uncharacterized FV3 proteins included in this study. Based on the results of the localization experiments, possible functions have been assigned to each protein, but these functions are speculative and extensive work is required to describe the actual functions of any of the proteins in an FV3 infection. The further characterization of FV3 88R, 91R, and 94L will provide insight into the cellular cycle of FV3 infection and maybe even certain cellular processes that are modified or commandeered by FV3 in order to successfully infect the host.

Author Contributions: Conceptualization, C.R.B. and E.P.; Methodology, C.R.B. and E.P.; Investigation, C.R.B. and E.P.; Resources, C.R.B.; Writing – Original Draft Preparation, E.P.; Writing – Review & Editing, C.R.B.; Visualization, C.R.B. and E.P.; Supervision, C.R.B.; Project Administration, C.R.B.; Funding Acquisition, C.R.B.

References

1. Goorha, R. Frog virus 3 DNA replication occurs in two stages. *J. Virol.* **1982**, *43*, 519–528. [PubMed]
2. He, J.G.; Deng, M.; Weng, S.P.; Li, Z.; Zhou, S.Y.; Long, Q.X.; Wang, X.Z.; Chan, S.M. Complete genome analysis of the mandarin fish infectious spleen and kidney necrosis iridovirus. *Virology* **2001**, *291*, 126–139. [CrossRef] [PubMed]
3. Goorha, R.; Murti, K.G. The genome of frog virus 3, an animal DNA virus, is circularly permuted and terminally redundant. *Proc. Natl. Acad. Sci. USA* **1982**, *79*, 248–252. [CrossRef] [PubMed]
4. Jakob, N.J.; Muller, K.; Bahr, U.; Darai, G. Analysis of the first complete DNA sequence of an invertebrate iridovirus: Coding strategy of the genome of chilo iridescent virus. *Virology* **2001**, *286*, 182–196. [CrossRef] [PubMed]
5. Williams, T. The iridoviruses. *Adv. Virus Res.* **1996**, *46*, 345–412. [PubMed]
6. Chinchar, V.G.; Waltzek, T.B.; Subramaniam, K. Ranaviruses and other members of the family iridoviridae: Their place in the virosphere. *Virology* **2017**, *511*, 259–271. [CrossRef] [PubMed]
7. Chinchar, V.G.; Hick, P.; Ince, I.A.; Jancovich, J.K.; Marschang, R.; Qin, Q.; Subramaniam, K.; Waltzek, T.B.; Whittington, R.; Williams, T.; et al. Ictv virus taxonomy profile: Iridoviridae. *J. Gen. Virol.* **2017**, *98*, 890–891. [CrossRef]
8. Chinchar, V.G.; Yu, K.H.; Jancovich, J.K. The molecular biology of frog virus 3 and other iridoviruses infecting cold-blooded vertebrates. *Viruses* **2011**, *3*, 1959–1985. [CrossRef]
9. Eaton, H.E.; Metcalf, J.; Penny, E.; Tcherepanov, V.; Upton, C.; Brunetti, C.R. Comparative genomic analysis of the family iridoviridae: Re-annotating and defining the core set of iridovirus genes. *Virol. J.* **2007**, *4*, 11. [CrossRef]
10. Delhon, G.; Tulman, E.R.; Afonso, C.L.; Lu, Z.; Becnel, J.J.; Moser, B.A.; Kutish, G.F.; Rock, D.L. Genome of invertebrate iridescent virus type 3 (mosquito iridescent virus). *J. Virol.* **2006**, *80*, 8439–8449. [CrossRef]
11. Gubser, C.; Hue, S.; Kellam, P.; Smith, G.L. Poxvirus genomes: A phylogenetic analysis. *J. Gen. Virol.* **2004**, *85*, 105–117. [CrossRef]
12. Iyer, L.M.; Aravind, L.; Koonin, E.V. Common origin of four diverse families of large eukaryotic DNA viruses. *J. Virol.* **2001**, *75*, 11720–11734. [CrossRef]
13. McGeoch, D.J. Molecular evolution of the gamma-herpesvirinae. *Philos. Trans. R. Soc. Lond B Biol. Sci.* **2001**, *356*, 421–435. [CrossRef]
14. Song, W.J.; Qin, Q.W.; Qiu, J.; Huang, C.H.; Wang, F.; Hew, C.L. Functional genomics analysis of singapore grouper iridovirus: Complete sequence determination and proteomic analysis. *J. Virol.* **2004**, *78*, 12576–12590. [CrossRef]
15. Tan, W.G.; Barkman, T.J.; Gregory Chinchar, V.; Essani, K. Comparative genomic analyses of frog virus 3, type species of the genus ranavirus (family iridoviridae). *Virology* **2004**, *323*, 70–84. [CrossRef]
16. Zhang, Q.; Gui, J.F. Virus genomes and virus-host interactions in aquaculture animals. *Sci. China Life Sci.* **2015**, *58*, 156–169. [CrossRef]
17. Long, F.W.; Liang, S.Y.; Liu, Z.J.; Chen, Y.; Tu, Z.D.; Shi, Y.J.; Bao, J.; Gong, J. Augmentor of liver regeneration ameliorates renal tubular epithelial cell injury after rat liver transplantation. *Transplant Proc.* **2008**, *40*, 2696–2699. [CrossRef]
18. Su, H.P.; Lin, D.Y.; Garboczi, D.N. The structure of g4, the poxvirus disulfide oxidoreductase essential for virus maturation and infectivity. *J. Virol.* **2006**, *80*, 7706–7713. [CrossRef]
19. Senkevich, T.G.; White, C.L.; Weisberg, A.; Granek, J.A.; Wolffe, E.J.; Koonin, E.V.; Moss, B. Expression of the vaccinia virus a2.5l redox protein is required for virion morphogenesis. *Virology* **2002**, *300*, 296–303. [CrossRef]
20. Senkevich, T.G.; White, C.L.; Koonin, E.V.; Moss, B. A viral member of the erv1/alr protein family participates in a cytoplasmic pathway of disulfide bond formation. *Proc. Natl. Acad. Sci. USA* **2000**, *97*, 12068–12073. [CrossRef]
21. Lotti, L.V.; Mottola, G.; Torrisi, M.R.; Bonatti, S. A different intracellular distribution of a single reporter protein is determined at steady state by kkxx or kdel retrieval signals. *J. Biol. Chem.* **1999**, *274*, 10413–10420. [CrossRef]

22. Akagi, S.; Yamamoto, A.; Yoshimori, T.; Masaki, R.; Ogawa, R.; Tashiro, Y. Distribution of protein disulfide isomerase in rat hepatocytes. *J. Histochem. Cytochem.* **1988**, *36*, 1533–1542. [CrossRef]
23. Senkevich, T.G.; Koonin, E.V.; Bugert, J.J.; Darai, G.; Moss, B. The genome of molluscum contagiosum virus: Analysis and comparison with other poxviruses. *Virology* **1997**, *233*, 19–42. [CrossRef]
24. White, C.L.; Senkevich, T.G.; Moss, B. Vaccinia virus g4l glutaredoxin is an essential intermediate of a cytoplasmic disulfide bond pathway required for virion assembly. *J. Virol.* **2002**, *76*, 467–472. [CrossRef]
25. Lewis, T.; Zsak, L.; Burrage, T.G.; Lu, Z.; Kutish, G.F.; Neilan, J.G.; Rock, D.L. An african swine fever virus erv1-alr homologue, 9gl, affects virion maturation and viral growth in macrophages and viral virulence in swine. *J. Virol.* **2000**, *74*, 1275–1285. [CrossRef]
26. Majji, S.; Thodima, V.; Sample, R.; Whitley, D.; Deng, Y.; Mao, J.; Chinchar, V.G. Transcriptome analysis of frog virus 3, the type species of the genus ranavirus, family iridoviridae. *Virology* **2009**, *391*, 293–303. [CrossRef]
27. Husain, M.; Weisberg, A.S.; Moss, B. Existence of an operative pathway from the endoplasmic reticulum to the immature poxvirus membrane. *Proc. Natl. Acad. Sci. USA* **2006**, *103*, 19506–19511. [CrossRef]
28. Netherton, C.; Rouiller, I.; Wileman, T. The subcellular distribution of multigene family 110 proteins of african swine fever virus is determined by differences in c-terminal kdel endoplasmic reticulum retention motifs. *J. Virol.* **2004**, *78*, 3710–3721. [CrossRef]
29. Tolonen, N.; Doglio, L.; Schleich, S.; Krijnse Locker, J. Vaccinia virus DNA replication occurs in endoplasmic reticulum-enclosed cytoplasmic mini-nuclei. *Mol. Biol. Cell* **2001**, *12*, 2031–2046. [CrossRef]
30. Ring, B.A.; Ferreira Lacerda, A.; Drummond, D.J.; Wangen, C.; Eaton, H.E.; Brunetti, C.R. Frog virus 3 open reading frame 97r localizes to the endoplasmic reticulum and induces nuclear invaginations. *J. Virol.* **2013**, *87*, 9199–9207. [CrossRef]
31. Eaton, H.E.; Ferreira Lacerda, A.; Desrochers, G.; Metcalf, J.; Angers, A.; Brunetti, C.R. Cellular litaf interacts with frog virus 3 75l protein and alters its subcellular localization. *J. Virol.* **2013**, *87*, 716–723. [CrossRef]
32. Eaton, H.E.; Metcalf, J.; Brunetti, C.R. Expression of frog virus 3 genes is impaired in mammalian cell lines. *Virol. J.* **2008**, *5*, 83. [CrossRef]

9

Ranaviruses Bind Cells from Different Species through Interaction with Heparan Sulfate

Fei Ke [1,2], Zi-Hao Wang [1], Cheng-Yue Ming [1] and Qi-Ya Zhang [1,2,*]

1 State Key Laboratory of Freshwater Ecology and Biotechnology, The Innovation Academy of Seed Design, Institute of Hydrobiology, Chinese Academy of Sciences, Wuhan 430072, China
2 College of Modern Agriculture Sciences, University of Chinese Academy of Sciences, Beijing 100049, China
* Correspondence: zhangqy@ihb.ac.cn

Abstract: Ranavirus cross-species infections have been documented, but the viral proteins involved in the interaction with cell receptors have not yet been identified. Here, viral cell-binding proteins and their cognate cellular receptors were investigated using two ranaviruses, *Andrias davidianus* ranavirus (ADRV) and *Rana grylio* virus (RGV), and two different cell lines, Chinese giant salamander thymus cells (GSTC) and *Epithelioma papulosum* cyprinid (EPC) cells. The heparan sulfate (HS) analog heparin inhibited plaque formation of ADRV and RGV in the two cell lines by more than 80% at a concentration of 5 μg/mL. In addition, enzymatic removal of cell surface HS by heparinase I markedly reduced plaque formation by both viruses and competition with heparin reduced virus-cell binding. These results indicate that cell surface HS is involved in ADRV and RGV cell binding and infection. Furthermore, recombinant viral envelope proteins ADRV-58L and RGV-53R bound heparin-Sepharose beads implying the potential that cell surface HS is involved in the initial interaction between ranaviruses and susceptible host cells. To our knowledge, this is the first report identifying cell surface HS as ranavirus binding factor and furthers understanding of interactions between ranaviruses and host cells.

Keywords: ranavirus; virus binding; heparan sulfate; *Andrias davidianus* ranavirus; *Rana grylio* virus; envelope protein

---

1. Introduction

Interspecies transmission and infection have been reported in several viruses that infect humans [1, 2]. Likewise, aquatic animal viruses can also infect and cause disease in a wide range of aquatic animals [3,4]. Specifically, the interspecies infection was reported following ranavirus infection [5,6]. However, the basis for the broad ranavirus host range is not clear [7,8]. Ranaviruses are large double-stranded DNA viruses within the family *Iridoviridae* [9]. Ranaviruses target aquatic animals globally and have been isolated from reptiles [10,11], amphibians [12–15], and bony fish [16,17]. Among them, several isolates represent great threats to the development of the aquaculture industry and wild animal populations [18]. Aside from infecting several species within a given taxonomic class (e.g., *frog virus 3* infects diverse amphibian species), ranaviruses may also infect members of different classes (e.g., *frog virus 3*-like agents have been isolated from both amphibians and fish) [19].

Binding susceptible host cells is the first step in viral infection and is one of the key factors responsible for determining host range [20]. Diverse cellular receptors and viral envelope and capsid proteins are involved in the process. For example, vaccinia virus utilizes four viral proteins that bind to glycosaminoglycans (GAGs) or laminin on the cell surface [21–25].

Cell surface GAGs consist of complex linear polysaccharides, which are ubiquitously expressed in most cell types [26]. Heparan sulfate (HS), chondroitin sulfate (CS), and dermatan sulfate comprise the main types of GAGs on the cell surface. Besides vaccinia virus, GAGs are involved in the binding of adenovirus [27], various alphaviruses [28,29], bunyaviruses [30,31], filoviruses [32], flavivirus [33], hepacivirus [34], herpesvirus [35], papillomaviruses [36], and rhabdoviruses [37]. It has been shown that cell surface GAGs, especially HS, serve as initial receptors in infections with these viruses. However, there is little information on the role of HS in ranavirus binding.

Rana grylio virus (RGV) and *Andrias davidianus* ranavirus (ADRV) are ranaviruses isolated from diseased pig frogs *R. grylio* (anura amphibian) and Chinese giant salamanders (CGS) *A. davidianus* (urodele amphibian), respectively [12,13]. The complete genomes of the two viruses have been sequenced, and several functional proteins have been characterized [38–45]. For example, RGV-53R, a homolog of ADRV-58L, was identified as an envelope protein [38]. Decreased expression of RGV-53R, or its *frog virus 3* (FV3) homolog FV3-53R impaired virus replication in cultured cells [45,46]. In this report, we examine the role of HS in ranavirus entry. Furthermore, since ranavirus viral envelope proteins [38–40,45–48] are likely involved in the initial interaction between virus and host, we examined the ability of RGV-53R and its ADRV homolog, ADVR-58L, to bind heparin-Sepharose beads.

2. Materials and Methods

2.1. Viruses and Cells

ADRV isolated from diseased Chinese giant salamanders (CGS) [13] and RGV isolated from the diseased pig frog *R. grylio* [12] were maintained in our laboratory and used in the present study. CGS thymus cells (GSTC) [49] and *Epithelioma papulosum* cyprinid (EPC) cells [39] were cultured in M199 medium supplemented with 10% bovine calf serum.

2.2. Virus Purification

ADRV and RGV were purified, as described previously [50]. Briefly, GSTC cells were infected with ADRV and RGV at a multiplicity of infection (MOI) of 0.1 PFU/cell, respectively, and incubated at 25 °C. The cell cultures were harvested when cytopathic effects (CPE) reached approximately 90%. The viral suspensions were frozen at −20 °C, thawed three times, and then centrifuged at 5000× *g* for 20 min. The resulting supernatants were ultracentrifuged at 110,000× *g* (Beckman, SW41, Brea, CA, USA) for 90 min. The pellets were resuspended in TE buffer (10 mM of Tris-HCl, 1 mM of EDTA, pH 7.4) and further purified in a discontinuous sucrose gradient (30%, 40%, 50%, and 60%) at 110,000× *g* for 60 min. The viral bands were collected, centrifuged to remove residual sucrose, and the resulting viral pellets were resuspended in TE buffer and stored at −80 °C.

2.3. Plaque Reduction Assay

Heparin is a structural homolog of highly sulfated HS and has been used as a surrogate for cell surface HS in research studies examining binding [51]. The effect of heparin on viral plaque formation was tested in assays using GSTC and EPC cells for the purpose of analyzing the effect of soluble glycosaminoglycans on viral infection. The indicated cells were seeded in 24-well plates 24 h prior to infection. Heparin (Sangon Biotech, Shanghai, China, from porcine intestinal mucosa, molecular weight range of 6–20 kDa) was diluted in cell culture medium and incubated with ADRV or RGV for 1 h at 4 °C. At a concentration of 10 μg/mL, the color of the growth medium (using phenol red as an indicator) did not change, indicating a stable pH. Cell culture media was removed and 100 μL of the virus-heparin solution, containing approximately 50 PFU of the indicated virus, was added to each well for 1 h at 25 °C. Three replicate wells were used in each treatment. After 1 h, the inoculum was removed, the cells were washed twice with fresh medium, and overlaid with culture media containing 0.7% agarose. Fresh medium was added to the well after agarose solidification. Plaques were counted after three days of incubation at 25 °C. Based on the results obtained from the heparin treatment, two other glycosaminoglycans, heparan sulfate and chondroitin sulfate (Sigma, St. Louis, MO, USA), were tested by the method described above.

2.4. Heparinase Treatment

To further investigate its role in virus binding, target cells were treated with heparinase to remove cell surface heparan sulfate. Heparinase I cleaves the linkages between hexosamines and the O-sulfated iduronic acids of heparin and HS. GSTC cells were seeded in 24-well plates 24 h prior to infection. Medium was removed before the assay, and the cells were incubated with different concentrations of heparinase I from *Flavobacterium heparinum* (Sigma) in 20 mM of Tris-HCl (pH 7.5), 4 mM of $CaCl_2$, 50 mM of NaCl, and 0.01% bovine serum albumin (BSA) for 1 h at 15 °C and then washed twice with fresh medium. One hundred microliters of medium containing ADRV or RGV (50 PFU) were added and incubated for another 1 h at 15 °C. The supernatant was removed, the cells washed twice with fresh medium, and overlaid with medium containing 0.7% agarose, as described above. After incubation for three days at 25 °C, the plaques were counted.

2.5. Cell Binding Assay

To investigate the effect of heparin on virus–cell binding, quantitative real-time PCR (qPCR) analysis that has been used in the detection of hepatitis C virus (HCV) genomes [34] was used to determine the relative quantity of virus bound to the cell surface. Both purified virions, isolated following separation on a sucrose step gradient, and a crude viral suspension, obtained by lysis of cells infected by the virus, were used in these assays. GSTC cells were seeded in 24-well plates 24 h prior to infection. Heparin was diluted in medium and mixed with the viral suspension or purified virions for 30 min at 4 °C, and then 100 μL of the mixture containing approximately 1000 PFU of the indicated virus was added to cells for 1 h at 4 °C, as described above. The inoculum was removed after incubation. The cells, washed twice with fresh medium, were collected by centrifugation, and DNA was extracted with the TakaRa MiniBEST Universal Genomic DNA Extraction Kit (TakaRa, Tokyo, Japan). Bound viral genomes, based on detection of the relative numbers of major capsid protein gene (MCP), were detected by qPCR, which was conducted using a StepOne Real-Time PCR system (The Applied Biosystems, Foster City, CA, USA). Each qPCR mixture contained 1 μL of DNA, 12.5 μL of SYBR Premix (2×), 0.5 μL of forward and reversed primers (for each primer, 5′-CACCTCCATCCCAGTCAGCA-3′/5′-AATCCCATCGAGCCGTTCA-3′), and 10.5 μL of ultrapure water. The qPCR conditions were as follows: 95 °C for 10 min; 40 cycles of 95 °C for 15 s and 60 °C for 1 min; and a melt curve analysis at 95 °C for 15 s, 60 °C for 1 min, and 95 °C for 15 s. The β-actin gene, used in a previous study [52], was used as a loading control. qPCR efficiency was evaluated with a standard curve, based on serially diluted DNA samples using the β-actin gene primers, which showed that there were no obvious differences on the qPCR efficiencies among samples treated with different concentrations of heparin. For the MCP detection, MCP levels were normalized to β-actin levels in each sample. The level of bound virus (MCP level) in the treated group versus that in the control group (no heparin) was calculated by the $2^{-\Delta\Delta CT}$ method [53].

2.6. Protein Expression and Purification

Considering the pivotal roles that viral envelope proteins play in virus attachment and entry, we determined whether viral envelope proteins interacted with HS by monitored the binding of purified recombinant proteins to heparin-Sepharose beads as described by Chung et al. [22]. Previous studies demonstrated that RGV-53R is 522 amino acid (aa) envelope protein with two predicted transmembrane (TM) domains (aa 193–211 and aa 218–237) [38]. ADRV-58L is the ADRV homolog of RGV-53R. The aa sequence identity between ADRV-58L and RGV-53R is 99.2% [13]. In the present assay, TM helices and topology of the two proteins were predicted with the online tools HMMTOP (http://www.enzim.hu/hmmtop/html/submit.html) and TMHMM (http://www.cbs.dtu.dk/services/TMHMM-2.0/) that presented on ExPASy. The primers 5′-GTAGAATTCATGGGAGCAGCGGAA-3′ and 5′-GATAAGCTTTATGTGGTGGGGTCCAGGCC-3′ were used for amplifying DNA sequences encoding the N-terminal region (1–192) of 58L and the N-terminal region (1–192) of 53R, respectively. The other pair of primers (5′-GGCGAATTCCCCAGGCCCGTCAAGA-3′/5′-CTATAAGCTTTTAACCCCTGTGGGC-3′) was used for amplifying the DNA sequences for the C-terminal region (238–522) of 58L. Because the amino acid sequence of the C-terminal regions of ADRV-58L and RGV-53R are identical, only the C-terminal region of ADRV-58L was expressed in the present study. The resulting fragments were digested using EcoR I and Hind III, and ligated into pET32a or pET28a vectors that had been digested with the same enzymes. Successful cloning was validated by DNA sequencing.

For protein expression and purification, the plasmids obtained above were used to transform *Escherichia coli* BL21 (DE3). Positive clones were cultured in LB medium and induced with 0.1 mM isopropyl-β-D-thiogalactopyranoside (IPTG) for 4 h at 24 °C. The bacterial pellets were lysed by sonication. The recombinant protein was purified using the HisBind Purification Kit (Novagen, Billerica, MA, USA) according to the manufacturer's instructions. The purified protein was dialyzed against PBS, the concentration determined using a BCA Protein Assay Kit (Beyotime, Wuhan, China), and stored at −80 °C.

2.7. Heparin-Sepharose Binding Assay

Heparin-Sepharose beads 6FF and control Sepharose beads (Purchased from SMART lifesciences, Changzhou, China, particle diameter range of 45–165 μm) were equilibrated with binding buffer (50 mM Tris-HCl, 10 mM sodium citrate, pH 7.4) before use. The purified recombinant protein (5 μg) was mixed with or without heparan sulfate (100 μg/mL) and incubated in binding buffer with 100 μL of beads for 1 h at 4 °C. The supernatant was collected after centrifugation for 1 min at 1800× *g*. The beads were washed with binding buffer (100 μL) five times, and bound protein was eluted with binding buffer containing 2 M NaCl. The samples were analyzed by 12% SDS-PAGE and subsequently transferred to a PVDF membrane (Millipore, Burlington, MA, USA). A monoclonal antibody against the His tag (Santa Cruz, Dallas, Texas, USA) was used as the primary antibody, horseradish peroxidase (HRP)-conjugated goat anti-mouse IgG (H + L) (Merck, Kenilworth, NJ, USA) as the secondary antibody, and antibody binding detected by chemiluminescence (Millipore).

3. Results

3.1. Heparin and HS Inhibit Infection by ADRV and RGV

The effect of heparin on virus infection was tested by monitoring viral plaque formation following incubation of ADRV and RGV in the presence of increasing concentrations of heparin. As shown in Figure 1, the number of plaques formed by either ADRV or RGV was reduced by heparin in a concentration-dependent manner. For ADRV, infectivity was reduced by approximately 70% by pre-exposure to heparin at 0.1 μg/mL and by more than 80% at 5 μg/mL. A similar phenomenon was observed in RGV-infected cells, which exhibited a 57% inhibition at 0.1 μg/mL and more than 75% at 5 μg/mL. Moreover, inhibition was detected regardless of the cell line used. The results indicate that heparin-like GAGs were involved in ADRV and RGV binding.

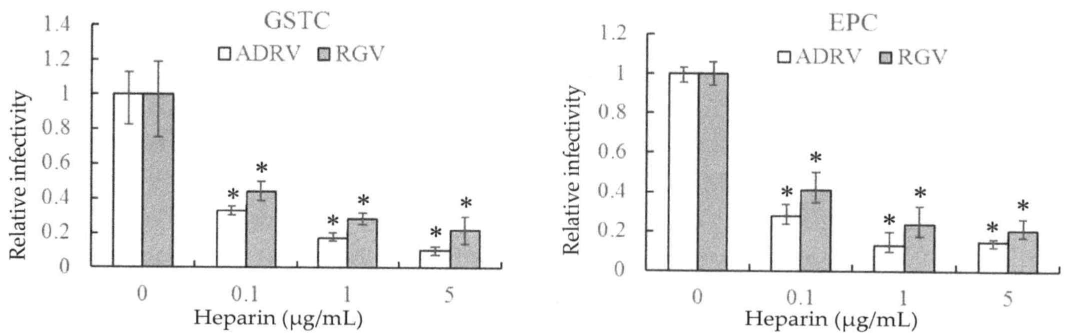

Figure 1. Soluble heparin inhibits *Andrias davidianus* ranavirus (ADRV) and *Rana grylio* virus (RGV) infection of giant salamander thymus cells (GSTC) and *Epithelioma papulosum* cyprinid (EPC) cells.

Cells were infected with ADRV or RGV that had been pre-incubated in the presence of different concentrations of heparin. The number of plaques obtained in the absence of heparin was set as 1. The data represent triplicate results and was analyzed with Student's *t*-test. Significant differences (versus virus without heparin) are marked with * ($p < 0.05$).

In a second experiment, HS and CS, linear polysaccharides that constitute two major classes of cell surface GAGs, were monitored for their ability to reduce plaque formation. As with heparin, HS reduced the number of plaques formed by the two viruses in GSTC and EPC cells in a concentration-dependent manner (Figure 2a). For both viruses, plaque formation was inhibited by more than 80% in GSTC cells and more than 60% in EPC cells at 5 μg/mL (Figure 2a). In contrast, a significant inhibitory effect was only seen at the highest concentration when CS was substituted for HS (Figure 2b). These results indicate that cell surface GAGs, including HS, likely play important roles in plaque formation by both ADRV and RGV.

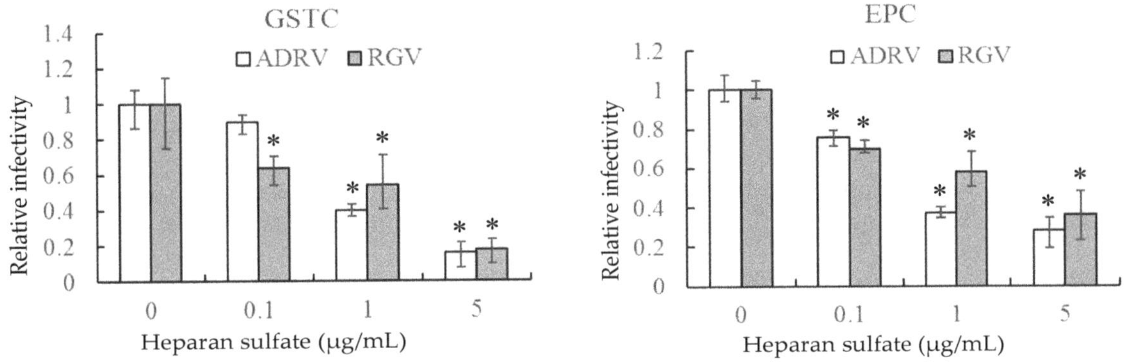

(**a**)

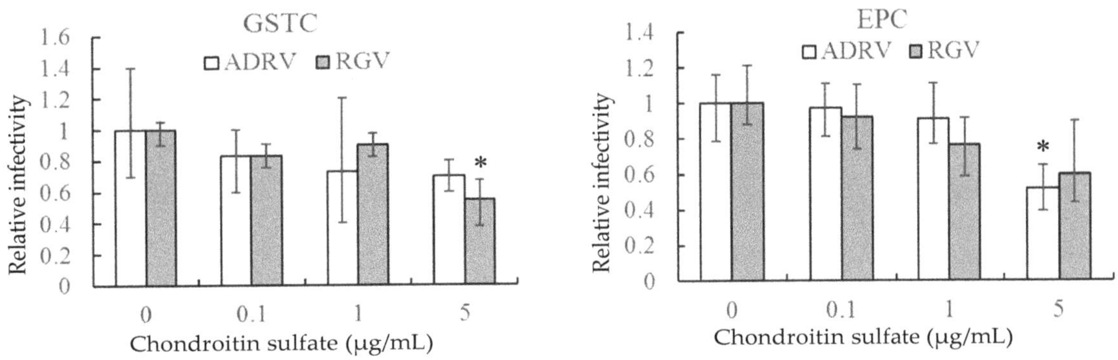

(**b**)

Figure 2. Heparan sulfate (**a**) and chondroitin sulfate (**b**) inhibit ADRV and RGV infection of GSTC and EPC cells. GSTC and EPC cells were infected with ADRV or RGV in the presence of different concentrations of heparan sulfate and chondroitin sulfate. The number of plaques obtained without glycosaminoglycans (GAGs) was set as 1. Experiments were conducted in triplicate and analyzed using Student's *t*-test. Significant differences (versus virus without exposure to GAGs) are marked with * ($p < 0.05$).

3.2. *Enzymatic Removal of Cell Surface HS Reduced Viral Infection*

If the interaction between heparin-like GAGs and viral particles is needed for viral infection, removal of GAGs should inhibit infection. To accomplish that task, heparinase I was used to remove cell surface HS. As shown in Figure 3, the number of plaques formed by the two viruses was markedly reduced in GSTC cells pretreated with heparinase I. The reduction was more than 50% at a heparinase concentration of 1.25 U/mL and slightly more at higher enzyme concentrations. Thus, the ability of heparinase treatment to reduce viral plaque formation supports the view that cell surface HS is a receptor for the two viruses.

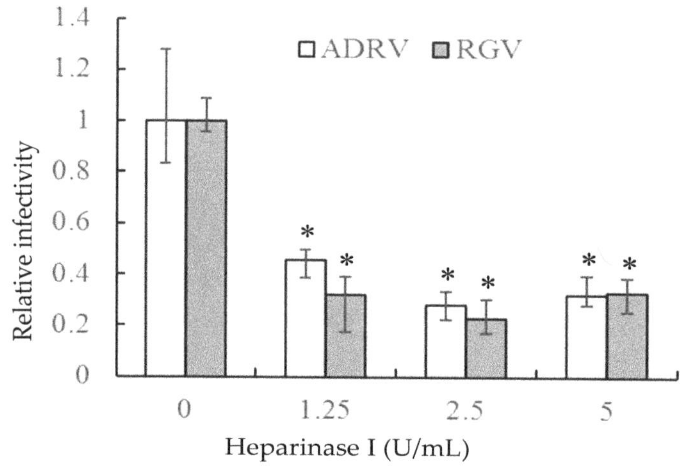

Figure 3. Heparinase I treatment reduced ADRV and RGV plaque formation in GSTC cells. Cells were infected with ADRV or RGV after treatment with different concentrations of heparinase I and plaque formation monitored. Plaque numbers obtained in the absence of heparinase treatment were set as 1. Triplicate results were analyzed by Student's t-test, and significant differences are marked with * ($p < 0.05$).

3.3. *Heparin Inhibits Virus–Cell Binding*

To further verify the role of cell surface HS on virus–cell binding, heparin was used to inhibit the binding of the two viruses competitively. The number of bound virions was determined by measurement of viral DNA copy number by qPCR to monitor binding. As shown in Figure 4, binding of either crude (viral suspension) or purified virions was inhibited by pre-exposure to heparin. Binding of crude suspensions of either ADRV or RGV were reduced to 20–30% of control levels at a concentration of 10 μg/mL. Similar results were obtained with purified virions. The inhibitory effect of heparin on virus-cell binding supports the finding that cell surface HS is a viral receptor.

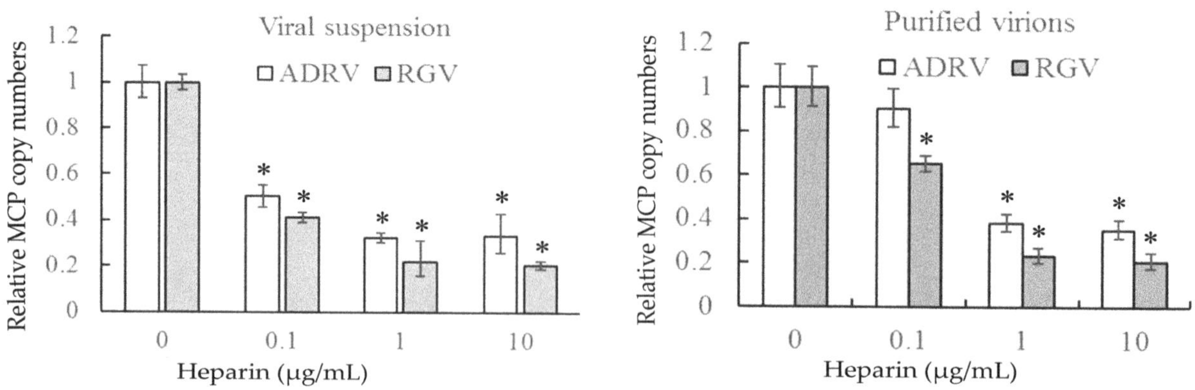

Figure 4. Soluble heparin inhibits virus binding to GSTC cells. Viral suspensions or purified virions were added to GSTC cells in the presence of different concentrations of heparin. After incubation, virion binding was assessed by determining the number of bound viral genomes by qPCR. DNA levels observed in the absence of heparin pre-treatment were set as 1. The data were obtained from three experiments and analyzed with Student's t-test. Significant differences are marked with * ($p < 0.05$).

3.4. *Recombinant Envelope Proteins Bind Heparin Beads In Vitro*

As shown above, cell surface HS is an important receptor for both ADRV and RGV. Here we examine the role that viral envelope proteins play in this process by monitoring the interaction of purified recombinant viral envelope protein with heparin-Sepharose beads. To accomplish this, the amino terminal 192 amino acids of ADRV-58L and RGV-53R and the C-terminal region of ADRV-58L (amino acids 238–522) were cloned into pET32a and pET28a and expressed in *E. coli* (Figure 5a). Because the amino acid sequences of the C-terminal regions of ADRV-58L and RGV-53R are identical, only the C-terminal region of ADRV-58L was expressed. Note the recombinant proteins were increased in size, due to a 17 kDa Trx-His-S tag in pET32a, and a 4 kDa His-T7 tag in pET28a. As shown in Figure 5b,c, recombinant proteins of the expected sizes were generated using the two expression systems. Subsequently, purified recombinant proteins were isolated and incubated with heparin-Sepharose beads or control beads lacking heparin. The three recombinant proteins (r58L-N, r53R-N, and r58L-C), expressed using either pET32a or pET28a, bound heparin-sepharose beads and were eluted in the presence of a high salt wash. In contrast, all three recombinant proteins failed to bind sepharose beads lacking heparin, and, as a result, were present in the unbound supernatant (S) fraction (Figure 5d). When considered together, these results support the view that HS is a cellular receptor for both ADRV and RGV and the binding fashion may be similar to the interaction between 53R/58L and heparin that occurred in vitro.

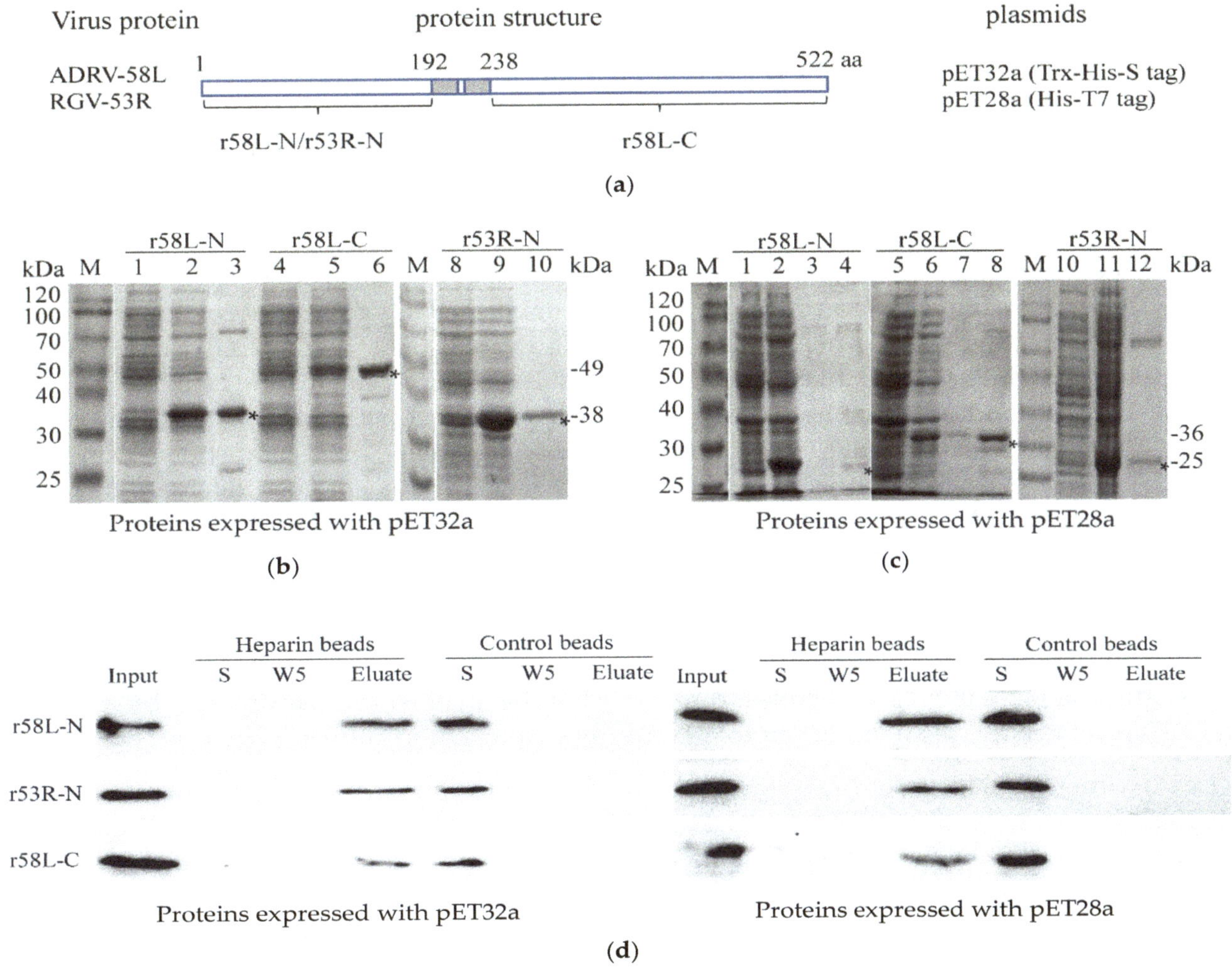

Figure 5. Recombinant proteins bind heparin-Sepharose beads. (**a**) Schematic diagram of the recombinant proteins: The N-terminal domain of ADRV-58L (r58L-N), the N-terminal domain of RGV-53R (r53R-N), and the C-terminal domain of ADRV-58L (r58L-C) were expressed using pET32a or pET28a. The predicted transmembrane region is shown in the grey box. (**b**) Expression and purification of the three proteins (r58L-N, r53R-N, and r58L-C) with pET32a vector. M: protein marker; 1, 4, 8: Bacteria without induction; 2, 5, 9: Bacteria with induction; 3, 6, 10: Purified proteins. The recombinant proteins are indicated with asterisks, and their predicted molecular weights are shown on the right.

(**c**) Expression and purification of the three proteins using pET28a. M: Protein marker; 1, 5, 10: Cacteria without induction; 2, 6, 11: Cacteria with induction; 3, 4, 7, 8, 12: Purified proteins. The recombinant proteins are indicated with asterisks, and their predicted molecular weights are shown on the right. (**d**) Binding of recombinant proteins and heparin-Sepharose beads. Recombinant proteins were incubated with heparin-Sepharose or Sepharose beads. The fractions of input (Input), supernatant after incubation (S), the fifth wash solution (W5), and the eluate (Eluate) were detected by Western blot with the anti-His antibody. Recombinant proteins expressed with pET32a or pET28a vectors were used. Recombinant proteins were observed in the Input and Elute fractions from heparin-Sepharose beads and S fraction from control beads.

4. Discussion

In this study, we tested the ability of heparin and two other GAGs (HS and CS) to inhibit plaque formation in fish and amphibian cell lines. These and other results showed that cell surface HS is an important receptor for the binding of ADRV and RGV to target cells. As far as we know, it is the first report describing the role of cell surface HS in iridovirus infection.

Previously we identified RGV-53R as an envelope protein [38]. Here we tested whether RGV-53R, or its ADRV homolog (58L), could bind HS. Our data showed that recombinant 53R and 58L proteins specifically bound heparin-Sepharose beads in vitro and that this binding could be inhibited by the presence of excess HS. The envelope protein 53R and its homologs among other members of the family constitute one of 26 core proteins and likely function in viral entry. Additional studies will be needed to identify the protein domains involved in 53R-heparan sulfate interaction.

Multiple steps are involved in virion entry and initiation of a successful infection. Binding to cell surface GAGs, including HS, has proved to be the initial event in the entry of several mammalian viruses [27–37]. Because GAGs are linked to proteins and usually exist as proteoglycans in vivo [26], viral envelope proteins may bind to cell surface heparan sulfate-linked proteins and facilitate attachment between the virus and potential host cells. Our results suggest that binding of cell surface HS is required for initiation of productive infection by ranaviruses. However, since competition by increasing concentrations of heparin or HS, or treatment with heparinase did not completely inhibit plaque formation, it appears that HS is not the sole cellular receptor for ranaviruses. A similar phenomenon has been observed in the binding of the vaccinia virus, which uses cell surface GAGs and cellular matrix laminin as receptors [22–25]. A recent study showed that class A scavenger receptors are utilized by *frog virus 3*, the type species of the genus *Ranavirus* [54]. Additional ranavirus binding factors and specific cellular protein receptors may be involved in viral attachment and entry. In addition, although the host range of ranaviruses could be determined by binding to other cellular proteins, binding to cell surface GAGs likely plays an important role in the initial interaction between virus and host cell.

It is worth noting that there are two types of virions for iridoviruses. One contains a central core surrounded by an internal membrane and a viral capsid. The other type has an outer viral envelope after the virions bud from the plasma membrane [9]. The two types of viral particles may have different cellular receptors when they infect cells. Thus, the existence of different virions that have an outer envelope or not could be another reason for the incomplete inhibition efficiency in the present study.

This is a report showing that iridoviruses bind cells of different species through the interaction between cell surface GAGs and envelope proteins. However, it remains to be determined whether binding to GAGs is both necessary and sufficient for subsequent virion entry, or whether virus—GAG interaction precedes binding to a second specific cellular receptor.

Author Contributions: Conceptualization, Q.-Y.Z.; methodology, F.K., Z.-H.W. and C.-Y.M.; writing—original draft preparation, F.K.; writing—review and editing, F.K. and Q.Y.Z.; supervision, funding acquisition, Q.-Y.Z.

Acknowledgments: We acknowledge the editorial assistance of V.G. Chinchar.

References

1. Long, J.S.; Mistry, B.; Haslam, S.M.; Barclay, W.S. Host and viral determinants of influenza A virus species specificity. *Nat. Rev. Microbiol.* **2018**. [CrossRef] [PubMed]
2. Li, W.; Hulswit, R.J.G.; Kenney, S.P.; Widjaja, I.; Jung, K.; Alhamo, M.A.; van Dieren, B.; van Kuppeveld, F.J.M.; Saif, L.J.; Bosch, B.J. Broad receptor engagement of an emerging global coronavirus may potentiate its diverse cross-species transmissibility. *Proc. Natl. Acad. Sci. USA* **2018**, *115*, E5135–E5143. [CrossRef] [PubMed]
3. Zhang, Q.Y.; Gui, J.F. Diversity, evolutionary contribution and ecological roles of aquatic viruses. *Sci. China Life Sci.* **2018**, *61*, 1486–1502. [CrossRef] [PubMed]
4. Gui, L.; Zhang, Q.Y. A brief review on aquatic animal virology researches in China. *J. Fish. China* **2019**, *42*, 1–20.
5. Bandin, I.; Dopazo, C.P. Host range, host specificity and hypothesized host shift events among viruses of lower vertebrates. *Vet. Res.* **2011**, *42*, 67. [CrossRef] [PubMed]
6. Price, S.J.; Ariel, E.; Maclaine, A.; Rosa, G.M.; Gray, M.J.; Brunner, J.L.; Garner, T.W.J. From fish to frogs and beyond: Impact and host range of emergent ranaviruses. *Virology* **2017**, *511*, 272–279. [CrossRef] [PubMed]
7. Gui, L.; Chinchar, V.G.; Zhang, Q.Y. Molecular basis of pathogenesis of emerging viruses infecting aquatic animals. *Aquac. Fish.* **2018**, *3*, 1–5. [CrossRef]
8. Gui, L.; Zhang, Q.Y. Disease prevention and control. In *Aquaculture in China: Success Stories and Modern Trends*; Gui, J.F., Tang, Q.S., Li, Z.J., Liu, J.S., Sena, S.S.D., Eds.; Wiley-Blackwell: Chichester, UK, 2018; pp. 577–598.
9. Chinchar, V.G.; Hick, P.; Ince, I.A.; Jancovich, J.K.; Marschang, R.; Qin, Q.; Subramaniam, K.; Waltzek, T.B.; Whittington, R.; Williams, T.; et al. ICTV virus taxonomy profile: *Iridoviridae*. *J. Gen. Virol.* **2017**, *98*, 890–891. [CrossRef]
10. Marschang, R.E.; Becher, P.; Posthaus, H.; Wild, P.; Thiel, H.J.; Müller-Doblies, U.; Kalet, E.F.; Bacciarini, L.N. Isolation and characterization of an iridovirus from Hermann's tortoises (*Testudo. hermanni*). *Arch. Virol.* **1999**, *144*, 1909–1922. [CrossRef]
11. Huang, Y.; Huang, X.; Liu, H.; Gong, J.; Ouyang, Z.; Cui, H.; Cao, J.; Zhao, Y.; Wang, X.; Jiang, Y.; et al. Complete sequence determination of a novel reptile iridovirus isolated from soft-shelled turtle and evolutionary analysis of *Iridoviridae*. *BMC Genom.* **2009**, *10*, 224. [CrossRef]
12. Zhang, Q.Y.; Xiao, F.; Li, Z.Q.; Gui, J.F.; Mao, J.; Chinchar, V.G. Characterization of an iridovirus from the cultured pig frog *Rana. grylio* with lethal syndrome. *Dis. Aquat. Organ.* **2001**, *48*, 27–36. [CrossRef]
13. Chen, Z.Y.; Gui, J.F.; Gao, X.C.; Pei, C.; Hong, Y.J.; Zhang, Q.Y. Genome architecture changes and major gene variations of *Andrias. davidianus* ranavirus (ADRV). *Vet. Res.* **2013**, *44*, 101. [CrossRef] [PubMed]
14. Chinchar, V.G.; Yu, K.H.; Jancovich, J.K. The molecular biology of frog virus 3 and other iridoviruses infecting cold-blooded vertebrates. *Viruses* **2011**, *3*, 1959–1985. [CrossRef] [PubMed]
15. Jancovich, J.K.; Mao, J.; Chinchar, V.G.; Wyatt, C.; Case, S.T.; Kumar, S.; Valente, G.; Subramanian, S.; Davidson, E.W.; Collins, J.P.; et al. Genomic sequence of a ranavirus (family *Iridoviridae*.) associated with salamander mortalities in North America. *Virology* **2003**, *316*, 90–103. [CrossRef] [PubMed]
16. Whittington, R.J.; Becker, J.A.; Dennis, M.M. Iridovirus infections in finfish—critical review with emphasis on ranaviruses. *J. Fish. Dis.* **2010**, *33*, 95–122. [CrossRef]
17. Qin, Q.W.; Chang, S.F.; Ngoh-Lim, G.H.; Gibson-Kueh, S.; Shi, C.; Lam, T.J. Characterization of a novel ranavirus isolated from grouper *Epinephelus. tauvina*. *Dis. Aquat. Org.* **2003**, *53*, 1–9. [CrossRef] [PubMed]
18. Zhang, Q.Y.; Gui, J.F. Virus genomes and virus-host interactions in aquaculture animals. *Sci. China Life Sci.* **2015**, *58*, 156–169. [CrossRef]
19. Chinchar, V.G.; Waltzek, T.B. Ranaviruses: not just for frogs. *PLoS Path.* **2014**, *10*, e1003850. [CrossRef]
20. Maginnis, M.S. Virus-receptor interactions: the key to cellular invasion. *J. Mol. Biol.* **2018**, *430*, 2590–2611. [CrossRef]
21. Moss, B. Poxvirus cell entry: How many proteins does it take? *Viruses* **2012**, *4*, 688–707. [CrossRef]
22. Chung, C.S.; Hsiao, J.C.; Chang, Y.S.; Chang, W. A27L protein mediates vaccinia virus interaction with cell surface heparan sulfate. *J. Virol.* **1998**, *72*, 1577–1585. [PubMed]
23. Chiu, W.L.; Lin, C.L.; Yang, M.H.; Tzou, D.L.; Chang, W. Vaccinia virus 4c (A26L) protein on intracellular mature virus binds to the extracellular cellular matrix laminin. *J. Virol.* **2007**, *81*, 2149–2157. [CrossRef] [PubMed]
24. Hsiao, J.C.; Chung, C.S.; Chang, W. Vaccinia virus envelope D8L protein binds to cell surface chondroitin sulfate and mediates the adsorption of intracellular mature virions to cells. *J. Virol.* **1999**, *73*, 8750–8761. [PubMed]

25. Lin, C.L.; Chung, C.S.; Heine, H.G.; Chang, W. Vaccinia virus envelope H3L protein binds to cell surface heparan sulfate and is important for intracellular mature virion morphogenesis and virus infection in vitro and in vivo. *J. Virol.* **2000**, *74*, 3353–3365. [CrossRef] [PubMed]
26. Aquino, R.S.; Park, P.W. Glycosaminoglycans and infection. *Front. Biosci. (Landmark Ed.)* **2016**, *21*, 1260–1277. [PubMed]
27. Dechecchi, M.C.; Tamanini, A.; Bonizzato, A.; Cabrini, G. Heparan sulfate glycosaminoglycans are involved in adenovirus type 5 and 2-host cell interactions. *Virology* **2000**, *268*, 382–390. [CrossRef] [PubMed]
28. Byrnes, A.P.; Griffin, D.E. Binding of Sindbis virus to cell surface heparan sulfate. *J. Virol.* **1998**, *72*, 7349–7356. [PubMed]
29. Ciano, K.A.; Saredy, J.J.; Bowers, D.F. Heparan sulfate proteoglycan: an arbovirus attachment factor integral to mosquito salivary gland ducts. *Viruses* **2014**, *6*, 5182–5197. [CrossRef]
30. Riblett, A.M.; Blomen, V.A.; Jae, L.T.; Altamura, L.A.; Doms, R.W.; Brummelkamp, T.R.; Wojcechowskyj, J.A. A haploid genetic screen identifies heparan sulfate proteoglycans supporting rift valley fever virus infection. *J. Virol.* **2016**, *90*, 1414–1423. [CrossRef]
31. Murakami, S.; Takenaka-Uema, A.; Kobayashi, T.; Kato, K.; Shimojima, M.; Palmarini, M.; Horimoto, T. Heparan sulfate proteoglycan is an important attachment factor for cell entry of akabane and schmallenberg viruses. *J. Virol.* **2017**, *91*, e00503-17. [CrossRef]
32. Salvador, B.; Sexton, N.R.; Carrion, R., Jr.; Nunneley, J.; Patterson, J.L.; Steffen, I.; Lu, K.; Muench, M.O.; Lembo, D.; Simmons, G. Filoviruses utilize glycosaminoglycans for their attachment to target cells. *J. Virol.* **2013**, *87*, 3295–3304. [CrossRef] [PubMed]
33. Kim, S.Y.; Zhao, J.; Liu, X.; Fraser, K.; Lin, L.; Zhang, X.; Zhang, F.; Dordick, J.S.; Linhardt, R.J. Interaction of zika virus envelope protein with glycosaminoglycans. *Biochemistry* **2017**, *56*, 1151–1162. [CrossRef] [PubMed]
34. Xu, Y.; Martinez, P.; Seron, K.; Luo, G.; Allain, F.; Dubuisson, J.; Belouzard, S. Characterization of hepatitis C virus interaction with heparan sulfate proteoglycans. *J. Virol.* **2015**, *89*, 3846–3858. [CrossRef] [PubMed]
35. Herold, B.C.; WuDunn, D.; Soltys, N.; Spear, P.G. Glycoprotein C of herpes simplex virus type 1 plays a principal role in the adsorption of virus to cells and in infectivity. *J. Virol.* **1991**, *65*, 1090–1098. [PubMed]
36. Giroglou, T.; Florin, L.; Schäfer, F.; Streeck, R.E.; Sapp, M. Human papillomavirus infection requires cell surface heparan sulfate. *J. Virol.* **2001**, *75*, 1565–1570. [CrossRef] [PubMed]
37. Sasaki, M.; Anindita, P.D.; Ito, N.; Sugiyama, M.; Carr, M.; Fukuhara, H.; Ose, T.; Maenaka, K.; Takada, A.; Hall, W.W.; et al. The role of heparan sulfate proteoglycans as an attachment factor for rabies virus entry and infection. *J. Infect. Dis.* **2018**, *217*, 1740–1749. [CrossRef] [PubMed]
38. Zhao, Z.; Ke, F.; Huang, Y.H.; Zhao, J.G.; Gui, J.F.; Zhang, Q.Y. Identification and characterization of a novel envelope protein in *Rana. grylio* virus. *J. Gen. Virol.* **2008**, *89*, 1866–1872. [CrossRef] [PubMed]
39. He, L.B.; Ke, F.; Wang, J.; Gao, X.C.; Zhang, Q.Y. *Rana. grylio* virus (RGV) envelope protein 2L: subcellular localization and essential roles in virus infectivity revealed by conditional lethal mutant. *J. Gen. Virol.* **2014**, *95*, 679–690. [CrossRef] [PubMed]
40. Zeng, X.T.; Gao, X.C.; Zhang, Q.Y. Rana grylio virus 43R encodes an envelope protein involved in virus entry. *Virus Genes* **2018**, *54*, 779–791. [CrossRef]
41. Lei, X.Y.; Ou, T.; Zhu, R.L.; Zhang, Q.Y. Sequencing and analysis of the complete genome of *Rana. grylio* virus (RGV). *Arch. Virol.* **2012**, *157*, 1559–1564. [CrossRef]
42. Sun, W.; Huang, Y.H.; Zhao, Z.; Gui, J.F.; Zhang, Q.Y. Characterization of the *Rana. grylio* virus 3beta-hydroxysteroid dehydrogenase and its novel role in suppressing virus-induced cytopathic effect. *Biochem. Biophys. Res. Commun.* **2006**, *351*, 44–50. [CrossRef] [PubMed]
43. Huang, X.; Fang, J.; Chen, Z.; Zhang, Q. *Rana. grylio* virus TK and DUT gene locus could be simultaneously used for foreign gene expression. *Virus Res.* **2016**, *214*, 33–38. [CrossRef] [PubMed]
44. Lei, X.Y.; Ou, T.; Zhang, Q.Y. *Rana. grylio* virus (RGV) 50L is associated with viral matrix and exhibited two distribution patterns. *PLoS ONE* **2012**, *7*, e43033. [CrossRef] [PubMed]
45. He, L.B.; Gao, X.C.; Ke, F.; Zhang, Q.Y. A conditional lethal mutation in *Rana grylio* virus ORF 53R resulted in a marked reduction in virion formation. *Virus Res.* **2013**, *177*, 194–200. [CrossRef] [PubMed]
46. Whitley, D.S.; Yu, K.; Sample, R.C.; Sinning, A.; Henegar, J.; Norcross, E.; Chinchar, V.G. Frog virus 3 ORF 53R, a putative myristoylated membrane protein, is essential for virus replication in vitro. *Virology* **2010**, *405*, 448–456. [CrossRef] [PubMed]
47. Zhou, S.; Wan, Q.; Huang, Y.; Huang, X.; Cao, J.; Ye, L.; Lim, T.K.; Lin, Q.; Qin, Q. Proteomic analysis of

Singapore grouper iridovirus envelope proteins and characterization of a novel envelope protein VP088. *Proteomics* **2011**, *11*, 2236–2248. [CrossRef] [PubMed]

48. Huang, X.; Gong, J.; Huang, Y.; Ouyang, Z.; Wang, S.; Chen, X.; Qin, Q. Characterization of an envelope gene VP19 from Singapore grouper iridovirus. *Virol. J.* **2013**, *10*, 354. [CrossRef] [PubMed]
49. Yuan, J.D.; Chen, Z.Y.; Huang, X.; Gao, X.C.; Zhang, Q.Y. Establishment of three cell lines from Chinese giant salamander and their sensitivities to the wild-type and recombinant ranavirus. *Vet. Res.* **2015**, *46*, 58. [CrossRef]
50. Zhang, Q.Y.; Zhao, Z.; Xiao, F.; Li, Z.Q.; Gui, J.F. Molecular characterization of three *Rana. grylio* virus (RGV) isolates and *Paralichthys. olivaceus* lymphocystis disease virus (LCDV-C) in iridoviruses. *Aquaculture* **2006**, *251*, 1–10. [CrossRef]
51. Rabenstein, D.L. Heparin and heparan sulfate: structure and function. *Nat. Prod. Rep.* **2002**, *19*, 312–331. [CrossRef]
52. Zhu, R.; Chen, Z.Y.; Wang, J.; Yuan, J.D.; Liao, X.Y.; Gui, J.F.; Zhang, Q.Y. Extensive diversification of MHC in Chinese giant salamanders *Andrias. davidianus* (Anda-MHC) reveals novel splice variants. *Dev. Comp. Immunol.* **2014**, *42*, 311–322. [CrossRef] [PubMed]
53. Livak, K.J.; Schmittgen, T.D. Analysis of relative gene expression data using real-time quantitative PCR and the $2^{-\Delta\Delta Ct}$ Method. *Methods* **2001**, *25*, 402–408. [CrossRef] [PubMed]
54. Vo, N.T.K.; Guerreiro, M.; Yaparla, A.; Grayfer, L.; DeWitte-Orr, S.J. Class A scavenger receptors are used by frog virus 3 during its cellular entry. *Viruses* **2019**, *11*, 93. [CrossRef] [PubMed]

IIV-6 Inhibits NF-κB Responses in *Drosophila*

Cara West [1], Florentina Rus [1], Ying Chen [2], Anni Kleino [1], Monique Gangloff [3], Don B. Gammon [4] and Neal Silverman [1,*]

1 Division of Infectious Diseases and Immunology, Department of Medicine, University of Massachusetts Medical School, Worcester, MA 01605, USA; Cara.West@umassmed.edu (C.W.); Florentina.Rus@umassmed.edu (F.R.); anni.kleino@aias.au.dk (A.K.)
2 RNA Therapeutics Institute, University of Massachusetts Medical School, Worcester, MA 01605, USA; cy4896@126.com
3 Department of Biochemistry, University of Cambridge, Cambridge CB2 1GA, UK; mg308@cam.ac.uk
4 Department of Microbiology, University of Texas Southwestern Medical Center, Dallas, TX T5390, USA; Don.Gammon@UTSouthwestern.edu
* Correspondence: neal.silverman@umassmed.edu

Abstract: The host immune response and virus-encoded immune evasion proteins pose constant, mutual selective pressure on each other. Virally encoded immune evasion proteins also indicate which host pathways must be inhibited to allow for viral replication. Here, we show that IIV-6 is capable of inhibiting the two *Drosophila* NF-κB signaling pathways, Imd and Toll. Antimicrobial peptide (AMP) gene induction downstream of either pathway is suppressed when cells infected with IIV-6 are also stimulated with Toll or Imd ligands. We find that cleavage of both Imd and Relish, as well as Relish nuclear translocation, three key points in Imd signal transduction, occur in IIV-6 infected cells, indicating that the mechanism of viral inhibition is farther downstream, at the level of Relish promoter binding or transcriptional activation. Additionally, flies co-infected with both IIV-6 and the Gram-negative bacterium, *Erwinia carotovora carotovora*, succumb to infection more rapidly than flies singly infected with either the virus or the bacterium. These findings demonstrate how pre-existing infections can have a dramatic and negative effect on secondary infections, and establish a *Drosophila* model to study confection susceptibility.

Keywords: viral immune evasion; immunomodulators; NF-κB; Imd; DNA virus; host-pathogen interactions; IIV-6

1. Introduction

The host immune system and the viruses that challenge it face constant, mutual, selective pressure for survival. This perpetual arms race, known as the Red Queen Hypothesis [1], has created a plethora of mechanisms that the cell uses to thwart viral replication, and also a plethora of novel immune evasion tactics that viruses use to evade the immune response, in some cases even stealing genes from their hosts to suit this purpose [2]. In order to evade the innate immune response, viruses—especially large DNA viruses—encode a variety of immune evasion proteins to facilitate their replication. For example, the model poxvirus Vaccinia encodes a variety NF-κB pathway inhibitors, and a recently identified invertebrate DNA virus, Kallithea virus, is able to inhibit *Drosophila* Toll signaling [3,4]. Invertebrate Iridescent Virus-6 (IIV-6) is a large DNA virus capable of infecting Drosophila [5,6], with an estimated 215 open reading frames (ORFs) [7]. The large genome of IIV-6 suggests it encodes for many evasion proteins in addition to the characterized suppressor of RNAi (340L) [8], and Inhibitor of Apoptosis Protein (IAP) (193R) [9,10]. The interplay between the virus and host, particularly which pathways the virus devotes its resources to inhibit, provides insight as to which defense mechanisms pose the most threat to viral replication. A virus devoting its resources to shutting down a particular host pathway indicates that this pathway has applied selective pressure against the virus.

Whether the Toll and Imd pathways play a role in antiviral defense has been a topic probed in several studies, often with mixed or conflicting results [11–14]. Of course, these two NF-κB signaling pathways are well-known for their roles in anti-bacterial and anti-fungal defense, through the regulation of antimicrobial peptides (AMPs) and other effectors [15–18]. While some studies have argued these two pathways also have some antiviral activities, the antiviral mechanisms have not been elucidated [2,12,19–21]. To address the question of the potential anti-DNA-viral role of the Toll and Imd pathways, we took the approach of using the virus to illuminate which pathways might be antiviral by examining whether IIV-6 inhibits the Toll and Imd pathways. We found that cells infected with IIV-6 have suppressed AMP production, while other immune genes (i.e., *Turandots*, c-Jun N-terminal kinase (JNK) targets) remain induced, suggesting that this is not a global suppression of host transcription [22]. Surprisingly, the cleavage of both Imd and Relish, key signaling events in the Imd pathway [23,24], remain intact in IIV-6 infected cells. Relish nuclear translocation is also at least partly intact, indicating the blockage in Imd response likely occurs in the nucleus, at the level of promoter binding or transcriptional activation. The inhibition of NF-κB transcription factor activity may be more general, as the Toll pathway was also blocked by IIV-6 infection. Consistent with the inhibition of these critical antibacterial defense responses, flies infected with IIV-6 were hyper-susceptible to infection with the Gram-negative bacteria *Erwinia carotovora carotovora* (*Ecc15*), establishing a system for studying co-infection morbidity in the *Drosophila* model.

2. Materials and Methods

2.1. RNA Isolation and qRT-PCR

Total RNA from flies or S2* cells was extracted using TRIzol (Invitrogen, Waltham, MA, USA). Samples were then DNase treated (RQ1, Promega, Madison, WI, USA) and RNA re-extracted by phenol-chloroform. The cDNA was synthesized using iScript cDNA Synthesis kit (BioRad, Hercules CA, USA). Alternatively, the gDNAclear cDNA synthesis kit (BioRad) was used following TRIzol purification. The quantitative reverse transcriptase–polymerase chain reaction (qRT-PCR) was analyzed by normalizing to the housekeeping gene *Rp49*.

2.2. The nCounter Analysis

The expression levels of 102 *Drosophila* immune genes (Supplemental Tables S1 and S2) were assayed from 100 nanograms of RNA via a customized Nanostring nCounter codeset. Two biological replicates of S2* cells were analyzed for each treatment and timepoint. The results were analyzed using nSolver 4.0 software according to the manufacturer's instructions (NanoString Technologies, Seattle, WA, USA), and the heatmap was created using nSolver 4.0 software and JavaTree. Shown are the AMPs, p38, JNK, and JAK-STAT targets with counts above background. Figure 3 analyzes data previously presented in West et al. [22].

2.3. Fly Stocks and In Vivo Studies

Three to five day old w^{1118} flies, of approximately equal proportions male and female, maintained at 22 °C, were used for all experiments. Flies were injected intrathoracically with 32.2 nL of virus (1×10^4 TCID50) or vehicle (PBS) using a Nanoject II (Drummond, Broomall, PA, USA). For survival assays, a minimum of fifty flies were used per treatment, per genotype, and the dead were counted daily. Kaplan-Meier curves are shown and significance was determined by Mantel-Cox log-rank using GraphPad Prism.

Erwinia carotovora carotovora 15, also known as *Pectobacterium carotovora*, was cultured overnight by shaking at 250 RPMs in LB broth, pelleted at 13,000× *g* for 10 min in a microfuge, washed in PBS, and re-pelleted. Infections were performed by dipping a microsurgery needle into the concentrated bacterial pellet, and pricking in the thorax. The IIV-6 injection site was identified using melanization, and bacterial pricking was performed on the opposite side of the thorax.

2.4. Cell Culture

S2* cells were cultured as previously described [25,26] and were treated with 1 μM 20-hydroxyecdysone, (EcD) for 18 h followed by infection with IIV-6 at a multiplicity of infection (MOI) of 2 for 6 h. Cells were then stimulated with 2 μg/mL PGN for 6 h to stimulate the Imd pathway, or stimulated with cleaved Spätzle for 18 h for Toll pathway stimulations, and were harvested for RNA isolation.

2.5. Immunoblots

S2* cells were cultured and treated as described above. For protein analysis, cells were then stimulated with 2 μg/mL PGN for 15 min to stimulate the Imd pathway, and harvested in lysis buffer consisting of 1% Triton X-100, 20 mM Tris pH 7.5, 150 mM NaCl, 2 mM EDTA, and 10% glycerol, with 1 mM DTT, 1 mM Sodium Vandate, 20 mM β-glycerolphosphate, and protease inhibitors. Imd or c-Rel antibodies were used as previously described [24,27].

2.6. Confocal Microscopy

YFP-Relish over-expressing cells were kept in selection media with hygromycin (200 μg/mL), and cultured as previously described [28]. Cells were treated with EcD for 2 h before being infected with mCherry-IIV-6 at an MOI of 2 for 18 h. Cells were then plated on Alcian Blue or ConA treated coverslips for 30 min and stimulated with PGN for 15 min. Cells were fixed in 4% paraformaldehyde, and stained with anti-lamin Dm0 (Developmental Studies Hybridoma Bank, ADL84.12) and Hoechst 33342.

2.7. Virus Preparation

IIV-6 was provided by Luis Teixeira. IIV-6 was propagated and purified on DL-1 cells as previously described [5], with a final resuspension in PBS, and quantified on DL-1 cells by TCID50. Cells were infected at an MOI of 2 unless otherwise noted, while flies were injected with 1×10^4 TCID50, as detailed above.

The ΔTS-MCP-mCherry-IIV-6 (mCherry-IIV-6) was created by inserting mCherry under the control of the major capsid promoter into the thymidylate synthase locus, using homologous recombination [29]. The major capsid promoter sequence was designed based on the analysis of IIV-6 MCP regulatory sequences by Nalçacioğlu et al. [30]

3. Results

3.1. AMP Production is Suppressed in the Presence of IIV-6

Given the scarce and somewhat conflicting data on the antiviral effects of the *Drosophila* NF-κB pathways in response to viral infections [2,12,19–21], we tested whether IIV-6 inhibits the Imd or Toll pathways. S2* cells were differentiated with 1 μM 20-hydroxyecdysone (EcD) for 18 h followed by IIV-6 infection for 6 h. Treatment with EcD is required to differentiate these cells and induce the expression of the receptor PGRP-LC, making these cells highly responsive to DAP-type peptidoglycan (PGN) stimulation [28]. After 6 h of IIV-6 infection, cells were then stimulated with 2 μg/mL of PGN for an additional 6 h, and gene expression analyzed by qRT-PCR. IIV-6 infection reduced transcriptional induction of the Imd-dependent AMP gene *Diptericin* to near background levels (Figure 1A).

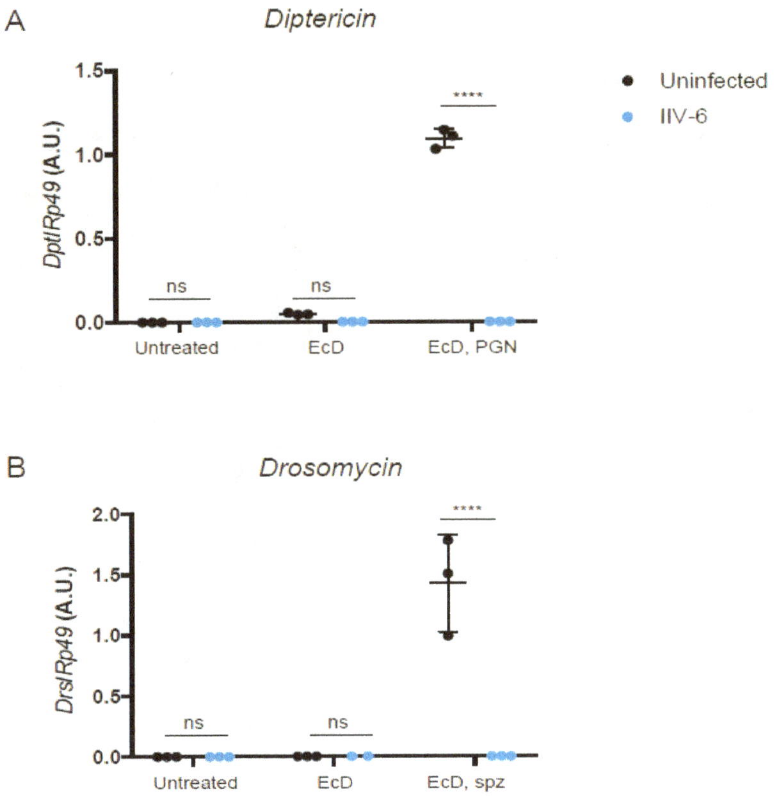

Figure 1. IIV-6 inhibits Imd and Toll Signaling. (**A**) S2* cells were treated with 20-hydroxyecdysone (EcD) as indicated for 18 hours and then infected with IIV-6 (blue circles) or uninfected (black circles) for six hours. Cells were then stimulated with DAP-type PGN for six hours, where indicated. *Diptericin* (*Dpt*) levels were monitored by qRT-PCR. (**B**) S2* cells were treated with 20-hydroxyecdysone (EcD) as indicated for 18 h and then infected with IIV-6 (blue circles) or uninfected (black circles) for six hours. Cells were then stimulated with cleaved Spätzle (spz) for 18 h, where indicated. *Drosomycin* (*Drs*) levels were monitored by qRT-PCR. (**A**,**B**) Black bars indicate mean and error bars indicate standard deviation. Statistics were determined using two-way ANOVA and Sidak's multiple comparisons test; ns, not significant; ****, $p < 0.0001$.

To determine if IIV-6 more generally blocks NF-κB signaling, we similarly analyzed Toll signaling. S2* cells were cultured in a similar manner, treated with EcD for 18 h, which upregulates expression of the Toll receptor [28], infected with IIV-6 for 6 h, and then stimulated with the Toll ligand, cleaved Spätzle, for 18 h. Induction of the Toll target gene *Drosomycin* was strongly inhibited by IIV-6 infection, as analyzed by qRT-PCR (Figure 1B). To examine both of these NF-κB pathways more broadly, we utilized NanoString nCounter Analysis, with a custom designed codeset probing over 100 immune-related genes, including most known AMP genes. All of the antimicrobial peptides were down-regulated in IIV-6 infected samples when the Imd pathway was stimulated with PGN (Figure 2A, Supplemental Table S1) or Toll pathway stimulated with cleaved Späztzle (Figure 2B, Supplemental Table S2), compared to the mock-infected controls stimulated with Imd or Toll ligands. Notably, other groups of genes were induced by virus infection, including the JNK targets *puckered* and *punch*, and the p38-dependent genes *upd3* and *Ddc* [22,31]. The antiviral gene *Ars2* was also induced by IIV-6 [32]. Since JNK signaling can be initiated through TAK1, the fact that JNK targets are being transcribed suggests that Imd signaling is being initiated, and signaling successfully occurs through at least TAK1 [33]. The induction of these genes by IIV-6 argues that the virus is not simply shutting down all host transcription, but is specifically targeting the two NF-κB signaling pathways, Imd and Toll. Additionally, we saw no suppression of *GAPDH1* levels (Supplemental Figure S1).

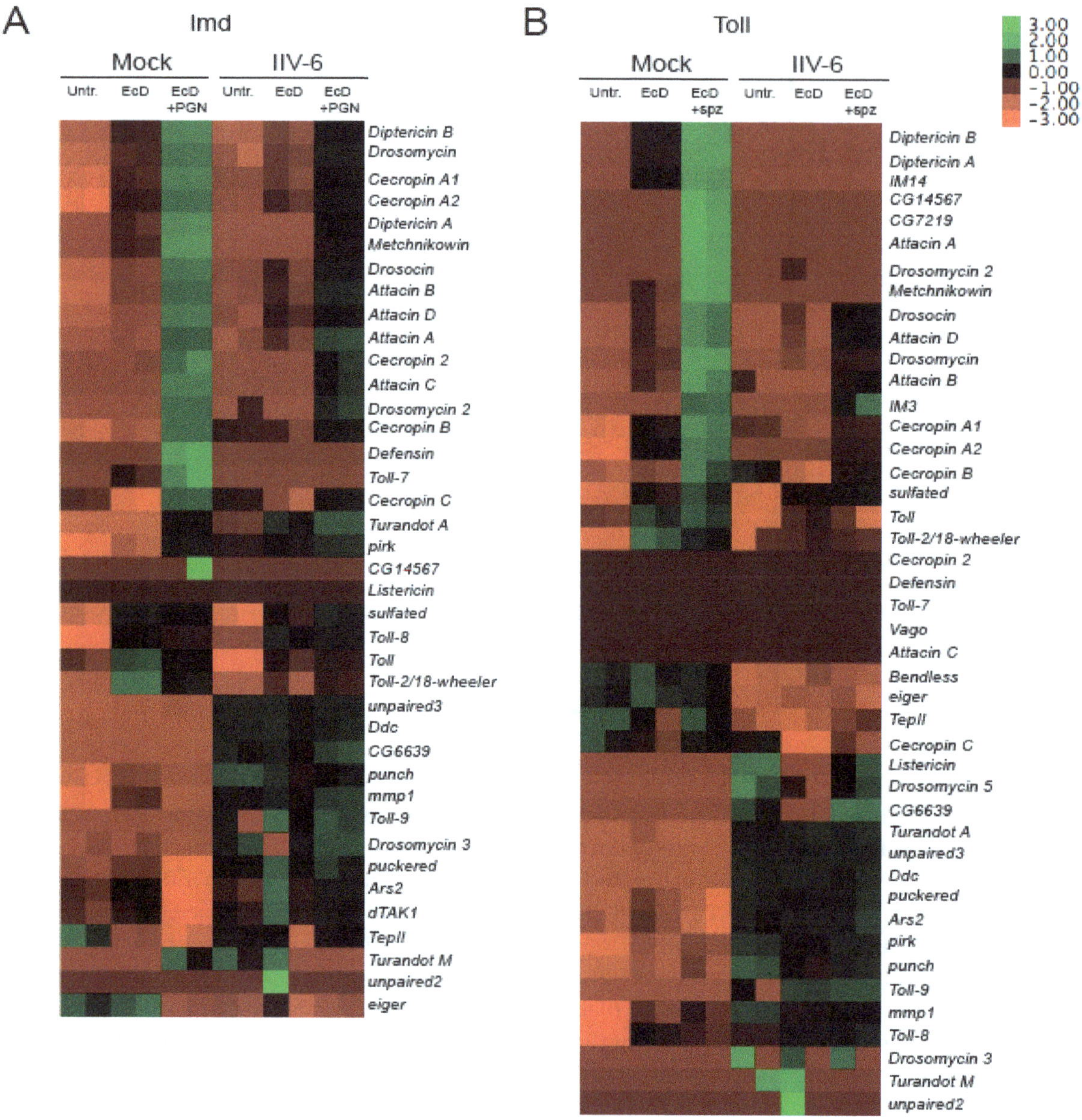

Figure 2. Both Imd and Toll regulated AMPs are suppressed by IIV-6 infection. S2* cells were treated with 20-hydroxyecdysone (EcD) for 18 h, and then infected with IIV-6 for 6 h. (**A**) Cells were stimulated with peptidoglycan (PGN) for 6 h prior to RNA isolation. (**B**) S2* cells were then stimulated with cleaved Spätzle (spz) for 18 h prior to RNA isolation, and then analyzed by Nanostring nCounter. Heatmaps display Z scores of mRNA levels of immune genes in the presence or absence of virus and pathway stimulation clustered by expression pattern. Biologically independent duplicates are shown. Untr., untreated cells. EcD, cells treated with 20-hydroxyecdysone. EcD + PGN, cells treated with 20-hydroxyecdysone and peptidoglycan. EcD + spz, cells treated with 20-hydroxyecdysone and Spätzle.

IIV-6 inhibition of both NF-κB pathways suggests that these pathways may function to limit viral replication. To examine this possibility, we analyzed an earlier Nanostring dataset from adult flies infected with IIV-6, or injected with a similar volume of PBS, to determine if virus infection induces any of the AMP genes [22]. Indeed, IIV-6 modestly induced AMP gene expression two-fold over PBS-injected controls 12 hours post-infection (Figure 3A,A'). However, by 24 h, this induction has returned to, or in some instances below, baseline levels. These findings are consistent with an early AMP response, possibly through NF-κB signaling pathways, which is quickly extinguished by viral inhibitors.

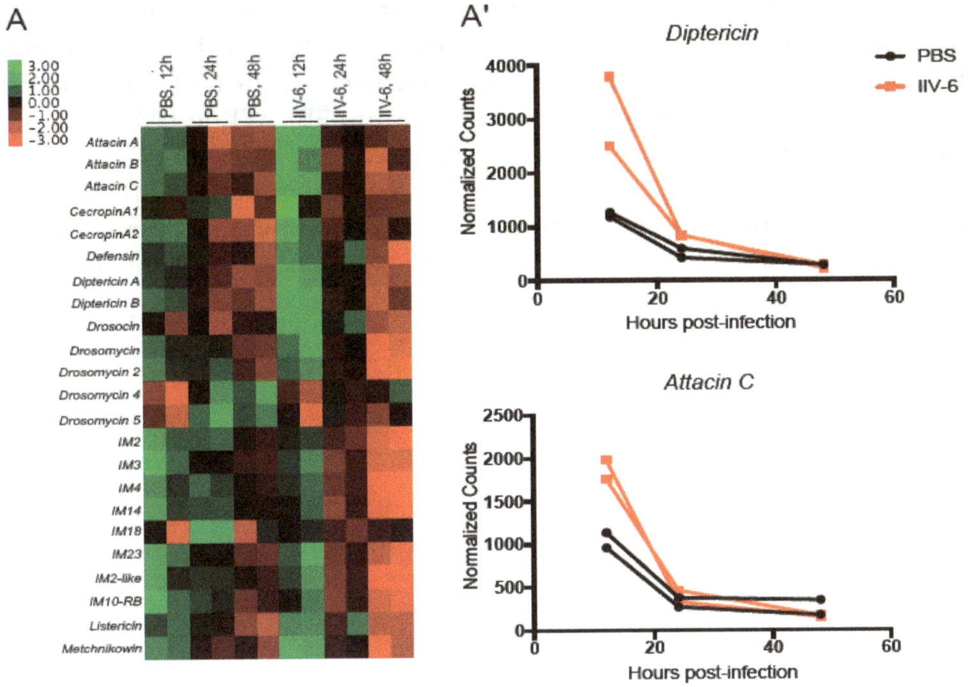

Figure 3. Some AMPs are elevated in vivo upon IIV-6 infection, before returning to baseline. (**A**) Heatmap of Z score transformed mRNA levels for AMP genes following IIV-6 infection of adult male w^{1118} flies for the indicated timepoints, assayed by Nanostring nCounter. RNA was isolated from PBS-injected flies at the same time points as a control. Biologically independent samples were analyzed in duplicate. (**A'**) Detailed comparison of mRNA levels for selected Imd-regulated AMP genes following IIV-6 infection of adult w^{1118} flies for 12, 24, and 48 h assayed by Nanostring nCounter.

3.2. NF-κB Inhibition is Downstream of Imd and Relish

Next, we focused on the Imd pathway to tease apart the mechanism of NF-κB inhibition by IIV-6. As Imd signaling requires cleavage of Imd by the caspase 8-like Dredd [24,34], we probed for cleaved Imd in IIV-6 infected or uninfected cell lysates. We found robust cleavage of Imd upon PGN stimulation in both the presence and absence of IIV-6, indicating that the blockage of Imd signaling is likely to occur downstream of Imd cleavage (Figure 4A). To probe further downstream, we examined Relish cleavage, another critical event in Imd signaling, utilizing a C-terminal Relish antibody. As expected, samples mock treated or infected with IIV-6 alone show a prominent 110 kD band indicating full-length Relish, and control samples treated with EcD and PGN show complete processing of full-length Relish. Likewise, upon PGN stimulation, Relish was fully cleaved in the presence of IIV-6, even at a relatively high MOI (Figure 4B, lanes 4–6). These results indicate the inhibition of Imd signaling by IIV-6 occurs downstream of Relish cleavage, the key event in the activation of this NF-κB precursor.

Since Relish is cleaved in the presence of IIV-6, we next sought to determine whether IIV-6 inhibits Imd signaling through blocking Relish nuclear translocation. To this end, we utilized a YFP-Relish over-expressing cell line, and a strain of IIV-6 expressing mCherry under the control of the major capsid protein (MCP). Even with these tools, scoring Relish nuclear translocation after virus infection presented several technical challenges. In particular, viral infected cells adhered poorly to treated coverslips and displayed reduced YFP-Relish signal. However, the levels of endogenous Relish protein are not altered by IIV-6 infection (Figure 4B). Additionally, IIV-6 undergoes massive DNA viral replication in the cytoplasmic viral factories, making it difficult to distinguish the nucleus. In order to identify the nucleus, cells were stained with both Hoechst 33342 and the nuclear envelope marker Lamin to accurately discern the nucleus from the cytosolic viral factory. In control conditions, cells stimulated with PGN showed 35% nuclear translocation, while Relish remained cytoplasmic in 20% of cells (Figure 5A, lower left panel, Figure 5B). The remaining 44% of cells contained a YFP signal in both the cytoplasm and nucleus or had a YFP signal too weak to determine localization. In IIV-6 infected cells, we found that approximately 17% showed nuclear translocation, with 33% remaining cytoplasmic, while nearly 50% displayed both cytoplasmic and nuclear Relish or had a YFP signal too weak to determine localization (Figure 5A lower right panel, Figure 5B). While IIV-6 reduces the detected level of PGN-induced Relish nuclear translocation compared to uninfected PGN-stimulated cells, this difference was not significant, suggesting PGN can still trigger some Relish nuclear translocation in virus infected cells. Interestingly, virus infection alone caused some Relish translocation (Figure 5A upper right panel, Figure 5B). These data suggest that IIV-6 does not completely inhibit Relish nuclear translocation, and that viral suppression of Imd signaling likely occurs downstream, either at the level of Relish promoter binding or transcriptional activation.

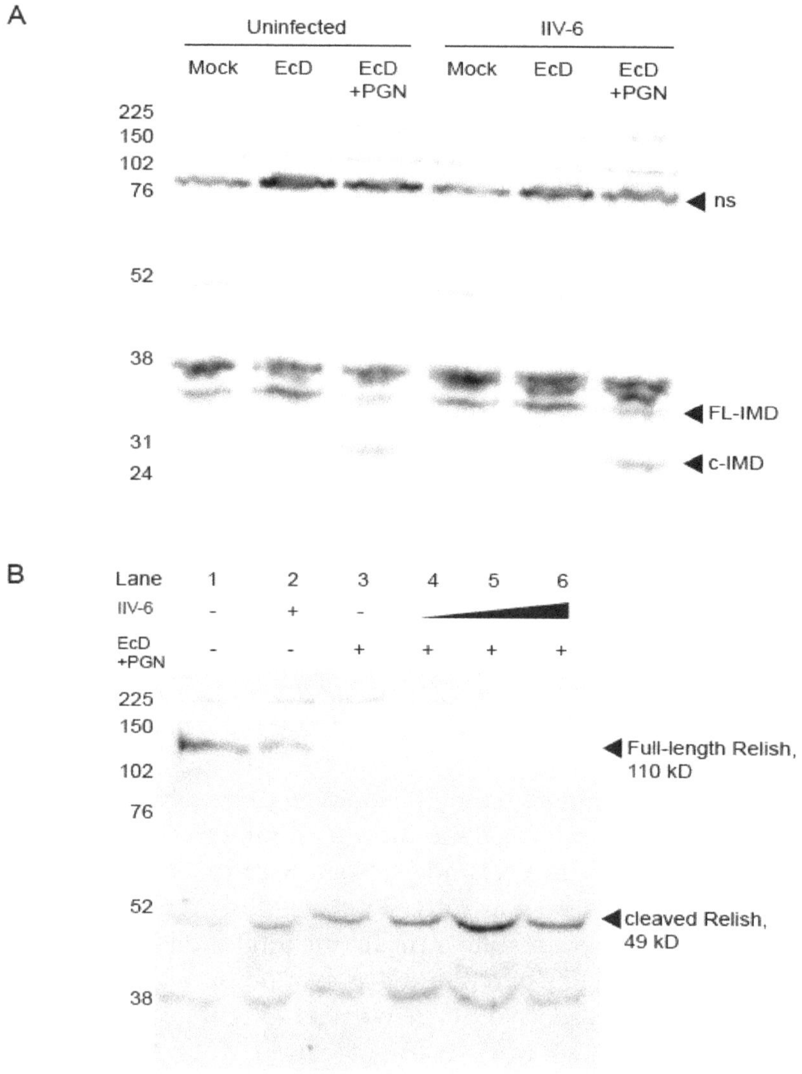

Figure 4. Imd signaling remains intact in the presence of IIV-6. (**A**) S2* cells were treated with 20-hydroxyecdysone (EcD) for 18 h. Cells were then infected with IIV-6, where indicated, for six hours. Samples were then stimulated with PGN for 15 min, where indicated, and lysed in standard lysis buffer. Endogenous Imd was monitored by immunoblotting; arrows indicate cleaved (c-Imd) or full-length (FL-Imd) Imd; ns, non-specific. (**B**) Cells were treated with 20-hydroxyecdysone (EcD), where indicated, for 18 h and infected with IIV-6, as indicated, for six hours. Samples stimulated with EcD were then stimulated with PGN for 15 min, and lysed in standard lysis buffer. Endogenous Relish was probed by immunoblotting using a C-terminal Relish antibody. MOI used was as follows. Lane 2: MOI =2; Lane 4: MOI = 0.2; Lane 5: MOI = 2; Lane 6: MOI = 5.

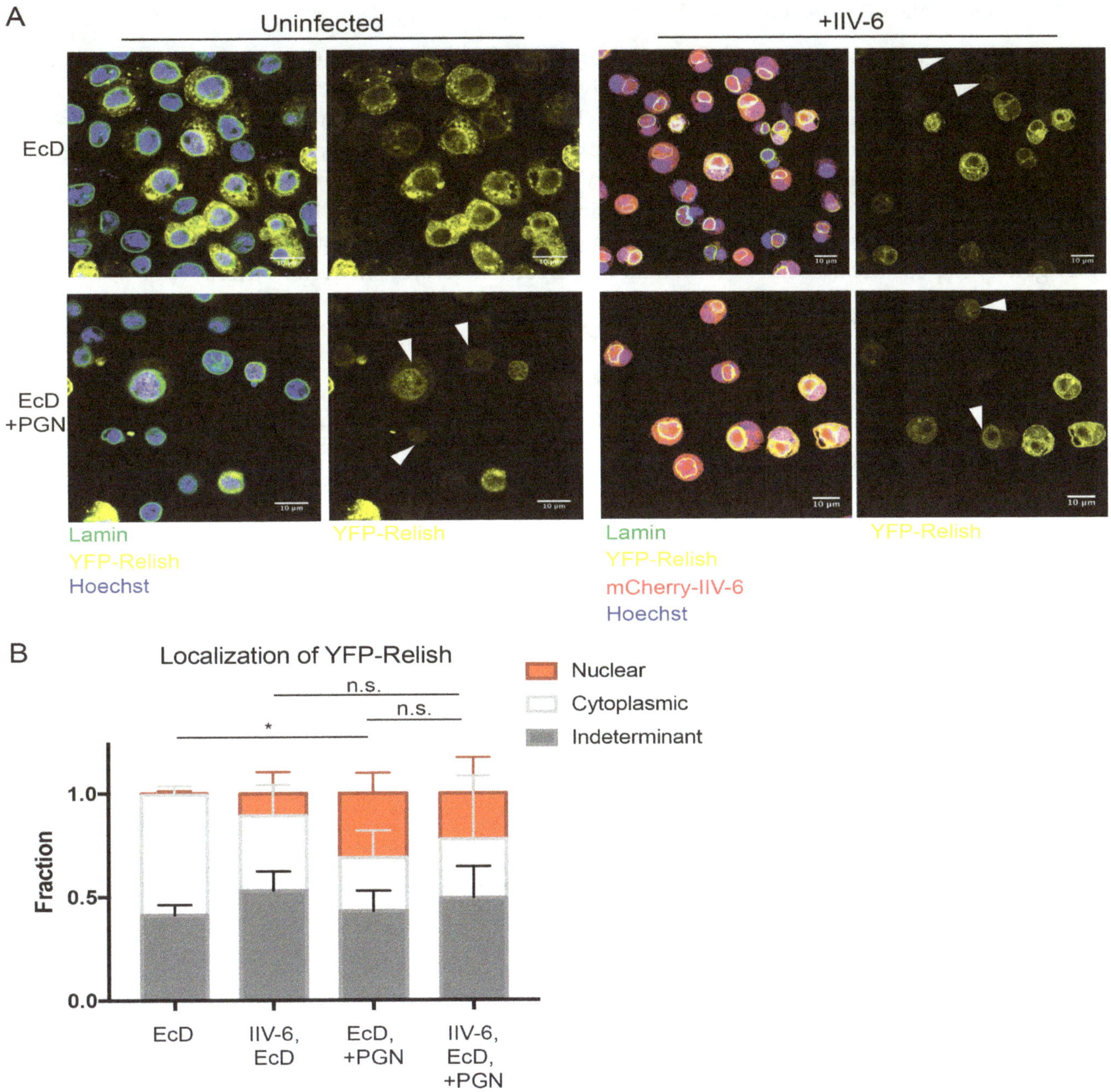

Figure 5. Nuclear localization of Relish remains intact in the presence of IIV-6. (**A**) S2* cells stably expressing YFP-Relish were treated with 20-hydroxyecdysone (EcD) for 2 h and then infected with mCherry-IIV-6 (right) or left uninfected (left) for 18 h. Cells were then stimulated with DAP-type PGN for 15 min (lower panels), prior to fixation and staining with anti-Lamin (shown in green) and Hoechst 33342 (shown in blue). Representative images from four biologically independent experiments are shown. Arrowheads mark cells with nuclear localized Relish. Left panels display the overlay of YFP-Relish, Hoeschst, Lamin, and mCherry-IIV-6 where applicable, while the right panels exhibit the YFP-Relish alone. Scale bars are 10 μm. (**B**) Quantification of Relish nuclear translocation. Between 400–1800 cells were scored for each condition as displaying Relish localization in the nucleus, in the cytoplasm, or as indeterminate when staining was diffuse throughout both compartments or the signal was too weak to discern. Statistics were determined using two-way ANOVA and Tukey's multiple comparisons test; *, $p < 0.05$; ns, not significant. Error bars represent standard deviation.

3.3. Inhibition of NF-κB Signaling is Mediated by an Immediate Early Gene

Most—but not all—immune evasion proteins are immediate early genes. Strategically, it is probably most effective, from the virus perspective, to shut down the host defense as quickly as possible and rapidly achieve high levels of replication. In order to determine if IIV6 inhibition of *Drosophila* NF-κB signaling also involves an early gene product, cells were treated with the viral polymerase inhibitor cidofovir, or infected with heat- or UV-inactivated virus, and then stimulated with PGN to probe Imd signaling. PGN-triggered *Diptericin* induction was blocked with all three virus-inactivating treatments (Figure 6), indicating that the IIV6-mediated NF-κB inhibition was the result of immediate early genes, or possibly associated with a factor directly delivered with the virion.

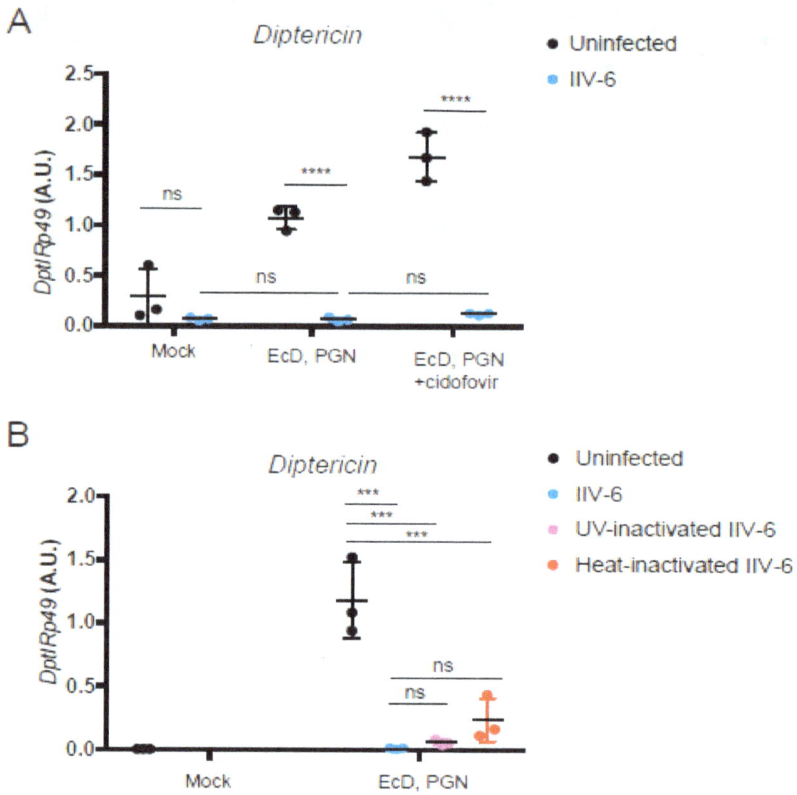

Figure 6. Viral replication is not needed for NF-κB inhibition. (**A**,**B**) S2* cells were treated with 20-hydroxyecdysone (EcD) as indicated for 18 h and then infected with IIV-6 (blue circles) or uninfected (black circles) for six hours. Cells were then stimulated with DAP-type PGN for six hours, where indicated. *Diptericin* levels were monitored by qRT-PCR. (**A**) Cells were treated with cidofovir, a viral polymerase inhibitor, where indicated. Mock cells untreated with EcD, PGN, or cidofovir are shown as a control. Each data point is a biologically independent replicate; $n = 3$. Error bars represent standard deviation. Statistics were determined using two-way ANOVA and Sidak's multiple comparisons test; ns, not significant; ****, $p < 0.0001$. (**B**) S2* cells were infected with IIV-6 (blue circles), or treated with UV- (purple circles) or heat- (red circles) inactivated IIV-6. Uninfected controls are shown in black. Each data point is a biologically independent replicate; $n = 3$. Error bars represent standard deviation. Statistics were determined using one-way ANOVA and Tukey's multiple comparisons test; ns, not significant; ***, $p < 0.001$.

3.4. Flies Infected with IIV-6 are more Susceptible to Bacterial Infection

Together, these results indicate that infection with IIV-6 results in the global suppression of NF-κB signaling in flies—a major component of the innate immune response to bacterial and fungal infections in flies and mammals. Consistent with our in vitro data showing universally suppressed AMPs, we found that flies infected with IIV-6 also had suppression of NF-κB signaling, as indicated by lower mRNA levels of the AMP genes *Diptericin* and *Drosomycin* compared to PBS-injected or unmanipulated flies at day 8 post-infection (Figure 7A, Supplemental Figure S2). This suggests that flies infected with IIV-6 should be highly susceptible to other microbial infections. To test this hypothesis, we infected *Drosophila* adults with IIV-6 for seven days. On day eight post-IIV-6 infection, we pricked one group with a sterile needle, and the other group of flies with a needle dipped in *Erwinia carotovora carotovora* (*Ecc15*), a Gram-negative pathogen with limited lethality in healthy, immunocompetent flies. Seven days after this secondary bacterial infection, 50% of the flies infected with both IIV-6 and *Ecc15* had succumbed to infection, with nearly 100% lethality by day 20 (Figure 7B). In contrast, flies that had been mock-injected with PBS prior to *Ecc15* infection had a median survival time of twenty-three days post-secondary (bacterial) infection. The animals injected with IIV-6 followed by a clean prick reached 50% lethality only at day 18, typical of the IIV-6 survival curve [22]. Together, these data show that an underlying IIV-6 infection can have a dramatic effect on a secondary bacterial infection, causing flies to be far more susceptible to *Ecc15* infection than unaffected animals. Given the critical role of the Imd pathway in defending against *Ecc15* infection, these results are consistent with the inhibition of Imd signaling observed in IIV-6 infected S2* cells and adult flies.

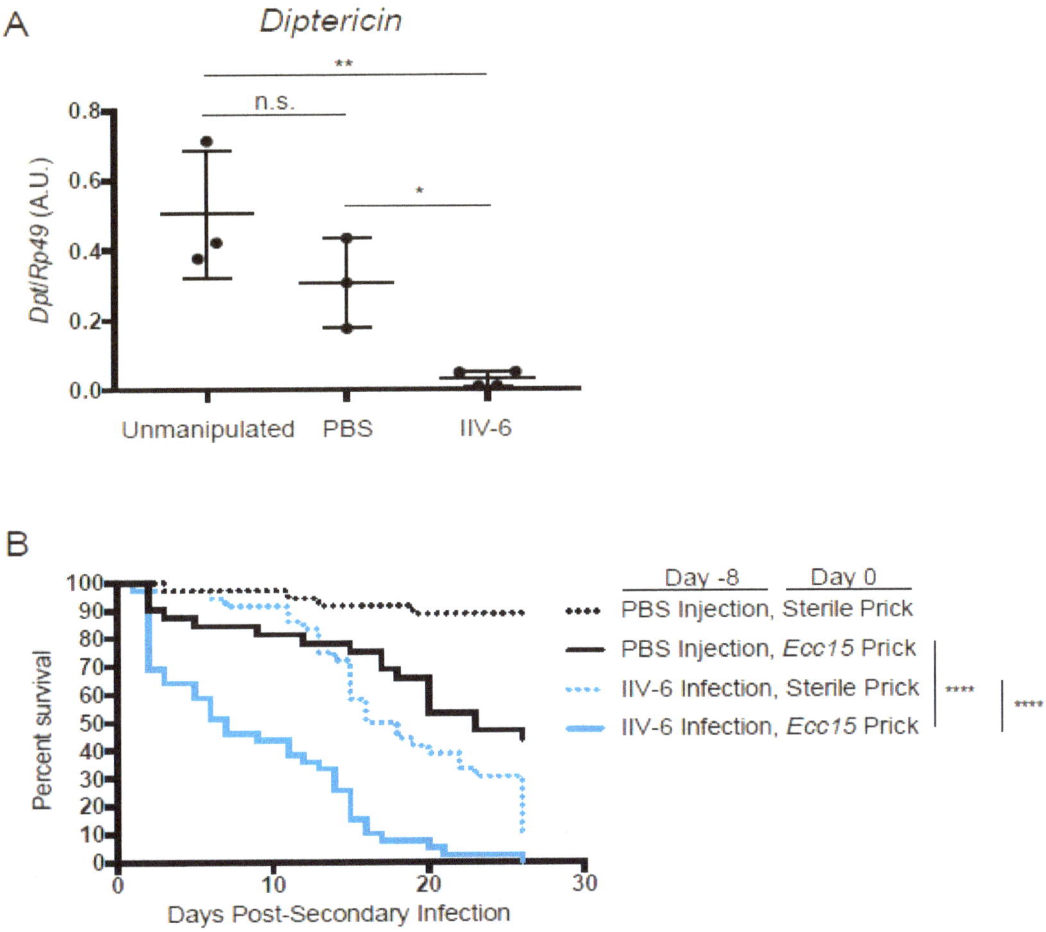

Figure 7. Flies infected with IIV-6 have lower AMP levels. (**A**) Adult w^{1118} flies were infected with IIV-6, injected with PBS, or left unmanipulated for 7 days, and then snap frozen for RNA isolation. *Dpt* levels were assayed by qRT-PCR. Black bars indicate mean and error bars indicate standard deviation. Statistics were determined using one-way ANOVA and Tukey's multiple comparisons test; ns, not significant; *, $p = 0.0463$; **, $p = 0.0033$. (**B**) Kaplan-Meier plots of adult w^{1118} flies infected with IIV-6 (blue lines) or PBS-injected (black lines) for 8 days prior to bacterial infection. On day 0, flies were pricked with *Ecc15* (solid lines) or were sterile pricked (dashed lines) with a microsurgery needle. Flies were counted daily for survivors. Statistics were determined using log-rank test; ****, $p < 0.0001$.

4. Discussion

Here, we show that IIV-6 infection interferes with the antimicrobial peptide response mediated through the Toll and Imd NF-κB pathways. In addition, we find that flies infected with IIV-6 succumb to an otherwise mildly pathogenic bacterial infection in vivo. Together, our data show that an underlying infection with this DNA virus can dampen the immune response and dramatically alter the outcome of a secondary bacterial infection, turning an otherwise innocuous infection into a lethal one.

While it appears that IIV-6 is suppressing the AMP response by actively inhibiting the NF-κB pathways, the mechanisms of inhibition remain unclear. It is possible that IIV-6 encodes an NF-κB inhibitor that prevents Relish binding to κB sites or recruitment of polymerase machinery. Several examples of viral-encoded NF-κB inhibitory proteins acting within the nucleus have been reported. NF-κB inhibitors functioning within the nucleus are encoded by vaccinia virus (VACV) [35], as well as African swine fever virus A238L [36]. VACV also encodes an inhibitor of Type 1 IFN, C6, that functions post-STAT1 and 2 nuclear translocation and DNA binding by binding the STAT2 transactivation domain [37]. Another VACV protein, N2 inhibits IRF3 within the nucleus [38]. Future studies will be aimed at examining the effect of IIV-6 on Relish binding at AMP gene κB sites, as well as the transcriptional activation process.

The timing of suppression also needs to be examined more closely. While our in vivo or in vitro data may suggest a difference in the suppressive effects of IIV-6, it is important to note that a more detailed in vivo time course is required. If there is a true shift in suppression, we hypothesize that this may be due to the fact that the in vitro infections occur at an MOI of 2, and we can more easily synchronize the timing of infection. In vivo, it may be that virus takes longer to diffuse through the hemolymph and replicate, resulting in a difference in response.

It should be noted that IIV-6 does not encode a homolog of Diedel, the putative Imd inhibitor, as other large insect DNA viruses do [2]. Given the role of RHIM-dependent amyloid fibrils in the Imd pathway [39], we also considered that IIV-6 might encode RHIM-inhibitor, similar to Murine cytomegalovirus M45 [40], but found no evidence for such a factor in the IIV-6 genome.

Future studies will be required to determine which viral gene(s) are responsible for NF-κB inhibition. Our work shows that Imd signaling is inhibited at a step downstream of Relish activation. However, the mechanisms of Toll inhibition have not been investigated yet. IIV-6 may encode more than one NF-κB inhibitor, targeting the Imd and Toll pathways at different points. Since other large DNA viruses, such as VACV, encode multiple inhibitors targeting NF-κB pathways at various points in mammalian NF-κB signaling pathways [3], elimination of single viral genes may not reveal any phenotype due to redundancies. Determining which viral genes are responsible for NF-κB inhibition will require both loss and gain of function approaches and will be the subject of future studies.

IIV-6 has a broad host range, infecting a large variety of insects with agricultural and economic importance. The slow replication cycle of this virus results in infected insects surviving for weeks, allowing it to spread amongst a population, and consequently, leaving that population with an increased susceptibility to a secondary bacterial infection that would otherwise be cleared. Given the current decline of honey bees, as well as the entire insect population [41], a persistent viral infection could further damage this already precarious population by increasing susceptibility to a range of other microbial pathogens. While previous studies have ruled out IIV-6 as the causative agent of colony collapse disorder [42], it should be noted that IIV-6 infection may expose other species to increased vulnerability to a secondary infection.

In summary, we have shown that IIV-6 infection results in an inability to mount an NF-κB mediated AMP response upon stimulation of either the Imd or Toll pathways. It appears that IIV-6 infection may briefly induce AMP expression before shutting it down. Suppression of Imd signaling occurs downstream of Relish activation, suggesting a viral inhibitor may act at the level of DNA binding and transcriptional activation. Nanostring data showing that IIV-6 induces transcription of JNK targets suggest that the Imd pathway is being successfully activated through at least the branch point of TAK1 [25]. These results are consistent with the data showing that Relish cleavage and some nuclear translocation remain intact and argue that the virally generated block occurs far downstream in this pathway. Strikingly, flies infected with IIV-6 subjected to subsequent infection with *Ecc15* died more rapidly than flies singly infected with virus or bacteria, indicating that this virus-mediated block in NF-κB signaling may have a profound effect on resistance to bacterial infections.

The involvement of the Drosophila NF-κB pathways in antiviral signaling is a controversial topic. Other studies have suggested an antiviral role for AMPs, examining a recombinant analog of the RNA virus Sindbis virus [21]. Whether or not some AMPs are also potent against DNA viruses, such as IIV-6, will be the focus of future studies, for example by examining the replication of mCherry -expressing virus or by quantifying viral loads from infected flies over-expressing AMPs. It is also possible that AMP induction is simply a by-product of Relish activation, similar to many ISGs that have no direct protective role against the pathogen eliciting their expression. In this case, other Relish target genes may be important for antiviral defense. Additionally, with some picrorna-like viruses, such as Drosophila C virus and Cricket Paralysis virus, some—but not all—Imd pathway components seem to be restrictive [12,13]. Identifying a viral-encoded Imd inhibitor would provide more evidence supporting that this pathway induces antiviral activities. However, how Imd signaling is initiated upon virus infection, when the established agonist for this pathway is bacterial peptidoglycan [43,44], and whether all Imd signaling components are necessary, need to be examined in greater detail. For example, as the key receptors in this pathway, PGRP-LC and PGRP-LE, directly bind to DAP-type PGN, it seems unlikely that they will be involved in viral recognition. Antiviral studies implicating Imd signaling should be reevaluated in light of recent work showing that the microbiome plays an important role activating Imd signaling in the gut to prime an antiviral response via ERK following an oral infection route [45]. Whether commensals can act to influence antiviral responses in other organ systems requires further exploration. For example, the Malpighian tubules are organs branching off from the *Drosophila* gut, which absorb waste from the hemolymph, but have also been shown to serve important roles in immune function, including induction and secretion of AMPs in response to infection [43,46–48]. Whether the Malpighian tubules provide crosstalk between the gut and hemolymph during viral infections should be explored.

Author Contributions: Conceptualization, C.W., D.B.G., and N.S.; formal analysis, C.W.; funding acquisition, N.S.; investigation, C.W., F.R., Y.C., and A.K.; methodology, C.W.; project administration, N.S.; resources, M.G. and D.B.G.; supervision, D.B.G. and N.S.; visualization, C.W.; writing – original draft, C.W.; writing – review and editing, C.W. and N.S.

Acknowledgments: The authors thank Luis Texeira for the IIV-6 strain, Deniz Ertürk-Hasdemir for generating the YFP-Relish cell lines, and Niamh Hallinan and Kendi Okuda for technical assistance.

References

1. Valen, L.V. A new evolutionary law. *Evol. Theory* **1973**, *1*, 1–30.
2. Lamiable, O.; Kellenberger, C.; Kemp, C.; Troxler, L.; Pelte, N.; Boutros, M.; Marques, J.T.; Daeffler, L.; Hoffmann, J.A.; Roussel, A.; et al. Cytokine diedel and a viral homologue suppress the imd pathway in drosophila. *Proc. Natl. Acad. Sci. USA* **2016**, *113*, 698–703. [CrossRef]
3. Smith, G.L.; Benfield, C.T.; Maluquer de Motes, C.; Mazzon, M.; Ember, S.W.; Ferguson, B.J.; Sumner, R.P. Vaccinia virus immune evasion: Mechanisms, virulence and immunogenicity. *J. Gen. Virol.* **2013**, *94*, 2367–2392. [CrossRef]
4. Palmer, W.H.; Joosten, J.; Overheul, G.J.; Jansen, P.W.; Vermeulen, M.; Obbard, D.J.; van Rij, R.P. Induction and suppression of nf-kappab signalling by a DNA virus of drosophila. *J. Virol.* **2019**, *93*. [CrossRef]

5. Teixeira, L.; Ferreira, Á.; Ashburner, M. The bacterial symbiont wolbachia induces resistance to rna viral infections in drosophila melanogaster. *PLoS Biol.* **2008**, *6*, e1000002. [CrossRef]
6. Bronkhorst, A.W.; Van Cleef, K.W.R.; Vodovar, N.; Ince, I.A.; Blanc, H.; Vlak, J.M.; Saleh, M.-C.; van Rij, R.P. The DNA virus invertebrate iridescent virus 6 is a target of the drosophila rnai machinery. *Proc. Natl. Acad. Sci. USA* **2012**, *109*, E3604–E3613. [CrossRef]
7. Ince, I.A.; Boeren, S.A.; van Oers, M.M.; Vervoort, J.J.M.; Vlak, J.M. Proteomic analysis of chilo iridescent virus. *Virology* **2010**, *405*, 253–258. [CrossRef]
8. Bronkhorst, A.W.; Van Cleef, K.W.R.; Venselaar, H.; Van Rij, R.P. A dsrna-binding protein of a complex invertebrate DNA virus suppresses the drosophila rnai response. *Nucleic Acids Res.* **2014**, *42*, 12237–12248. [CrossRef]
9. Chitnis, N.S.; D'Costa, S.M.; Paul, E.R.; Bilimoria, S.L. Modulation of iridovirus-induced apoptosis by endocytosis, early expression, jnk, and apical caspase. *Virology* **2008**, *370*, 333–342. [CrossRef]
10. Chitnis, N.S.; Paul, E.R.; Lawrence, P.K.; Henderson, C.W.; Ganapathy, S.; Taylor, P.V.; Virdi, K.S.; D'Costa, S.M.; May, A.R.; Bilimoria, S.L. A virion-associated protein kinase induces apoptosis. *J. Virol.* **2011**, *85*, 13144–13152. [CrossRef]
11. Avadhanula, V.; Weasner, B.P.; Hardy, G.G.; Kumar, J.P.; Hardy, R.W. A novel system for the launch of alphavirus rna synthesis reveals a role for the imd pathway in arthropod antiviral response. *PLoS Pathog.* **2009**, *5*, e1000582. [CrossRef]
12. Costa, A.; Jan, E.; Sarnow, P.; Schneider, D. The imd pathway is involved in antiviral immune responses in drosophila. *PLoS ONE* **2009**, *4*, e7436. [CrossRef]
13. Goto, A.; Okado, K.; Martins, N.; Cai, H.; Barbier, V.; Lamiable, O.; Troxler, L.; Santiago, E.; Kuhn, L.; Paik, D.; et al. The kinase ikkbeta regulates a sting- and nf-kappab-dependent antiviral response pathway in drosophila. *Immunity* **2018**, *49*, 225–234 e224. [CrossRef]
14. Liu, Y.; Gordesky-Gold, B.; Leney-Greene, M.; Weinbren, N.L.; Tudor, M.; Cherry, S. Inflammation-induced, sting-dependent autophagy restricts zika virus infection in the drosophila brain. *Cell Host Microbe* **2018**, *24*, 57–68 e53. [CrossRef]
15. Lemaitre, B.; Hoffmann, J. The host defense of drosophila melanogaster. *Annu. Rev. Immunol.* **2007**, *25*, 697–743. [CrossRef]
16. Clemmons, A.W.; Lindsay, S.A.; Wasserman, S.A. An effector peptide family required for drosophila toll-mediated immunity. *PLoS Pathog.* **2015**, *11*, e1004876. [CrossRef]
17. Lindsay, S.A.; Lin, S.J.H.; Wasserman, S.A. Short-form bomanins mediate humoral immunity in drosophila. *J. Innate Immun.* **2018**, *10*, 306–314. [CrossRef]
18. Hanson, M.A.; Dostalova, A.; Ceroni, C.; Poidevin, M.; Kondo, S.; Lemaitre, B. Synergy and remarkable specificity of antimicrobial peptides in vivo using a systematic knockout approach. *Elife* **2019**, *8*. [CrossRef]
19. Zambon, R.A.; Nandakumar, M.; Vakharia, V.N.; Wu, L.P. The toll pathway is important for an antiviral response in drosophila. *Proc. Natl. Acad. Sci. USA* **2005**, *102*, 7257–7262. [CrossRef]
20. Tsai, C.W.; McGraw, E.A.; Ammar, E.D.; Dietzgen, R.G.; Hogenhout, S.A. Drosophila melanogaster mounts a unique immune response to the rhabdovirus sigma virus. *Appl. Environ. Microbiol.* **2008**, *74*, 3251–3256. [CrossRef]
21. Huang, Z.; Kingsolver, M.B.; Avadhanula, V.; Hardy, R.W. An antiviral role for antimicrobial peptides during the arthropod response to alphavirus replication. *J. Virol.* **2013**, *87*, 4272–4280. [CrossRef]
22. West, C.; Silverman, N. P38b and jak-stat signaling protect against invertebrate iridescent virus 6 infection in drosophila. *PLoS Pathog.* **2018**, *14*, e1007020. [CrossRef]
23. Stoven, S.; Silverman, N.; Junell, A.; Hedengren-Olcott, M.; Erturk, D.; Engstrom, Y.; Maniatis, T.; Hultmark, D. Caspase-mediated processing of the drosophila nf-kappab factor relish. *Proc. Natl. Acad. Sci. USA* **2003**, *100*, 5991–5996. [CrossRef]
24. Paquette, N.; Broemer, M.; Aggarwal, K.; Chen, L.; Husson, M.; Ertürk-Hasdemir, D.; Reichhart, J.-M.; Meier, P.; Silverman, N. Caspase-mediated cleavage, iap binding, and ubiquitination: Linking three mechanisms crucial for drosophila nf-kappab signaling. *Mol. Cell* **2010**, *37*, 172–182. [CrossRef]
25. Samakovlis, C.; Åsling, B.; Boman, H.G.; Gateff, E.; Hultmark, D. In vitro induction of cecropin genes–an immune response in a drosophila blood cell line. *Biochem. Biophys. Res. Commun.* **1992**, *188*, 1169–1175. [CrossRef]

26. Silverman, N.; Zhou, R.; Stoven, S.; Pandey, N.; Hultmark, D.; Maniatis, T. A drosophila ikappab kinase complex required for relish cleavage and antibacterial immunity. *Genes Dev.* **2000**, *14*, 2461–2471. [CrossRef]
27. Ertürk-Hasdemir, D.; Broemer, M.; Leulier, F.; Lane, W.S.; Paquette, N.; Hwang, D.; Kim, C.-H.; Stöven, S.; Meier, P.; Silverman, N. Two roles for the drosophila ikk complex in the activation of relish and the induction of antimicrobial peptide genes. *Proc. Natl. Acad.Sci. USA* **2009**, *106*, 9779–9784. [CrossRef] [PubMed]
28. Rus, F.; Flatt, T.; Tong, M.; Aggarwal, K.; Okuda, K.; Kleino, A.; Yates, E.; Tatar, M.; Silverman, N. Ecdysone triggered pgrp-lc expression controls drosophila innate immunity. *EMBO J.* **2013**, *32*, 1626–1638. [CrossRef]
29. Ozgen, A.; Muratoglu, H.; Demirbağ, Z.; Vlak, J.M.; van Oers, M.M.; Nalçacioğlu, R. Construction and characterization of a recombinant invertebrate iridovirus. *Virus Res.* **2014**. [CrossRef]
30. Nalçacioğlu, R.; Marks, H.; Vlak, J.M.; Demirbağ, Z.; van Oers, M.M. Promoter analysis of the chilo iridescent virus DNA polymerase and major capsid protein genes. *Virology* **2003**, *317*, 321–329. [CrossRef]
31. Davis, M.M.; Primrose, D.A.; Hodgetts, R.B. A member of the p38 mitogen-activated protein kinase family is responsible for transcriptional induction of dopa decarboxylase in the epidermis of drosophila melanogaster during the innate immune response. *Mol. Cell. Biol.* **2008**, *28*, 4883–4895. [CrossRef]
32. Sabin, L.R.; Zhou, R.; Gruber, J.J.; Lukinova, N.; Bambina, S.; Berman, A.; Lau, C.-K.; Thompson, C.B.; Cherry, S. Ars2 regulates both mirna- and sirna- dependent silencing and suppresses rna virus infection in drosophila. *Cell* **2009**, *138*, 340–351. [CrossRef] [PubMed]
33. Silverman, N.; Zhou, R.; Erlich, R.L.; Hunter, M.; Bernstein, E.; Schneider, D.; Maniatis, T. Immune activation of nf-kappab and jnk requires drosophila tak1. *J. Biol. Chem.* **2003**, *278*, 48928–48934. [CrossRef]
34. Kim, C.H.; Paik, D.; Rus, F.; Silverman, N. The caspase-8 homolog dredd cleaves imd and relish but is not inhibited by p35. *J. Biol. Chem.* **2014**, *289*, 20092–20101. [CrossRef]
35. Sumner, R.P.; Maluquer de Motes, C.; Veyer, D.L.; Smith, G.L. Vaccinia virus inhibits nf- b-dependent gene expression downstream of p65 translocation. *J. Virol.* **2014**, *88*, 3092–3102. [CrossRef] [PubMed]
36. Revilla, Y.; Callejo, M.; Rodriguez, J.M.; Culebras, E.; Nogal, M.L.; Salas, M.L.; Vinuela, E.; Fresno, M. Inhibition of nuclear factor kappab activation by a virus-encoded ikappab-like protein. *J. Biol. Chem.* **1998**, *273*, 5405–5411. [CrossRef]
37. Stuart, J.H.; Sumner, R.P.; Lu, Y.; Snowden, J.S.; Smith, G.L. Vaccinia virus protein c6 inhibits type i ifn signalling in the nucleus and binds to the transactivation domain of stat2. *PLoS Pathog.* **2016**, *12*, e1005955. [CrossRef]
38. Ferguson, B.J.; Benfield, C.T.; Ren, H.; Lee, V.H.; Frazer, G.L.; Strnadova, P.; Sumner, R.P.; Smith, G.L. Vaccinia virus protein n2 is a nuclear irf3 inhibitor that promotes virulence. *J. Gen. Virol.* **2013**, *94*, 2070–2081. [CrossRef]
39. Kleino, A.; Ramia, N.F.; Bozkurt, G.; Shen, Y.; Nailwal, H.; Huang, J.; Napetschnig, J.; Gangloff, M.; Chan, F.K.; Wu, H.; et al. Peptidoglycan-sensing receptors trigger the formation of functional amyloids of the adaptor protein imd to initiate drosophila nf-kappab signaling. *Immunity* **2017**, *47*, 635–647 e636. [CrossRef] [PubMed]
40. Upton, J.W.; Kaiser, W.J.; Mocarski, E.S. Cytomegalovirus m45 cell death suppression requires receptor-interacting protein (rip) homotypic interaction motif (rhim)-dependent interaction with rip1. *J. Biol. Chem.* **2008**, *283*, 16966–16970. [CrossRef]
41. Hallmann, C.A.; Sorg, M.; Jongejans, E.; Siepel, H.; Hofland, N.; Schwan, H.; Stenmans, W.; Muller, A.; Sumser, H.; Horren, T.; et al. More than 75 percent decline over 27 years in total flying insect biomass in protected areas. *PLoS ONE* **2017**, *12*, e0185809. [CrossRef]
42. Tokarz, R.; Firth, C.; Street, C.; Cox-Foster, D.L.; Lipkin, W.I. Lack of evidence for an association between iridovirus and colony collapse disorder. *PLoS ONE* **2011**, *6*, e21844. [CrossRef]
43. Kaneko, T.; Goldman, W.E.; Mellroth, P.; Steiner, H.; Fukase, K.; Kusumoto, S.; Harley, W.; Fox, A.; Golenbock, D.; Silverman, N. Monomeric and polymeric gram-negative peptidoglycan but not purified lps stimulate the drosophila imd pathway. *Immunity* **2004**, *20*, 637–649. [CrossRef]
44. Kaneko, T.; Yano, T.; Aggarwal, K.; Lim, J.-H.; Ueda, K.; Oshima, Y.; Peach, C.; Erturk-Hasdemir, D.; Goldman, W.E.; Oh, B.-H.; et al. Pgrp-lc and pgrp-le have essential yet distinct functions in the drosophila immune response to monomeric dap-type peptidoglycan. *Nat. Immunol.* **2006**, *7*, 715–723. [CrossRef]
45. Sansone, C.L.; Cohen, J.; Yasunaga, A.; Xu, J.; Osborn, G.; Subramanian, H.; Gold, B.; Buchon, N.; Cherry, S. Microbiota-dependent priming of antiviral intestinal immunity in drosophila. *Cell Host Microbe* **2015**, *18*, 571–581. [CrossRef]

46. McGettigan, J.; McLennan, R.K.; Broderick, K.E.; Kean, L.; Allan, A.K.; Cabrero, P.; Regulski, M.R.; Pollock, V.P.; Gould, G.W.; Davies, S.A.; et al. Insect renal tubules constitute a cell-autonomous immune system that protects the organism against bacterial infection. *Insect Biochem. Mol. Biol.* **2005**, *35*, 741–754. [CrossRef]
47. Verma, P.; Tapadia, M.G. Immune response and anti-microbial peptides expression in malpighian tubules of drosophila melanogaster is under developmental regulation. *PLoS ONE* **2012**, *7*, e40714. [CrossRef]
48. Paik, D.; Monahan, A.; Caffrey, D.R.; Elling, R.; Goldman, W.E.; Silverman, N. Slc46 family transporters facilitate cytosolic innate immune recognition of monomeric peptidoglycans. *J. Immunol.* **2017**, *199*, 263–270. [CrossRef]

11

Description of a Natural Infection with Decapod Iridescent Virus 1 in Farmed Giant Freshwater Prawn, *Macrobrachium rosenbergii*

Liang Qiu [1,†], Xing Chen [1,2,†], Ruo-Heng Zhao [1,3], Chen Li [1], Wen Gao [1,2], Qing-Li Zhang [1,2] and Jie Huang [1,2,3,*]

1 Yellow Sea Fisheries Research Institute, Chinese Academy of Fishery Sciences; Laboratory for Marine Fisheries Science and Food Production Processes, National Laboratory for Marine Science and Technology (Qingdao); Key Laboratory of Maricultural Organism Disease Control, Ministry of Agriculture and Rural Affairs; Qingdao Key Laboratory of Mariculture Epidemiology and Biosecurity, Qingdao 266071, China; qiuliang@ysfri.ac.cn (L.Q.); chenxing910520@163.com (X.C.); ruohengzhao@126.com (R.-H.Z.); lichen@ysfri.ac.cn (C.L.); gaowen1994@163.com (W.G.); zhangql@ysfri.ac.cn (Q.-L.Z.)

2 Shanghai Ocean University, Shanghai 201306, China

3 Dalian Ocean University, Dalian 116023, China

* Correspondence: huangjie@ysfri.ac.cn

† These authors contributed to the work equally and should be regarded as co-first authors.

Abstract: *Macrobrachium rosenbergii* is a valuable freshwater prawn in Asian aquaculture. In recent years, a new symptom that was generally called "white head" has caused high mortality in *M. rosenbergii* farms in China. Samples of *M. rosenbergii*, *M. nipponense*, *Procambarus clarkii*, *M. superbum*, *Penaeus vannamei*, and Cladocera from a farm suffering from white head in Jiangsu Province were collected and analyzed in this study. Pathogen detection showed that all samples were positive for Decapod iridescent virus 1 (DIV1). Histopathological examination revealed dark eosinophilic inclusions and pyknosis in hematopoietic tissue, hepatopancreas, and gills of *M. rosenbergii* and *M. nipponense*. Blue signals of in situ digoxigenin-labeled loop-mediated isothermal amplification appeared in hematopoietic tissue, hemocytes, hepatopancreatic sinus, and antennal gland. Transmission electron microscopy of ultrathin sections showed a large number of DIV1 particles with a mean diameter about 157.9 nm. The virogenic stromata and budding virions were observed in hematopoietic cells. Quantitative detection with TaqMan probe based real-time PCR of different tissues in naturally infected *M. rosenbergii* showed that hematopoietic tissue contained the highest DIV1 load with a relative abundance of 25.4 ± 16.9%. Hepatopancreas and muscle contained the lowest DIV1 loads with relative abundances of 2.44 ± 1.24% and 2.44 ± 2.16%, respectively. The above results verified that DIV1 is the pathogen causing white head in *M. rosenbergii*. *M. nipponense* and *Pr. clarkii* are also species susceptible to DIV1.

Keywords: DIV1; SHIV; CQIV; *Macrobrachium rosenbergii*; *Macrobrachium nipponense*; *Procambarus clarkii*; white head; susceptible species; viral load

---

1. Introduction

Globally, viral diseases have been acknowledged as a huge threat to the shrimp aquaculture industry. Among the viruses reported for crustaceans, *Cherax quadricarinatus* iridovirus (CQIV) and Shrimp hemocyte iridescent virus (SHIV) are two newly found viruses isolated from red claw crayfish *C. quadricarinatus* [1] and white leg shrimp *Penaeus vannamei* [2], respectively. CQIV and SHIV both have a typical icosahedral structure with a mean diameter of about 150 nm [1,2]. Evidence from histopathological study, transmission electron microscope (TEM) of ultrathin sections, and in situ hybridization (ISH) indicated that SHIV may mainly infect the hematopoietic tissue and hemocytes in gills, hepatopancreas, pereiopods, and muscle of *P. vannamei* [2]. Similarly, the TEM also showed that CQIV could infect hematopoietic tissue and gills in *C. quadricarinatus* and *P. vannameii* [1]. Phylogenetic analysis supported that SHIV and CQIV belong to a new genus, which was originally proposed to be named *Xiairidovirus* or *Cheraxvirus*, in the family *Iridoviridae* [2–4]. Alignment of the complete genomic sequences revealed that SHIV and CQIV might be different strains or genotypes of the same viral species [5]. In March 2019, the Executive Committee of the International Committee on Taxonomy of Viruses (ICTV) approved the proposal made by Chinchar et al. [6] that a new species of Decapod iridescent virus 1 (DIV1) in a new genus *Decapodiridovirus* to include SHIV 20141215 and CQIV CN01 as two isolates. We follow the ICTV's decision to use the formally recognized name for general indication of the virus or newly identified strains, and SHIV 20141215 and CQIV CN01 for individualization of the original isolations. To date, DIV1 has been detected in farmed *P. vannamei*, *P. chinensis*, *P. japonicus*, *C. quadricarinatus*, *Procambarus clarkii*, *Macrobrachium nipponense*, and *M. rosenbergii* in China since 2014 [1,4,7], indicating that DIV1 is a new threat to the shrimp farming industry.

The giant freshwater prawn, *M. rosenbergii*, a valuable crustacean species in Asian aquaculture, are widely cultured in tropical and subtropical areas. *M. rosenbergii* is native to Malaysia and other Asian countries, including Vietnam, Cambodia, Thailand, Myanmar, Bangladesh, India, Sri Lanka, and the Philippines [8,9]. Being popular for its delicious flesh and high nutritional value, the global production of this species has increased from about 3,000 tons in 1980 to more than 220,000 tons in 2014 [10,11]. Generally, *M. rosenbergii* is considered less prone to some viral diseases in aquaculture when compared to penaeid shrimps [12]. Some viral pathogens, such as *Macrobrachium* hepatopancreatic parvo-like virus (MHPV), *Macrobrachium* muscle virus (MMV), Infectious hypodermal and hematopoietic necrosis virus (IHHNV), White spot syndrome virus (WSSV), *Macrobrachium rosenbergii* nodavirus (MrNV), and Extra small virus like particle (XSV), have been reported in prawns [13]. Recently, results of RT-LAMP and histopathological examination indicated that *M. rosenbergii* could be infected with Covert mortality nodavirus (CMNV) [14]. To date, only PCR results showed that cultured *M. rosenbergii* were DIV1 positive [2,7] and more pathological information is not available.

In recent years, a new symptom that occurred in *M. rosenbergii* farms in China has been commonly called "white head" or "white spot", due to the diseased prawn exhibited a typical white triangle under the carapace at the base of rostrum [15]. Moribund prawns resting on the bottom in deep water and dead prawns can be found every day, with a cumulative mortality up to 80%. It is noteworthy that many *M. rosenbergii* populations suffering from white head were polycultured with *P. vannamei* [16]. In the present study, we investigated a diseased polyculture pond with *M. rosenbergii* and *Pr. clarkii*. In the pond, most of the *M. rosenbergii* exhibited typical the white triangle under the carapace at the base of rostrum and appeared moribund or died. One month before we arrived, all of the *P. vannamei* in an adjacent pond had died. Samples were collected and analyzed in this study.

2. Materials and Methods

All the protocols of animal handling and sampling were approved by the Animal Care and Ethics Committee, Yellow Sea Fisheries Research Institute, Chinese Academy of Fishery Sciences, and all efforts were made to minimize the suffering of animals according to recommendations proposed by the European Commission (1997). The study was carried out in accordance with the approved protocol. All the methods were applied in accordance with relevant guidelines.

2.1. Samples

Samples of farmed *M. rosenbergii* (4–6 cm) and *Pr. clarkii* (5–7 cm) were collected from the pond with high mortality in a farm in Jiangsu Province on 20 June 2018. In the same pond, some wild crustaceans, including *M. nipponense*, *M. superbum*, and some species of Cladocera, were also sampled for further analysis. Dead and dry bodies of *P. vannamei* (5–7 cm) were collected on the drained bottom of the adjacent pond, which suffered from a severe disease one month before in the farm.

2.2. DNA and RNA Extraction

Total DNA and RNA were extracted from 30 mg individual cephalothorax tissue of prawns, shrimp, or crayfish, or 30 mg multiple individuals of Cladocera by TIANamp Marine Animal DNA Kit and RNAprep pure Tissue Kit (TIANGEN Biotech, Beijing, China), respectively, according to the manufacturer's instructions.

2.3. Pathogen Detection

DNA samples of *M. rosenbergii*, *P. vannamei*, *Pr. clarkii*, *M. nipponense*, *M. superbum*, and Cladocera were tested for White spot syndrome virus (WSSV), IHHNV, acute hepatopancreas necrosis disease-causing *Vibrio parahaemolyticus* (Vp_{AHPND}), and DIV1 by real-time PCR methods. The RNA samples were tested for Yellow head virus (YHV), Infectious myonecrosis virus (IMNV), and CMNV by RT-real-time PCR methods. All the detection methods were recommended by the World Organization for Animal Health (OIE) [17] or developed before [5,18].

2.4. Histopathological Sections

Samples were fixed in Davidson's alcohol-formalin-acetic acid fixative (DAFA) [19] for 24 h and then changed to 70% ethanol. Paraffin sections were prepared and stained with hematoxylin and eosin (H&E) staining according to the procedures of Bell and Lightner [19].

2.5. In Situ Digoxigenin-Labeled Loop-Mediated Isothermal Amplification (ISDL)

Samples were fixed in DAFA for 24 h and changed to 70% ethanol. The paraffin sections were then prepared and subjected to ISDL assays targeting the gene of the second largest subunit of DNA-directed RNA polymerase II of DIV1, according to the method adapted for DIV1 infection by Chen et al. [20].

2.6. Transmission Electron Microscopy (TEM)

Ultrathin sections of the white triangle tissue under cuticle at the base of rostrum (hematopoietic tissue) from diseased *M. rosenbergii* were prepared for observation with TEM. Small pieces of the hematopoietic tissue in <1 mm^3 of sampled animals were fixed in TEM fixative (2% paraformaldehyde, 2.5% glutaraldehyde, 160 mM NaCl, and 4 mM $CaCl_2$ in 200 mM PBS) (pH 7.2) for 24 h at 4 °C. Before ultrathin sectioning, the fixed tissues were secondarily fixed with 1% osmium tetroxide for 2 h, then embedded in Spurr's resin and stained with uranyl acetate and lead citrate. Ultrathin sections were laid on collodion-coated grids and examined under a JEOL JEM-1200 electron microscope (Jeol Solutions for Innovation, Japan) operated at 80–100 kV in the Equipment Center of the Medical College of Qingdao University.

2.7. Quantitative Detection of DIV1 in Different Tissues of Naturally Infected M. rosenbergii

Total 15 moribund *M. rosenbergii* samples frozen at −80 °C were chosen to be defrosted and different tissues were separated, including hematopoietic tissue, antenna, uropods, pleopods, gills, pereiopods, muscle, and hepatopancreas. Total DNA was extracted from different tissues using TIANamp Marine Animal DNA Kit. The DIV1 loads in different tissues were detected by TaqMan probe-based quantitative real-time PCR (TaqMan qPCR) following our published method [5].

2.8. Relative Abundance of DIV1 in Different Tissues

In order to evaluate the distribution of DIV1 in different tissues, the relative abundance (RA_i) of DIV1 in different tissues was calculated with the DIV1 load in each tissue (L_i) to compare with the total load of DIV1 in the whole body, which were resulted from the sum of the DIV1 loads in all tested tissues ($\sum L_i$). The calculation is based on the following formula:

$$RA_i = \frac{L_i}{\sum L_i}$$

Significance analysis of relative abundance data between each of the two tissues was carried out using the t-test for heteroscedasticity hypothesis of two groups of samples with the add-in tool Data Analysis in Microsoft® Excel® 2016 MSO 64-bit.

3. Results

3.1. Observation of Clinical Signs of Diseaesd *M. rosenbergii*

According to on-farm inquiry, the investigated pond, about 1.5 ha in size was stocked with 45 postlarva/m^2 of prawn *M. rosenbergii* in the middle of May and some juveniles of crayfish *Pr. clarkii* one week before. The adjacent pond stocked with shrimp *P. vannamei* suffered from an unknown disease and died out one month before. No disinfection or other effective control measure was taken for the ponds except drainage of the diseased shrimp pond. Symptoms of "white head" and "yellow gills" in the prawn population were observed two weeks before our arrival. During the first week, the disease developed slowly, but it became more and more severe in the second week. The moribund prawns lost their swimming ability and sank to the bottom of water and were rarely observed in shallow water. Moribund and dead prawns could be found every day in the diseased pond. In the following inquiry, we were told that the cumulative mortality was higher than 80%.

While the samples were taken and processed, it was observed that most of the caught *M. rosenbergii* prawns from the diseased pond exhibited obvious clinical signs, including a distinct white triangle area under the carapace at the base of rostrum, hepatopancreatic atrophy with color fading and yellowing in the section, empty stomach and guts (Figure 1A,B), and some moribund prawns were accompanied by slightly whitish muscle and mutilated antenna.

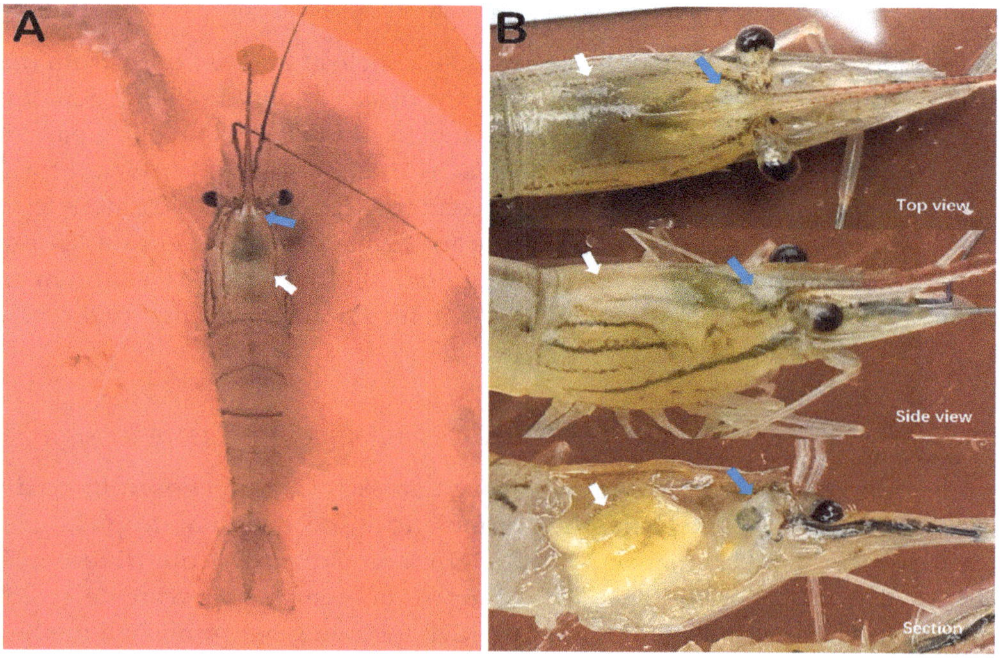

Figure 1. Clinical symptoms of *M. rosenbergii* (20180620) naturally infected with DIV1. (**A**) Overall appearance of a diseased prawn in water. (**B**) Close-up of cephalothoraxes. Blue arrows show white area under the cuticle at the base of rostrum. White arrows indicate hepatopancreas atrophy, color fading, and yellowing.

3.2. Pathogen Detection of Samples

A total of 20 DNA samples extracted from cephalothorax tissues of individual shrimp or multiple Cladocera were tested by real-time PCR or RT-real-time PCR. All samples were negative for WSSV, IHHNV, Vp_{AHPND}, YHV, IMNV, and CMNV, but positive for DIV1. Samples of *M. rosenbergii* contained the highest DIV1 load range from 3.16×10^8 to 9.83×10^8 copies/μg-DNA. Samples of *M. superbum* contained the lowest DIV1 load (Table 1).

Table 1. Samples detected with the real time PCR for DIV1.

Species	Positive Samples	Total Samples	Geometric Mean (Copies/μg-DNA)	DIV1 Range (Copies/μg-DNA)
M. rosenbergii	5	5	$10^{(8.65 \pm 0.21)}$	3.16×10^8–9.83×10^8
P. vannamei	3	3	$10^{(5.96 \pm 0.79)}$	4.56×10^5–7.19×10^6
M. nipponense	3	3	$10^{(4.17 \pm 1.68)}$	1.30×10^3–1.30×10^6
Pr. clarkii	5	5	$10^{(3.82 \pm 0.36)}$	2.20×10^3–1.57×10^4
Cladocera	3	3	$10^{(1.10 \pm 0.06)}$	1.09×10^1–1.43×10^1
M. superbum	1	1	$10^{(1.04 \pm 0.05)}$	1.00×10^1–1.18×10^1

3.3. Histopathology

Histological examination of DAFA fixed samples showed that dark eosinophilic inclusions mixed with basophilic tiny staining and karyopyknosis existed in hematopoietic tissue (Figure 2A) and hemocytes in hepatopancreatic sinus (Figure 2B) and in gills (Figure 2C) of *M. rosenbergii*. For *M. nipponense*, dark eosinophilic inclusions and karyopyknosis were also observed in the hepatopancreas (Figure 2D). No typical histopathological feature was found in the tissue sections of *Pr. Clarkii* and Cladocera.

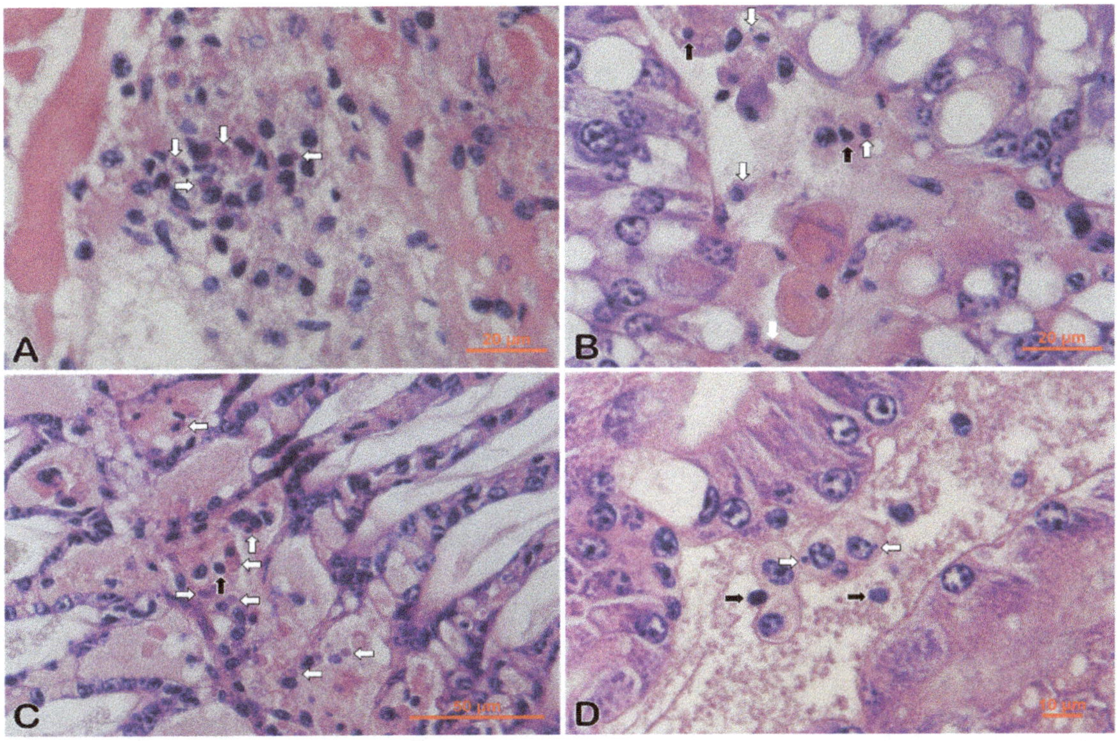

Figure 2. Histopathological features of Davidson's alcohol-formalin-acetic acid fixative (DAFA) fixed *M. rosenbergii* (**A**–**C**) and *M. nipponense* (**D**) samples 20180620. White arrows show the eosinophilic inclusions and black arrows show the karyopyknotic nuclei. (**A**) Hematoxylin and eosin (H&E) staining of the hematopoietic tissue. (**B**,**D**) H&E staining of hepatopancreas. (**C**) H&E staining of gills. Bar, 20 μm (**A**,**B**), 50 μm (**C**), and 10 μm (**D**), respectively.

3.4. ISDL

ISDL results of *M. rosenbergii* showed that blue signals existed in hematopoietic tissue (Figure 3A), hemocytes in the hepatopancreatic sinus and in gills (Figure 3B,C), some R-cells, and myoepithelial fibers of the hepatopancreas (Figure 3D), coelomosac epithelium of antennal gland (Figure 3E), and epithelium of ovaries (Figure 3F). In addition, similar distribution of positive signals was also observed in the hepatopancreas of *M. nipponense* (Figure 3G) and hepatopancreas and hematopoietic tissue of *Pr. clarkii* (Figure 3H,I). No positive signals were observed in sections prepared from Cladocera, and uninfected prawn or crayfish.

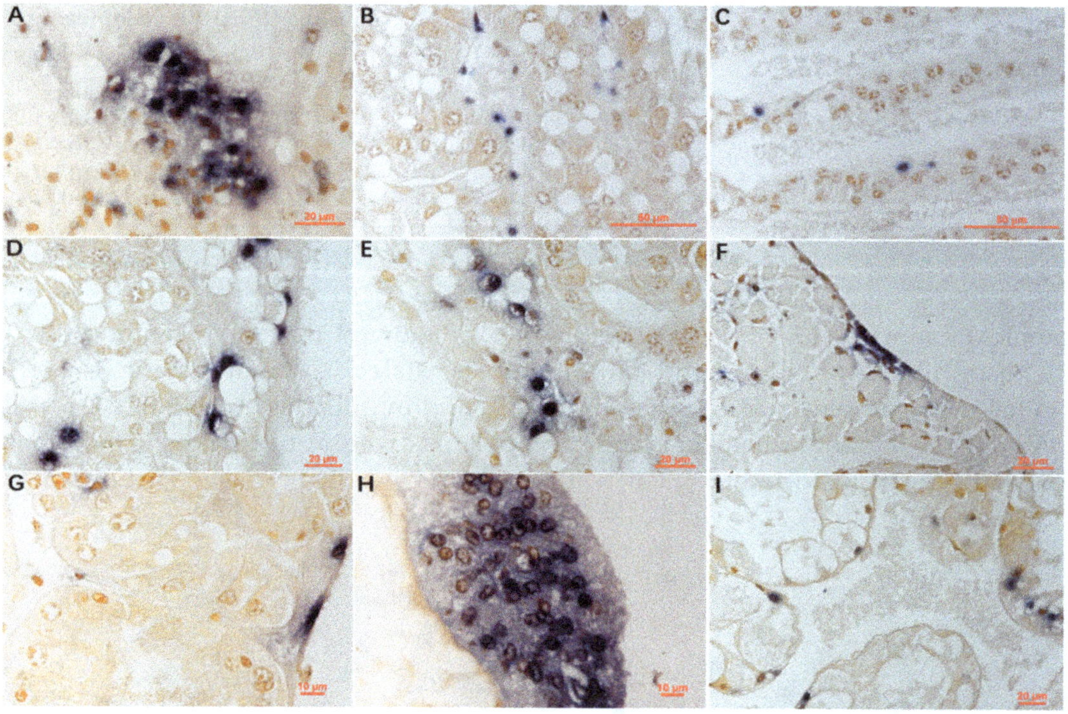

Figure 3. In situ digoxigenin-labeled loop-mediated isothermal amplification (ISDL) targeting the gene of the second largest subunit of DNA-directed RNA polymerase II of DIV1 on histological sections of *M. rosenbergii* (**A–F**), *M. nipponense* (**G**), and *Pr. clarkii* (**H**,**I**) samples 20180620. (**A**,**H**) Hematopoietic tissue; (**B**,**D**,**G**,**I**) hepatopancreas; (**C**) gills; (**E**) antennal gland; (**F**) ovaries. In (**A–C**,**H**), blue signals were observed in hematopoietic tissue, hemocytes in the sinus of the hepatopancreas, and in gills. In (**D**,**G**,**I**), blue signals exist in some hepatopancreatic R-cells and myoepithelial fibers. In (**E**), blue signals exist in the coelomosac epithelium. In (**F**), blue signals exist in the epithelium. Bar, 20 μm (**A**,**D–F**,**I**), 50 μm (**B**,**C**), and 10 μm (**G**,**H**), respectively.

3.5. TEM of Ultrathin Sections

Visualization with TEM of ultrathin sections of the naturally infected *M. rosenbergii* revealed the presence of a large number of icosahedral particles with typical iridescent viral structure, both inside and outside hematopoietic cells in the tissue (Figure 4A). Non-enveloped virions were 166.3 ± 14.8 nm (N = 39) vertex to vertex (v-v), 149.4 ± 13.8 nm (n = 39) from face to face (f-f), and about 157.9 nm of an average equivalent diameter, with a nucleoid at 93.5 ± 9.9 nm (n = 39). At the margin of the cytoplasm, many virions were budding from the plasma membrane, and in budded virions, an outer viral envelope was acquired from the plasma membrane (Figure 4B). Virion formation took place in the cytoplasmic morphologically distinct regions, termed virogenic stromata, which were electron-lucent areas containing numerous immature and empty capsids, few mature virions, and were devoid of cellular organelles, with paracrystalline array of viral particles and budding virions in the same cell (Figure 4C). Assembly of nucleocapsid can be described in three progressive stages (Figure 4D), during which crescent-shaped structures of early capsid complexes subsequently assembled into spherical intermediates at stage 1 (Figure 4E–G, and indicated with 1 on Figure 4D), followed by formation of icosahedral capsids with a small opening at one vertex at stage 2 (Figure 4H, and indicated with 2 on Figure 4D) and recruitment of electron-dense nucleic acid at stage 3 (Figure 4I and indicated with 3 on Figure 4D). Complete nucleocapsids were observed in a fully filled state (Figure 4J, and indicated with 4 on Figure 4D).

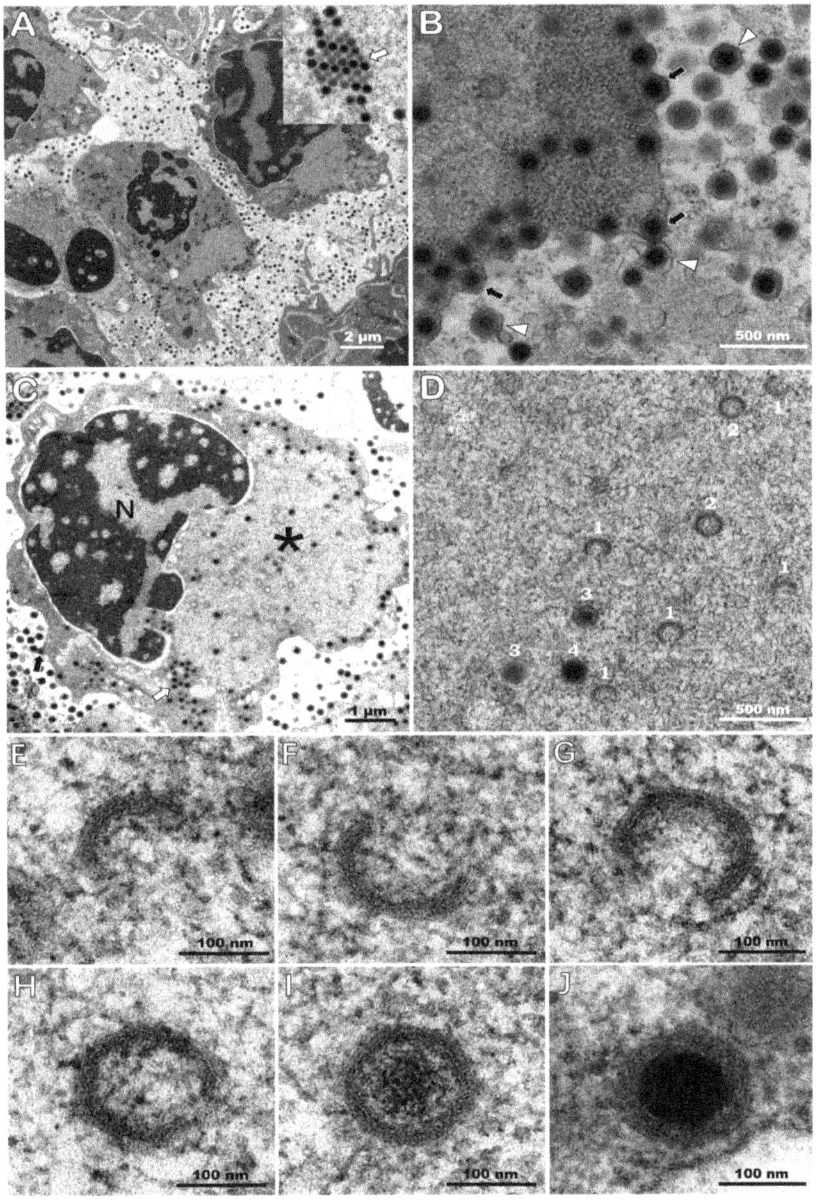

Figure 4. TEM of hematopoietic tissue of naturally infected *M. rosenbergii* samples 20180620. (**A**) A large numbers of virions in hematopoietic tissue. (**B**) DIV1 budded and acquired an envelope from the plasma membrane. (**C**) DIV1 replication and assembly in hematopoietic cells. (**D**) The stages of nucleocapsid assembly, which are indicated with numbers 1–3, and a complete nucleocapsid is indicated with number 4. The capsids at stage 2 and 3 should have a small opening at one vertex but may not be visible in the picture due to the ultrathin section. (**E**) Crescent-shaped structures. (**F–I**) As the assembling process continues, the crescent-shaped structure curves to form icosahedral capsids. (**J**) A mature virion with a dense core was eventually formed. N: nucleus; *: a large electron-lucent virogenic stroma; white arrows: paracrystalline array of viral particles; black arrows: budding virions; and white triangles: budded virions that acquired an envelope.

3.6. Quantitative Detection of DIV1 in Different Tissues of Naturally Infected *M. rosenbergii*

A total of 120 tissues were separated from 15 naturally infected *M. rosenbergii* prawns. DIV1 loads in tissues of different prawns were examined with TaqMan qPCR. Copies of DIV1 per μg tissue DNA sample were converted to their logarithms for calculation of the geometric means and standard deviations of each tissue. The results showed that hematopoietic tissue samples contained an average DIV1 load of $10^{(7.92 \pm 0.91)}$ copies/μg-DNA, which was the highest load of DIV1 in tissues tested. Antenna had a mean DIV1 load of $10^{(7.84 \pm 0.70)}$ copies/μg-DNA, which approached hematopoietic tissue. Uropods, pleopods, gills, and pereiopods also had high loads of DIV1 above $10^{7.6}$ copies/μg-DNA of geometric means. Moreover, muscles and the hepatopancreas, as the vast majorities of cephalothorax and abdominal segments of the prawn, contained the lowest load of DIV1, which were at $10^{(6.96 \pm 0.57)}$ and $10^{(6.85 \pm 0.72)}$ copies/μg-DNA, respectively (Table 2). It was also noted that there were big differences of DIV1 loads by 3–4 orders of magnitude in all the tissues among different individuals, shown in the range column of Table 2.

Table 2. DIV1 copies in different tissues detected in DIV1-positive prawns *Macrobrachium rosenbergii*.

Samples	n	Geometric Mean (Copies/μg-DNA)	Range (Copies/μg-DNA)
Hematopoietic tissue	15	$10^{(7.92 \pm 0.91)}$	8.11×10^4–6.33×10^8
Antenna	15	$10^{(7.84 \pm 0.70)}$	1.44×10^6–1.03×10^9
Uropods	15	$10^{(7.74 \pm 0.71)}$	6.45×10^5–6.04×10^8
Pleopods	15	$10^{(7.77 \pm 0.69)}$	4.77×10^5–5.30×10^8
Gills	15	$10^{(7.69 \pm 0.66)}$	4.45×10^5–4.10×10^8
Pereiopod	15	$10^{(7.62 \pm 0.69)}$	6.10×10^5–1.05×10^9
Muscle	15	$10^{(6.96 \pm 0.57)}$	1.87×10^5–4.36×10^7
Hepatopancreas	15	$10^{(6.85 \pm 0.72)}$	2.35×10^4–3.18×10^7

3.7. Relative Abundance of DIV1 in Different Tissues

As the differences of DIV1 loads in tissues among different individuals reached 3–4 orders of magnitude, direct comparison on DIV1 loads in a specific tissue could be easily upset by a single sample with a very high DIV1 load. Therefore, relative abundance of DIV1 load based on a uniformization of the sum of DIV1 loads in all detected tissues was introduced to evaluate the distribution of DIV1 in different tissues. The relative abundance results showed that more than one-quarter of DIV1 relative load was in hematopoietic tissue, which contained the highest relative abundance at 25.4 ± 16.9%. Antenna, pleopods, and uropods contained significantly lower DIV1 relative abundances at 18.7 ± 10.3%, 14.5 ± 5.1%, and 13.8 ± 4.7%, respectively, more than hematopoietic tissue did ($p < 0.05$). Gills and pereiopods contained very significant lower DIV1 relative abundances at 12.1 ± 4.1% to 10.7 ± 5.5%, respectively, more than hematopoietic tissue did ($p < 0.01$). Hepatopancreas and muscle contained the very significantly lowest levels ($p < 0.01$) of the relative abundances of DIV1 at 2.44 ± 2.16% and 2.44 ± 1.24%, compared with all other tissues, respectively (Figure 5). When the total DIV1 load in all tissues fell between $10^{8.24}$ to $10^{9.25}$ copies/μg-DNA, most of the relative abundances of DIV1 in hematopoietic tissue fell into 12.2% to 35.9%, and even 71.7%; however, when the total DIV1 load in all tissues was lower than 10^7 or higher than $10^{9.5}$ copies/μg-DNA, the relative abundances of DIV1 in hematopoietic tissue dropped to about 5% or lower, while the most DIV1 existed in antenna and pleopods.

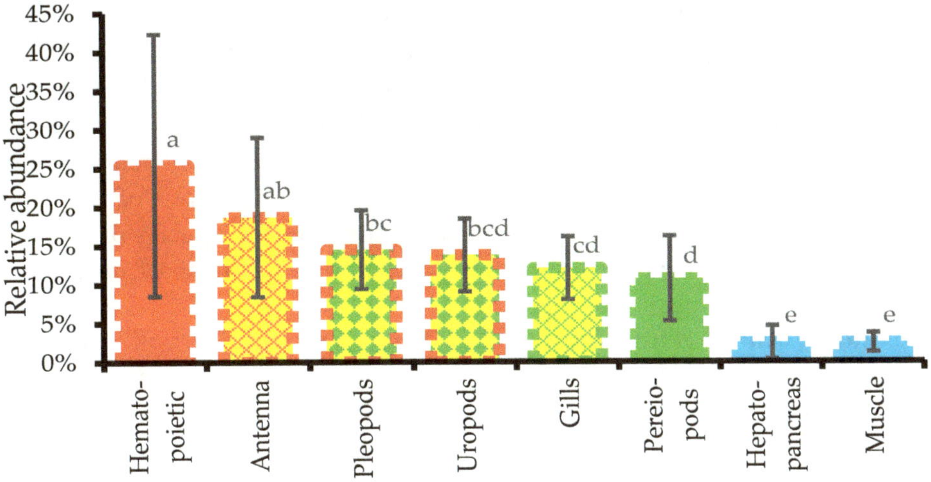

Figure 5. Relative abundance of DIV1 for different tissues of fifteen DIV1-infected *M. rosenbergii* samples. Columns without sharing of a same letter indicate significant difference of $p < 0.05$; columns without a same color indicate a highly significant difference of $p < 0.01$.

4. Discussion

DIV1 belongs to a novel genus *Decapodiridovirus* accepted by ICTV, found independently in *C. quadricarinatus* and *P. vannamei* as CQIV and SHIV. Target surveillance was started in China in 2017 and revealed that the virus has been detected in 6 provinces out of the 13 surveyed provinces and caused massive economic losses [7].

As far as we know, there was no reported pathological information of *M. rosenbergii* infected with DIV1, but a PCR detection result showed that 5 of 33 cultured *M. rosenbergii* samples were DIV1 positive from 2014 to 2016 [2]. This study reported farmed *M. rosenbergii* and *Pr. clarkii*, cohabitating with some wild crustaceans, *M. nipponense*, *M. superbum*, and Cladocera in a pond in Jiangsu Province, suffered from severe mortality in June 2018, following the death of the *P. vannamei* population in the adjacent pond. Diseased giant freshwater prawns have exhibited typical symptom commonly, known as "white head" or "white spot", since 2015 [15]. Real-time PCR results showed that all samples were negative for WSSV, IHHNV, Vp_{AHPND}, YHV, IMNV, and CMNV, but positive for DIV1. Samples of *M. rosenbergii* contained the highest DIV1 loads, ranging from 3.16×10^8 to 9.83×10^8 copies/μg-DNA, which were higher than any other naturally infected species in this and earlier studies [5], indicating that the disease of *M. rosenbergii* in this case was caused by a severe infection with DIV1. This is the first confirmation of the causative agent of "white head" symptoms in farmed *M. rosenbergii*. In addition, in this case, DIV1 was also proved to be a natural pathogen to *P. clarkii* and *M. nipponense*. The disease firstly broke out and caused death of farmed *P. vannamei* population in the adjacent pond two weeks before clinical signs for prawns were found in the pond stocked with *M. rosenbergii*. This provided evidence that the transmitted disease cross ponds and species due to lack of biosecurity in the farm management.

Notably, susceptibilities of *M. rosenbergii*, *M. nipponense*, *P. vannamei*, and *Pr. clarkii* to infection with DIV1 and infection with WSSV are different. An earlier study revealed that *M. rosenbergii* and *M. nipponense* resisted infection with WSSV via intermuscular injection, but *Pr. clarkii* had high mortality at 94% [21]. Further challenge studies showed *M. rosenbergii* could be infected by injection with WSSV stains Thai-1 and Viet, however, the infectious dose to reach 50% endpoint in *M. rosenbergii* needed 20-fold and 400-fold more than that in *P. vannamei*, respectively [22]. The half lethal dose (LD50) of WSSV to *M. nipponense* by injection was $10^{(3.84 \pm 0.06)}$ copies/g, which was about 1780-fold higher than that of *P. vannamei* at $10^{(0.59 \pm 0.22)}$ copies/g [23]. It was demonstrated that *M. rosenbergii* could clear infectious WSSV after 5 to 50 days post-injection [24]. There is no report for the quantitative comparison on the virulence of DIV1 to different species yet. Based on the time course of mortality in intramuscular challenge with DIV1 (CQIV CN01), *P. vannamei* may be slightly more susceptible to DIV1 infection than *C. quadricarinatus* and *Pr. clarkii* [1]. The disease course observed in the farm for this study indicated that *M. rosenbergii* and *M. nipponense* have no tolerance to infection with DIV1. Many farms in Jiangsu, Guangdong, and Zhejiang Provinces [16,25], as well as in Southeast Asia [26,27] and Africa [28], have stocked ponds in polyculture mode with *M. rosenbergii* and *P. vannamei* or *P. monodon*. As *M. rosenbergii* has tolerance to infection with WSSV [21], the polyculture mode provides a profitable approach for farmers under the threat of WSSV. However, the emergence of DIV1 has shattered the vision and verified our earlier warning that polyculture with different species of crustaceans may bring risks for spread of disease, increase of susceptible species, and evolution of pathogens, based on our early surveillance on shrimp epidemiology [29].

M. rosenbergii and *P. vannamei* infected with DIV1 both exhibited hepatopancreatic atrophy with color fading on the surface and in the section, empty stomach, and gut. However, these symptoms are not distinctive, because empty stomach and gut also occurred in some other diseases, such as infection with WSSV [17,30], infection with Taura syndrome virus (TSV) and AHPND [17], and loss color of hepatopancreas is similar to the clinical feature of AHPND [17]. It is worth noting that "white head" is a typical clinical sign for on-site diagnosis of *M. rosenbergii* infected with DIV1. Xu et al. [1] reported that experimentally-challenged individuals of *Pr. clarkii* showed gross signs, such as cessation of feeding and flaccidity, at day-5 post-infection. In this case, cultured *Pr. clarkii* naturally infected with DIV1 contained a lower viral load (2.20×10^3 to 1.57×10^4 copies/μg-DNA) and suffered from an asymptomatic infection, because the juveniles of crayfish were stocked in the pond for only one week and some of them may have just been infected with the virus for a few days. Almost all of the *P. vannamei* shrimp in the adjacent pond died out in two weeks and the pond was abandoned by drainage. It was one month later when we arrived at the farm, and only several dried bodies of *P. vannamei* could be collected from the bottom of a drained pond, which could not be used for observation of gross signs.

Histopathological examination showed the existence of dark eosinophilic inclusions mixed or surrounded with basophilic staining and karyopyknosis in the hematopoietic tissues and hemocytes of gills and hepatopancreatic sinus in *M. rosenbergii*. Similar pathological features also existed in *M. nipponense* collected in this study and *Exopalaemon carinicauda* experimentally challenged with DIV1 (SHIV 20141215) [20]. The inclusions of DIV1 infection found in *P. vannamei* in previous research were described as basophilic, but the color of the inclusions on the published pictures was dark eosinophilic mixed with tiny basophilic staining, the same as in this study [2]. The eosinophilic inclusions presented in the cytoplasm of hemocytes or hematopoietic cells, which is very similar to some shrimp cases caused by the iridescent virus reported earlier [31,32], and the karyopyknosis is similar to some fish cases [33,34]. Compared with ISH in *P. vannamei* [2], ISDL also specifically indicated the existence and location of DIV1 in histologic sections. Blue signals were observed in hematopoietic tissue and hemocytes in the sinus of the hepatopancreas and gills in *M. rosenbergii*, *M. nipponense*, and *Pr. clarkii*, which is consistent with the ISH result in sections of infected *P. vannamei* [2]. However, differing from the results of ISH, positive signals resulted from ISDL were observed in some R-cells and myoepithelial fibers of the hepatopancreas, coelomosac epithelium of antennal glands, and the epithelium of ovaries, which indicated that besides hematopoietic tissues and hemocytes, DIV1 may also infect some other tissues at a lower level. More remarkably, blue signals of ISDL appeared in both the nucleus and cytoplasm of hematopoietic cells, hemocytes, and other infected cells. It signified that DIV1 may employ a replication strategy to include both nuclear and cytoplasmic stages, as Frog virus 3 (FV3), the typical species of genus *Ranavirus*, does [35–37]. It should be kept in mind that the amplification of target DNA in ISDL results in a similar climax quantity of amplified products, so that the density of positive signals does not relatively reflect the original quantity of the DNA target, and it can only indicate the presence and location of target viral nucleic acid in the cells of various tissues and organs [38].

TEM evidence also proved the diseased prawns *M. rosenbergii* were infected with DIV1 by viral morphology and cytopathology. The icosahedral morphology and intracytoplasmic location of virions are consistent with reports of DIV1 infections in *P. vannamei* [2] or in *C. quadricarinatus*, *Pr. clarkii*, and *P. vannamei* [4]. Non-enveloped virions showed values 166.3 ± 14.8 nm (v-v) and 149.4 ± 13.8 nm (f-f), with a nucleoid at 93.5 ± 9.9 nm, which were 7.7 nm, 5.8 nm, and 7.7 nm larger than previous study in *P. vannamei* [2]. The different size of virions may result from shrinkage of virions caused by the duration time of tissue samples stored in the TEM fixative and the dehydration and embedding procedures before ultrathin sectioning. Typical electron-lucent virogenic stroma was observed in hematopoietic cells of infected *M. rosenbergii*, containing immature, empty, and mature virions, with paracrystalline array of viral particles and budding virions in the same cell. The appearance and location of virogenic stroma were consistent with the dark eosinophilic inclusions observed on the H&E stained histopathological slides. Viral nucleic acid, intensive mature virions, and paracrystalline array embedded in and surrounding the virogenic stroma resulted in the dark eosinophilic and tiny basophilic staining. The progressive assembling stages, from the crescent shape complex to filled hexagonal nucleocapsids, are very similar to the forming process of Singapore grouper iridovirus (SGIV) nucleocapsid [39]. This observation was not reported for DIV1 in previous studies.

TaqMan qPCR specifically detected the highest loads of DIV1 in the lesion of whitish hematopoietic tissue at the base of rostrum of diseased *M. rosenbergii*. This data fills the gap in the previous study, which lacked the DIV1 load in hematopoietic tissue [5]. Unlike *M. rosenbergii*, *P. vannamei* has several very small hematopoietic tissues, which makes it hard to be seen and collected for quantitative detection of DIV1. Decapods have separate hematopoietic tissues located above the stomach and at the base of antennas, pereiopods, and gills as appendages of the cephalothorax [19]. That is why appendages under the cephalothorax also contained relatively higher DIV1 load than the main body. It is notable that the high loads of DIV1 were also detected in the appendages of the abdominal segments, including uropods and pleopods, which hinted that these appendages of abdominal segments may be attached with some hematopoietic cells. Muscle and hepatopancreas contained the lowest DIV1 loads, which may be due to the virus in hemocytes and hemolymph in these tissues. The relative abundances of DIV1 load in different tissues calculated based on the results of TaqMan qPCR could be used to quantitatively estimate the virus distribution for a tissue tropism study in comparison with different hosts. Interestingly, the naturally dried bodies of shrimp *P. vannamei* that had been dead for more than one month could still be detected as positive up to 7.19×10^6 copies/μg-DNA. It indicated that dried shrimp bodies are still available for DIV1 detection.

5. Conclusions

This is a report of a natural occurrence of DIV1 in farmed giant freshwater prawn *M. rosenbergii*. All if the evidence resulting from symptom description, detection of known pathogens, histopathological and cytopathological observation, in situ DIG-labeled LAMP location, and quantitative detection of tissues consistently confirmed that the "white head" symptom of *M. rosenbergii* is the typical sign caused by infection with DIV1. Additionally, this study also provided evidence to add *Pr. clarkii* and *M. nipponense* as susceptible species to DIV1. The disease was likely transmitted from the adjacent pond stocked with *P. vannamei*, which had died out during the outbreak of infection with DIV1 due to lock of biosecurity in the farm management. The study provides a typical example of how DIV1 threaten the freshwater polyculture modes with different species of crustaceans, which we discourage.

Author Contributions: Conceptualization, L.Q., J.H., and Q.-L.Z.; formal analysis, L.Q. and X.C.; funding acquisition, J.H.; sampling, L.Q., X.C., J.H., and R.-H.Z.; methodology, X.C., R.-H.Z., C.L., and W.G.; sample testing, X.C., R.-H.Z., and W.G.; project administration, J.H. and L.Q.; resources, L.Q., X.C., and R.-H.Z.; supervision, J.H.; writing—original draft, L.Q; revision, J.H. and L.Q.

Acknowledgments: We are grateful to Guo-Liang Yang of Jiangsu Shufeng Aquatic Seed Industry Co., Ltd. for connecting the diseased farm, the agent serving the farm, and the farmers in Jiangsu Province for their help during our sampling, as well as all of our laboratory members for their technical advice and helpful discussions.

References

1. Xu, L.; Wang, T.; Li, F.; Yang, F. Isolation and preliminary characterization of a new pathogenic iridovirus from redclaw crayfish *Cherax quadricarinatus*. *Dis. Aquat. Organ.* **2016**, *120*, 17–26. [CrossRef]
2. Qiu, L.; Chen, M.M.; Wan, X.Y.; Li, C.; Zhang, Q.L.; Wang, R.Y.; Cheng, D.Y.; Dong, X.; Yang, B.; Wang, X.H.; et al. Characterization of a new member of Iridoviridae, Shrimp hemocyte iridescent virus (SHIV), found in white leg shrimp (*Litopenaeus vannamei*). *Sci. Rep.* **2017**, *19*, 11834. [CrossRef]
3. Qiu, L.; Chen, M.M.; Wang, R.Y.; Wan, X.Y.; Li, C.; Zhang, Q.L.; Dong, X.; Yang, B.; Xiang, J.H.; Huang, J. Complete genome sequence of shrimp hemocyte iridescent virus (SHIV) isolated from white leg shrimp, *Litopenaeus vannamei*. *Arch. Virol.* **2018**, *163*, 781–785. [CrossRef] [PubMed]
4. Li, F.; Xu, L.; Yang, F. Genomic characterization of a novel iridovirus from redclaw crayfish *Cherax quadricarinatus*: Evidence for a new genus within the family *Iridoviridae*. *J. Gen. Virol.* **2017**, *98*, 2589–2595. [CrossRef] [PubMed]
5. Qiu, L.; Chen, M.M.; Wan, X.Y.; Zhang, Q.L.; Li, C.; Dong, X.; Yang, B.; Huang, J. Detection and quantification of shrimp hemocyte iridescent virus by TaqMan probe based real-time PCR. *J. Invertebr. Pathol.* **2018**, *154*, 95–101. [CrossRef] [PubMed]
6. One New Genus with One New Species in the Subfamily *Betairidovirinae*. Available online: https://talk.ictvonline.org/files/ictv_official_taxonomy_updates_since_the_8th_report/m/animal-dna-viruses-and-retroviruses/8051 (accessed on 10 March 2019).
7. Qiu, L.; Dong, X.; Wan, X.Y.; Huang, J. Analysis of iridescent viral disease of shrimp (SHID) in 2017. In *Analysis of Important Diseases of Aquatic Animals in China in 2017*; Fishery and Fishery Administration Bureau under the Ministry of Agriculture and Rural Affairs, National Fishery Technical Extension Center, Eds.; China Agriculture Press: Beijing, China, 2018; pp. 187–204, ISBN 978-7-109-24522-8. (In Chinese)
8. Rao, R.; Bhassu, S.; Bing, R.Z.; Alinejad, T.; Hassan, S.S.; Wang, J. A transcriptome study on *Macrobrachium rosenbergii* hepatopancreas experimentally challenged with white spot syndrome virus (WSSV). *J. Invertebr. Pathol.* **2016**, *136*, 10–22. [CrossRef]
9. Hameed, A.S.S. Viral infections of *Macrobrachium* spp.: Global status of outbreaks, diagnosis, surveillance, and research. *Isr. J. Aquac.* **2009**, *61*, 240–247.
10. Ho, K.L.; Gabrielsen, M.; Beh, P.L.; Kueh, C.L.; Thong, Q.X.; Streetley, J.; Tan, W.S.; Bhella, D. Structure of the *Macrobrachium rosenbergii* nodavirus: A new genus within the Nodaviridae? *PLoS Biol.* **2018**, *16*, e3000038. [CrossRef]
11. *Macrobrachium rosenbergii* (De Man, 1879). Available online: http://www.fao.org/fishery/culturedspecies/Macrobrachium_rosenbergii/en (accessed on 9 January 2019).
12. Bonami, J.R.; Sri Widada, J. Viral diseases of the giant fresh water prawn *Macrobrachium rosenbergii*: A review. *J. Invertebr. Pathol.* **2011**, *106*, 131–142. [CrossRef]
13. Sahul Hameed, A.S.; Bonami, J.R. White tail disease of freshwater prawn, *Macrobrachium rosenbergii*. *Indian J. Virol.* **2012**, *23*, 134–140. [CrossRef]
14. Zhang, Q.L.; Xu, T.T.; Wan, X.Y.; Liu, S.; Wang, X.H.; Li, X.P.; Dong, X.; Yang, B.; Huang, J. Prevalence and distribution of Covert mortality nodavirus (CMNV) in cultured crustacean. *Virus. Res.* **2017**, *233*, 113–119. [CrossRef] [PubMed]
15. Who Is the Ringleader of the "White Spot" Disease in *Macrobrachium rosenbergii*. Available online: http: //www.shuichan.cc/article_view-41757.html (accessed on 9 January 2019). (In Chinese)
16. Huang, S. Polyculture of *Macrobrachium rosenbergii* and *Litopenaeus vannamei* Will Become the Main Mixed Culture Mode in Chaozhou and Shantou of Guangdong Province. Chinese Aquaculture Gateway Web. 2015. Available online: http://www.bbwfish.com/article.asp?artid=173520 (accessed on 9 January 2019). (In Chinese)
17. Manual of Diagnostic Tests for Aquatic Animals. Available online: http://www.oie.int/en/standard-setting/aquatic-manual/access-online/ (accessed on 11 September 2018).

18. Zhang, Q.L.; Liu, Q.; Liu, S.; Yang, H.L.; Liu, S.; Zhu, L.L.; Yang, B.; Jin, J.T.; Ding, L.X.; Wang, X.H.; et al. A new nodavirus is associated with covert mortality disease of shrimp. *J. Gen. Virol.* **2014**, *95*, 2700–2709. [CrossRef] [PubMed]
19. Bell, T.A.; Lightner, D.V. *A Handbook of Normal Penaeid Shrimp Histology*; World Aquaculture Society: Tucson, AZ, USA, 1988.
20. Chen, X.; Qiu, L.; Wang, H.L.; Zou, P.Z.; Dong, X.; Li, F.H.; Huang, J. Susceptibility of *Exopalaemon carinicauda* to the infection with Shrimp hemocyte iridescent virus (SHIV 20141215), a strain of Decapod iridescent virus 1 (DIV1). *Viruses* 2019, accepted.
21. Huang, C.H.; Shi, Z.L.; Zhang, J.H.; Zhang, L.R.; Chen, D.H.; Bonami, J.R. Establishment of a model for proliferating White spot syndrome virus in vivo. *Virol. Sin.* **1999**, *14*, 358–363.
22. Corteel, M.; Dantas-Lima, J.J.; Tuan, V.V.; Thuong, K.V.; Wille, M.; Alday-Sanz, V.; Pensaert, M.B.; Sorgeloos, P.; Nauwynck, H.J. Susceptibility of juvenile *Macrobrachium rosenbergii* to different doses of high and low virulence strains of white spot syndrome virus (WSSV). *Dis. Aquat. Org.* **2012**, *100*, 211–218. [CrossRef]
23. Zhao, C.Y.; Fu, H.T.; Sun, S.M.; Qiao, H.; Zhang, W.Y.; Jin, S.B.; Jiang, S.F.; Xiong, Y.W.; Gong, Y.S. Experimental inoculation of oriental river prawn *Macrobrachium nipponense* with white spot syndrome virus (WSSV). *Dis. Aquat. Org.* **2017**, *126*, 125–134. [CrossRef]
24. Sarathi, M.; Nazeer Basha, A.; Ravi, M.; Venkatesan, C.; Senthil Kumar, B.; Sahul Hameed, A.S. Clearance of white spot syndrome virus (WSSV) and immunological changes in experimentally WSSV-injected *Macrobrachium rosenbergii*. *Fish Shellfish Immunol.* **2008**, *25*, 222–230. [CrossRef]
25. Shen, Y.J.; Zhu, B.K.; Xu, K.C.; Xu, J.F. Pond culture experiment of polyculture with white leg shrimp and *Macrobrachium rosenbergii*. *Xiandai Nongye Keji* **2017**, *17*, 231–241. (In Chinese)
26. Azim, M.E.; Mazid, M.A.; Alam, M.J.; Nurullah, M. The potential of mixed culture of freshwater giant prawn *Macrobrachium rosenbergii* de Man and tiger shrimp *Penaeus monodon* Fab. at Khulna region, Bangladesh. *Bangladesh J. Fish. Res.* **2001**, *5*, 67–74.
27. Ali, H.; Rahman, M.M.; Rico, A.; Jaman, A.; Basak, S.K.; Islam, M.M.; Khan, N.; Keus, H.J.; Mohan, C.V. An assessment of health management practices and occupational health hazards in tiger shrimp (*Penaeus monodon*) and freshwater prawn (*Macrobrachium rosenbergii*) aquaculture in Bangladesh. *Vet. Ani. Sci.* **2018**, *5*, 10–19. [CrossRef]
28. Sadek, S.; Moreau, J. Performance of *Macrobrachium rosenbergii* and *Penaeus semisulcatus* under mono and mixed culture systems, and their suitability for polyculture with Florida Red Tilapia, in Egypt. *J. Aquacult. Trop.* **2000**, *15*, 97–107.
29. Diseases of Farmed Shrimp and Their Biosecurity Control Technologies. Available online: http://www.shuichan.cc/news_view-268040.html (accessed on 9 February 2019). (In Chinese)
30. Escobedo-Bonilla, C.M.; Alday-Sanz, V.; Wille, M.; Sorgeloos, P.; Pensaert, M.B.; Nauwynck, H.J. A review on the morphology, molecular characterization, morphogenesis and pathogenesis of White spot syndrome virus. *J. Fish. Dis.* **2008**, *31*, 1–18. [CrossRef]
31. Lightner, D.V.; Redman, R.M. A putative iridovirus from the penaeid shrimp *Protrachypene precipua* Burkenroad (Crustacea: Decapoda). *J. Invertebr. Pathol.* **1993**, *62*, 107–109. [CrossRef]
32. Tang, K.F.J.; Redman, R.M.; Pantoja, C.R.; Groumellec, M.L.; Duraisamy, P.; Lightner, D.V. Identification of an iridovirus in *Acetes erythraeus*, (Sergestidae) and the development of *in situ*, hybridization and PCR method for its detection. *J. Invertebr. Pathol.* **2007**, *96*, 255–260. [CrossRef]
33. Mahardika, K.; Yamamoto, A.; Miyazaki, T. Susceptibility of juvenile humpback grouper *Cromileptes altivelis* to grouper sleepy disease iridovirus (GSDIV). *Dis. Aquat. Org.* **2004**, *59*, 1–9. [CrossRef] [PubMed]
34. Sudthongkong, C.; Miyata, M.; Miyazaki, T. Iridovirus disease in two ornamental tropical freshwater fishes: African lampeye and dwarf gourami. *Dis. Aquat. Org.* **2002**, *48*, 163–173. [CrossRef] [PubMed]
35. Chinchar, V.G.; Hick, P.; Ince, I.A.; Jancovich, J.K.; Marschang, R.; Qin, Q.; Subramaniam, K.; Waltzek, T.B.; Whittington, R.; Williams, T.; et al. ICTV Virus Taxonomy Profile: *Iridoviridae*. *J. Gen. Virol.* **2017**, *98*, 890–891. [CrossRef] [PubMed]
36. Chinchar, V.G.; Hyatt, A.; Miyazaki, T.; Williams, T. Family *Iridoviridae*: Poor viral relations no longer. *Curr. Top. Microbiol. Immunol.* **2009**, *328*, 123–170. [CrossRef]

37. Williams, T.; Barbosa-Solomieu, V.; Chinchar, V.G. A decade of advances in iridovirus research. *Adv. Virus Res.* **2005**, *65*, 173–248. [CrossRef]
38. Jitrakorn, S.; Arunrut, N.; Sanguanrut, P.; Flegel, T.W.; Kiatpathomchai, W.; Saksmerprome, V. In situ DIG-labeling, loop-mediated DNA amplification (ISDL) for highly sensitive detection of infectious hypodermal and hematopoietic necrosis virus (IHHNV). *Aquaculture* **2016**, *456*, 36–43. [CrossRef]
39. Liu, Y.; Tran, B.N.; Wang, F.; Ounjai, P.; Wu, J.; Hew, C.L. Visualization of assembly intermediates and budding vacuoles of Singapore grouper iridovirus in grouper embryonic cells. *Sci. Rep.* **2016**, *6*, 18696. [CrossRef] [PubMed]

12

Pathogen Risk Analysis for Wild Amphibian Populations Following the First Report of a Ranavirus Outbreak in Farmed American Bullfrogs (*Lithobates catesbeianus*) from Northern Mexico

Bernardo Saucedo [1], José M. Serrano [2,3], Mónica Jacinto-Maldonado [4,5], Rob S. E. W. Leuven [6,7], Abraham A. Rocha García [8], Adriana Méndez Bernal [8], Andrea Gröne [1], Steven J. van Beurden [1] and César M. Escobedo-Bonilla [9,*]

1 Department of Pathobiology, Faculty of Veterinary Medicine, Utrecht University, 3584 CL Utrecht, The Netherlands; b.saucedogarnica@uu.nl (B.S.); a.groene@uu.nl (A.G.); steven.vanbeurden@gupta-strategists.nl (S.J.v.B.)

2 Laboratorio de Genética y Evolución, Departamento de Ciencias Ecológicas, Facultad de Ciencias, Universidad de Chile, Las Palmeras, Santiago 3425, Chile; jose.rano@gmail.com

3 Programa de Fisiología y Biofísica, Facultad de Medicina, Universidad de Chile, Santiago 8380453, Chile

4 Wildlife and Laboratory Animals, Department of Ethology, Faculty of Veterinary Medicine, National Autonomous University of Mexico, Mexico City 045010, Mexico; monica.jacinto@c3.unam.mx

5 C3-Complexity sciences Center, Autonomous University of Mexico, Mexico City 045010, Mexico

6 Department of Animal Ecology and Physiology, Institute for Water and Wetland Research, Radboud University, 6500 GL Nijmegen, The Netherlands; r.leuven@science.ru.nl

7 Netherlands Centre of Expertise on Exotic Species, 6500 GL Nijmegen, The Netherlands

8 Department of Pathology 04510, Faculty of Veterinary Medicine, National Autonomous University of Mexico, Mexico City 045010, Mexico; ab_abraham88@hotmail.com (A.A.R.G.); mvzadrimb@gmail.com (A.M.B.)

9 Department of Aquaculture, Instituto Politécnico, Nacional-CIIDIR Unidad Sinaloa, Guasave Sinaloa 81101, Mexico

* Correspondence: cescobe@ipn.mx

Abstract: Ranaviruses are the second deadliest pathogens for amphibian populations throughout the world. Despite their wide distribution in America, these viruses have never been reported in Mexico, the country with the fifth highest amphibian diversity in the world. This paper is the first to address an outbreak of ranavirus in captive American bullfrogs (*Lithobates catesbeianus*) from Sinaloa, Mexico. The farm experienced high mortality in an undetermined number of juveniles and sub-adult bullfrogs. Affected animals displayed clinical signs and gross lesions such as lethargy, edema, skin ulcers, and hemorrhages consistent with ranavirus infection. The main microscopic lesions included mild renal tubular necrosis and moderate congestion in several organs. Immunohistochemical analyses revealed scant infected hepatocytes and renal tubular epithelial cells. Phylogenetic analysis of five partial ranavirus genes showed that the causative agent clustered within the Frog virus 3 clade. Risk assessment with the Pandora$^+$ protocol demonstrated a high risk for the pathogen to affect amphibians from neighboring regions (overall Pandora risk score: 0.619). Given the risk of American bullfrogs escaping and spreading the disease to wild amphibians, efforts should focus on implementing effective containment strategies and surveillance programs for ranavirus at facilities undertaking intensive farming of amphibians.

Keywords: amphibians; histopathology; immunohistochemistry; Mexico; outbreak; ranavirus; risk assessment

1. Introduction

Mexico is the fifth ranking nation in terms of amphibian biodiversity with a total of 252 endemic species [1]. Moreover, the country ranks second in number of threatened amphibian species ($n = 164$) [1]. Main causes for amphibian population declines are anthropogenic activities such as deforestation and habitat fragmentation, and infectious diseases [2,3]. The two deadliest amphibian pathogens worldwide are the fungus *Batrachochytrium dendrobatidis* (Bd) and ranavirus [4]. The fungus Bd has caused epidemics in amphibian populations from mountainous regions of central and southern Mexico since the 1970s [5]. In contrast, ranaviruses (double-stranded DNA viruses from the family *Iridoviridae*) [4] have never been reported in Mexico, despite the wide distribution of these pathogens throughout the United States and Central and South America [6]. Insufficient surveillance is suspected to be the main reason for the lack of ranavirus records in Mexico [6].

Ranaviruses have been associated with die-offs of wild amphibians in America, where *Ambystoma tigrinum* virus (ATV) and frog virus 3 (FV3) are most prevalent [7,8]. In Europe, the common midwife toad viruses (CMTV) prevail in the wild [9–11], whereas in Asia, the tiger frog virus (TFV) and FV3 virus are present in the wild [12,13] and the CMTV strains have only been found in captive populations [14]. In many of these cases, ranavirus outbreaks have affected zoo collections [15], fisheries [16], or laboratory facilities [17].

The present study describes an outbreak of ranavirus disease in a captive colony of American bullfrogs (*Lithobates catesbeianus*) in the Mexican province of Sinaloa. Sinaloa is located in a transitional zone in the Pacific coast of Mexico and is home to at least 40 amphibian species, among them, salamanders of the families Ambystomatidae (1 species) and Plethodontidae (1 species), and anurans belonging to Bufonidae (9 species), Craugastoridae (4 species), Eleutherodactylidae (4 species), Hylidae (11 species), Leptodactylidae (1 species), Mycrohylidae (3 species), Pelobatidae (1 species), and Ranidae (5 species) [18]. Around 10% of the species from these families show evidence of population declines, 80% are considered stable and for the remaining 10% no information is available [1].

Commercial farming of American bullfrogs (*Lithobates catesbeianus*) started in Mexico during the 1950s when the species was introduced from the United States [19,20]. This species was first noticed in the wild in the northwestern part of Mexico in 1969 and has since become invasive and a major threat for ecological niches of endemic Mexican amphibians [19,21].

This work describes, for the first time, a ranavirus outbreak in farmed American bullfrogs (*Lithobates catesbeianus*) from Mexico. In late March of 2017, the rana-culture facility experienced a die-off involving an undetermined number of dead animals at various life stages, particularly juveniles and sub-adults. Macroscopic and microscopic lesions of affected animals, along with immunohistochemical and molecular analyses of the samples, revealed that the die-off was associated with an FV3-like ranavirus.

The aims of this study were to investigate the first ranavirus disease outbreak in Mexico and to analyze the potential risk that this pathogen poses for wild amphibian populations in neighboring areas.

2. Materials and Methods

2.1. General Information on the Rana-Culture Facility

The farm is located in the city of Guasave, province of Sinaloa (Figure 1). It started operations around 2009. The bullfrogs were originally brought in from a facility in central Mexico (state of Mexico) and were subsequently bred. Afterwards, a batch of frogs raised in the farm was bred to produce their own eggs and tadpoles, closing the culture system. The culture cycle normally lasts four to six months from tadpole to harvest size (120–250 g). The animals are used mainly for educational purposes and not for human consumption.

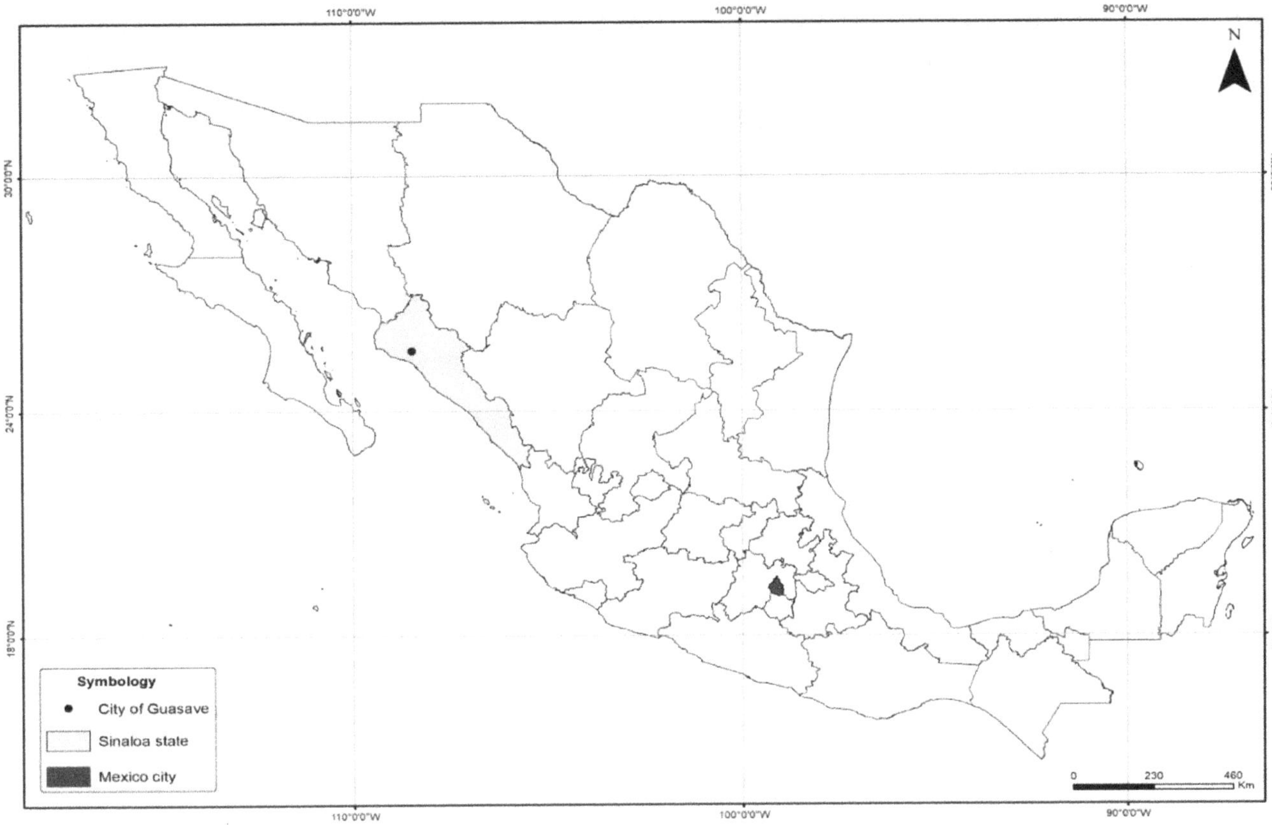

Figure 1. Location of city of the outbreak in Mexico. The city of Guasave is indicated by a black circle within the province of Sinaloa (grey). Mexico City, the place from which the bullfrogs were imported from, is shown in black.

2.2. Necropsies and Sample Collection

Animals that reached humane end-points manifested as severe lesions and clinical signs (ulcerations and skin erythema, lethargy,) were euthanized with an overdose of barbiturates (by a licensed veterinarian). Necropsies of five bullfrogs were performed in situ and spleen, kidney, and liver were sampled for histopathology and PCR.

2.3. Histopathology and Immunohistochemistry

Fragments of spleen, kidney, and liver were stored in 10% formalin for 24 h. Subsequently, samples were dehydrated with ascending grades of ethyl alcohol (ETOH), embedded in paraffin blocks (Paraplast, Tissue Embedding Medium, MacCormick Scientific, Chicago, IL, USA), cut at 3 μm thickness and stained with hematoxylin and eosin (H&E) for microscopic examination [22]. Immunohistochemistry was performed based on an established protocol with a polyclonal rabbit anti-European catfish virus antibody (kindly donated by Dr. Anna Toffan, Instituto Zooprofilattico Sperimentale delle Venezie, Italy) with slight modifications [23]. Sections were dewaxed and hydrated through decreasing concentrations of alcohol, and gently washed twice with PBS, pH 7.2, for 5 min. Endogenous peroxidase activity was inhibited by immersion of the slides in 4% hydrogen peroxide for 30 min. Blocking was done by placing the slides on PBS buffer with 5% bovine serum albumin (BSA) for 30 min at room temperature. Antigen retrieval was achieved by incubating the slides with trypsin (0.1% working solution) at 37 °C for 20 min. Afterwards, slides were incubated with the polyclonal rabbit anti-European catfish virus antibody diluted to 1:1800 in PBS buffer with 2.5% BSA for 1 h at room temperature. After washing three times with PBS-Tween 0.1%, a biotinylated goat anti-rabbit IgG antibody (Vector, Burlingame, CA, USA) diluted to 1:250 in TBS buffer with 2.5% BSA was added and followed by a 30-minute incubation period. The slides were then washed with PBS-Tween three times, incubated with Avidin-Biotin Complex (Dako, Palo Alto, CA, USA) diluted to 1:250 in PBS buffer for 30 min at room temperature, washed with PBS for 5 min and a chromogen substrate (AEC) was added (Dako, Palo Alto, CA, USA) for 8 min. Afterwards, the slides were placed in running tap water for 5 min and counterstained with hematoxylin for 1 min. Finally, after a 10 min bath in running tap water, the slides were mounted with Aquatex (Merck, Kenilworth, NJ, USA) and examined under a light microscope. Negative controls included tissues from non ranavirus-infected amphibians, tissues omitting primary or secondary antibody and an isotype control.

2.4. DNA Extraction, PCR and Sequencing

DNA extraction was performed at the Laboratory of Pathology and Molecular Diagnosis (IPN) in Sinaloa, Mexico. This was done using 3% acetyl trimethyl ammonium bromide (CTAB) buffer (Tris HCl 100 mM pH 8.0, EDTA 20 mM, NaCl 1.4 M CTAB 3% *w*/*v*) added with 0.2% (*v*/*v*) β-mercaptoethanol [24].

Each sample was separately ground with a pestle in a 1.5 mL microtube and incubated at 60 °C for 30 min. Then, a mix (24:1) of chloroform-isoamyl alcohol was added to denature proteins and centrifuged (13,000 rpm, 10 min) to separate the aqueous phase containing nucleic acids. This was transferred to a new tube and nucleic acids were precipitated with cold isopropyl alcohol and centrifuged as above. The pellet was washed with cold 70% ethanol, centrifuged and supernatant decanted. Nucleic acids were air dried and dissolved in 70 µL nuclease-free water. Concentration was measured by nano-spectrophotometry (nanodrop 2000c, Thermo Scientific, Waltham, MA, USA).

For the PCR analyses, the primers used were those designed by Mao et al., which target a portion of the viral major capsid protein gene [25]. The PCR volume was 50 µL. A PCR mix was done using a DreamTaq DNA polymerase (Thermo Fisher Scientific, Waltham, MA, USA) containing 5 µL of DreamTaq PCR Green buffer (10×), the respective forward (1 µL) and reverse (1 µL) primer pairs and 40.5 µL nuclease-free water. One µL of DNA template was added to each reaction. A positive and a negative control (distilled water) were included.

Amplification conditions were: an initial DNA denaturation step at 95 °C for 3 min, followed by 35 cycles at 95 °C for 30 s, annealing at 49 °C for all primers used and extension at 72 °C for 60 s. A final extension at 72 °C for min was included. PCR products were resolved in a 1% agarose gel in tris HCl, acetic acid EDTA (TAE) buffer and visualized in a Gel Doc XR+ documentation system (BioRad, Inc., Hercules, CA, USA).

The resulting PCR products were purified using a ExoSAP-IT® PCR Product Cleanup kit (GE Healthcare, Santa Clara, CA, USA) and sent for Sanger sequencing at (Macrogren, Amsterdam, NL, USA). The sequence was used for multiple sequence alignment and phylogenetic analyses.

2.5. Phylogenetic Analysis

For phylogenetic analysis, five partial ranavirus genes were selected (13R, 16L/MCP, 22L, 59R and 82L). The partial gene sequences of the Mexican ranavirus were compared to those of 20 other members of the family *Iridoviridae*. The DNA sequences were aligned using the software CLUSTAL Omega with default settings (https://www.ebi.ac.uk/Tools/msa/clustalo/) [26]. Maximum likelihood phylogeny was re-constructed with the software MEGA 6.0 using 1000 bootstrap replicates. The best-fit model Kimura 2 parameter + Gamma distribution (K2 + G) was chosen based on the lowest Bayesian Information Criterion score.

2.6. Risk Classification Using the PANDORA + Protocol

To assess the risks that ranavirus would pose for the wild endemic amphibians in Sinaloa, Mexico, the Pandora$^+$ protocol was used [27]. This protocol is specific for invasive microbes and complies with EU regulations for risk assessment that supports the listing of invasive alien species. This program originates from the Invasive Species Environmental Impact Assessment protocol (ISEIA) [28]. The ISEIA protocol was previously used to perform a risk analysis on ranaviruses in The Netherlands [29]. The protocol consists of a questionnaire (20 questions) covering distinct modules which include entry, exposure, and consequences for various targets including environment, plants, domestic animals, and humans. Each question classifies the risk as low, medium, or high with a corresponding low, medium, or high level of confidence. Default settings of the Pandora^{+} program were used. In the default settings, each alternative answer to a question from a module is ranked 0–1 (lowest to highest) and then the arithmetic mean of the questions are calculated to obtain a value for each module. An equal weight value of 1 was allocated to each of the questions. The exposure/entry score is calculated by taking the geometric mean of the emerging/entry and emerging/exposure modules. The scores from the different impact/consequence modules are similarly aggregated to calculate a general consequence score by using the geometric mean from each module. Finally, the overall risk score is calculated by multiplying the exposure/entry score by the consequence score. The overall risk score may be: low (0–0.25), medium (0.25–0.50), high (0.50–0.75), and very high (0.75–1.00). Four of the coauthors, each with a particular area of expertise (biology, virology, veterinary pathology, and ecology) performed the risk assessment separately and later unified criteria to provide a consensus risk score.

2.7. Level of Wild Species Threat

The level of ranavirus infection threat from each of the amphibian species within the province of Sinaloa was evaluated by consulting the species status in the IUCN Red List (International Union for Conservation of Nature) [1] and the Environmental Vulnerability Score (EVS), which is an algorithm comprised of three scales of added values which consist of geographical distribution (score range 1–6), ecological distribution based on types of forest formations (score range 1–8), and reproductive mode which takes into account type of wetlands where larvae develop and in which egg masses are laid (score range 1–5). Once these elements are added, the score range can be from 3 to 19, this allows to categorize the threat as low (3–9), medium (10–13), and high (14–19) [30].

3. Results

External macroscopic lesions of affected individuals included skin hemorrhages and ulcerations (Figure 2A), while internal inspection of the coelomic cavity revealed multiple lesions including hepato-splenomegaly, hepatic necrosis, and epicardial hemorrhages (Figure 2B).

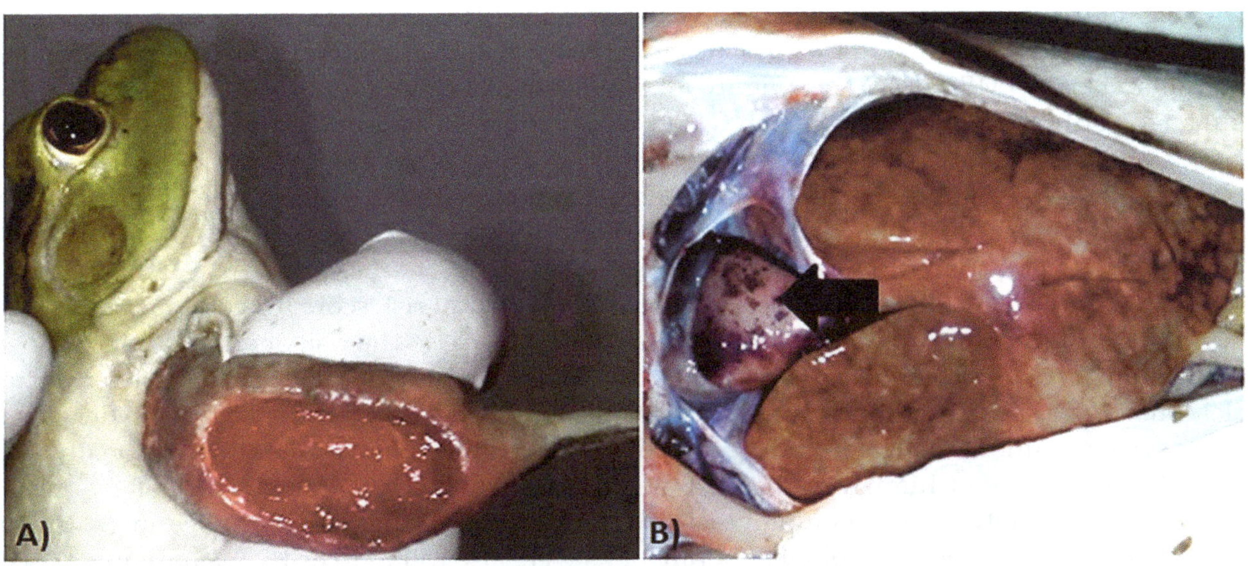

Figure 2. Macroscopic lesions of ranavirus-infected bullfrogs. (**A**) Extensive ulceration of the forelimb. (**B**) Extensive hepatic necrosis and epicardial pallor and hemorrhages (black arrow).

Most of the microscopic lesions in organs from examined specimens were moderate and consisted primarily of congestive changes (Figure 3A) with the exception of the kidney which had multifocal areas of mild tubular necrosis and intraluminal eosinophilic protein casts (Figure 3B).

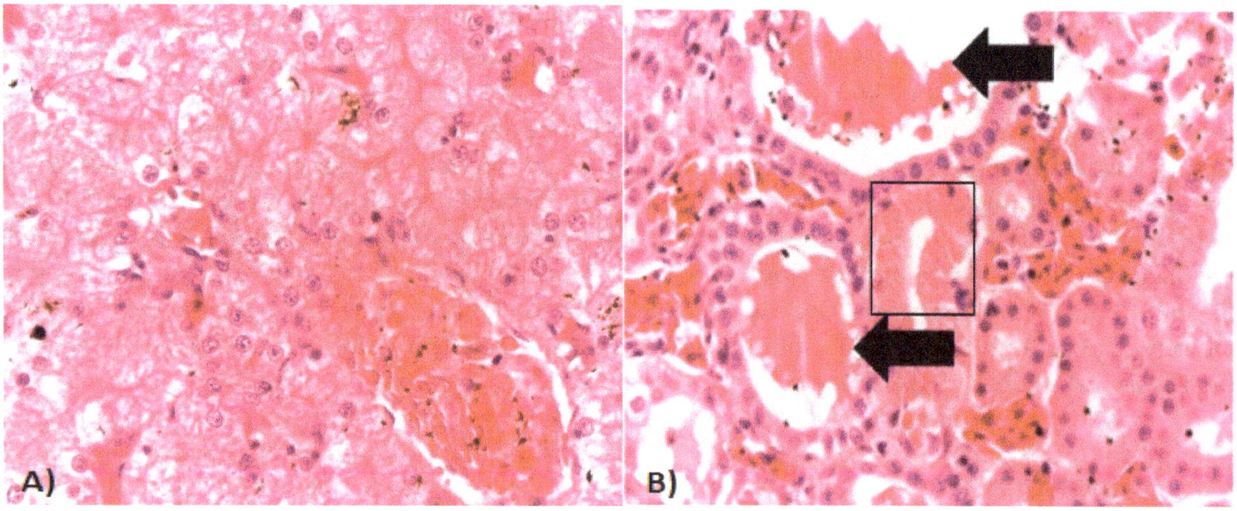

Figure 3. Histopathology of ranavirus-infected American bullfrogs. (**A**) Liver section from an affected bullfrog with mild congestion of sinusoids. (**B**) Kidney section shows an area of mild tubular necrosis characterized by cytoplasmic hyper-eosinophilia, loss of nuclei, and mild nuclear pyknosis (black square). Intraluminal protein casts are also seen (black arrows). Hematoxylin/Eosin staining. Magnification for all photos 20×/100 μm.

Intracytoplasmic inclusion bodies, which are considered the microscopic hallmark of ranavirus disease [9,10], were not observed in any of the specimens examined. Areas of positive immunohistochemical staining were scant and were only found in a few Kupffer cells of the liver (Figure 4A) and in the cortical interstitium of the kidney (Figure 4B). Initial PCR on samples from five frogs showed mild to strong positive signal in both liver and kidney of three frogs, mild positive signal in the liver of one frog, and one frog negative in all organs PCR bands for all five partial genes were from similar size to an FV3 isolate used as a positive control (GenBank no. KT003504) [31] (Figure 5). Phylogeny of five partial ranavirus genes revealed that the Mexican isolate clustered within the FV3 group and was closely related to USA isolates FV3 (AY548484) [32], FV3 SSME (KJ175144) [33], and a Nicaraguan isolate (MF360246) [31] (Figure 6). Sanger sequencing of a portion of the major capsid protein (462 bp) [25] showed that the ranavirus involved in this outbreak shared a 100% nucleotide homology with the original frog virus 3 isolate from the United States (AY 548484) [31]. The GenBank accession numbers for the Mexican virus sequences and the genomic identity to the ranavirus strains used for phylogeny are available as Supplementary Table S1.

The results of the Pandora$^+$ protocol showed a high overall risk score for endemic amphibians in neighboring areas (overall risk score 0.619, confidence level 1). The risk of entry of the pathogen through animals harboring subclinical infections or through virions in water was found to be medium (risk score 0.5, confidence level 1), given that waste water disposal of the farm is through the general sewage system without the water undergoing any prior disinfection treatment. Additionally, there is also the possibility of human-mediated transport and subsequent introduction of the virus if virions persist in footwear from employees. The possibilities of the pathogen to emerge, become established, and affect native amphibian populations were also found to be high (consequence score 0.875, confidence level 1) due to susceptible amphibian species present in the area. Since the pathogen does not affect plants or warm-blooded domestic animals or humans, the risk score for all these factors was classified as inapplicable.

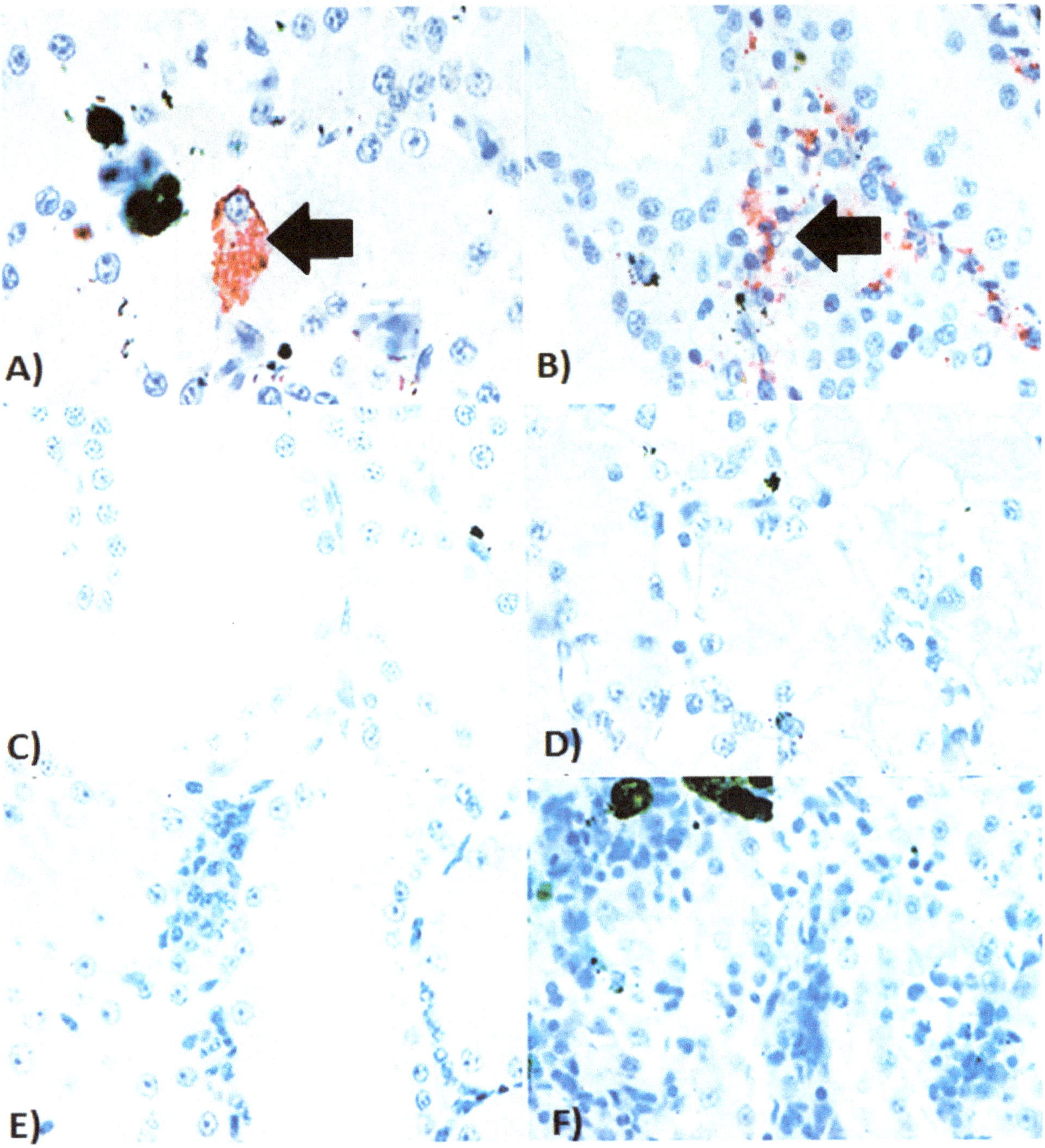

Figure 4. Immunohistochemistry of ranavirus-infected American bullfrogs (*Lithobates catesbeianus*). (**A**) Immunohistochemical staining of kidney, in which positive immunolabelling (red staining) is present in the renal cortical interstitium along with areas of necrosis and scant inflammation. (**B**) Immunohistochemical staining of a liver from an affected bullfrog with a focal area of positive immunolabelling (red staining) in the cytoplasm of a Kupffer cell. (**C**) Serial section of the kidney from the same animal without secondary antibody (control), (**D**) Serial section of liver from the same animal without secondary antibody (control). (**E**) Histological kidney section from an anuran which was PCR negative for ranavirus infection (control). (**F**) Histological liver section from an anuran which was PCR negative for ranavirus infection (control). Magnification for all photos 40×/50 μm.

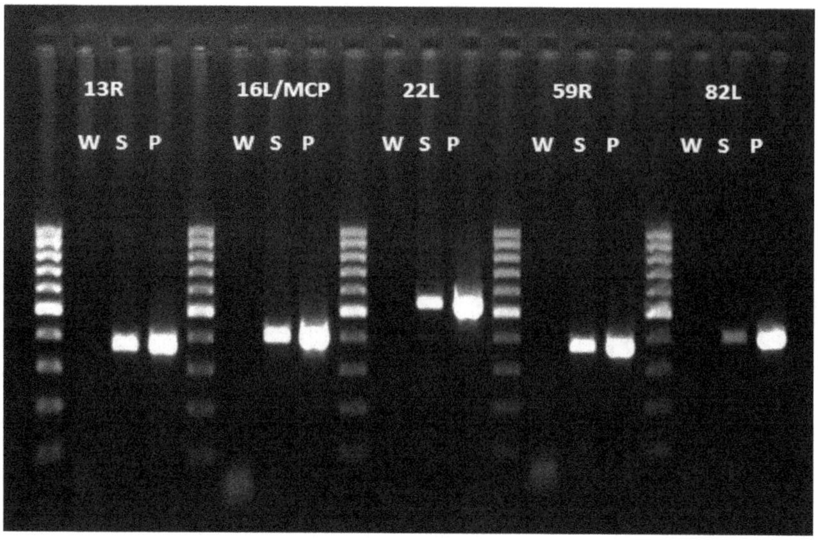

Figure 5. PCR products of five ranavirus genes. All products of the Mexican virus samples (S) are of similar size to those of the FV3 virus positive control (P). Abbreviations: W (water), S (Mexican virus sample), P (Positive control frog virus 3 isolate *Oophaga pumilio*/2015/Netherlands/UU3150324001 (GenBank no. MF360246)). 100 bp ladder.

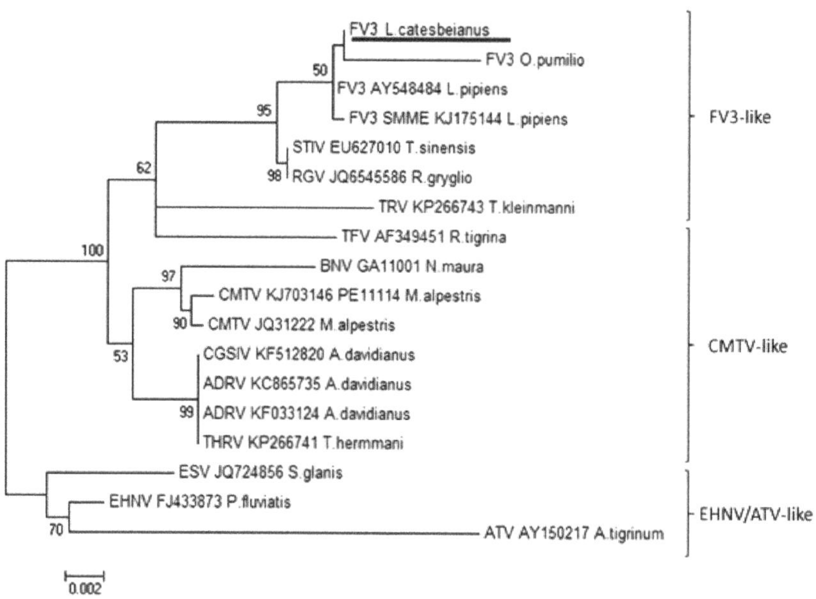

Figure 6. Phylogeny of five partial ranavirus gene sequences (1000 bootstrap values). The Mexican isolate FV3 *L. catesbeianus* (underlined) clusters closely within the FV3 ranavirus clade. Isolates and Genbank numbers used: Frog virus 3/*Lithobates catesbeianus*/2018/Mexico (GenBank accession numbers for partial sequences available in Supplementary Table S1), frog virus 3 (AY548484), frog virus 3 isolate SMME (KJ175144), tortoise ranavirus isolate 1 (882/96) (KP266743), Bosca's newt virus isolate GA11001 (KJ703118), frog virus 3 isolate *Oophaga pumilio*/2015/Netherlands/UU3150324001 (MF360246), common midwife toad ranavirus isolate *Mesotriton alpestris*/2008/E (JQ231222), common midwife toad virus isolate P11114 (KJ703146), *Testudo hermanni* ranavirus isolate CH8/96 (KP266741), tiger frog virus (AF389451), soft-shelled turtle iridovirus (EU627010), European sheatfish virus (JQ724856), epizootic hematopoietic necrosis virus (FJ433873), *Ambystoma tigrinum stebbensi* virus (AY150217), *Andrias davidianus* ranavirus isolate (KF033124), *Andrias davidianus* ranavirus isolate 1201 (KC865735), Chinese giant salamander iridovirus, isolate CGSIV-HN1104, (KF512820), and *Rana grylio* virus (JQ654586). Only bootstrap values higher than 50 are shown.

Finally, given the potential effects of ranavirus on the ecosystem, like the disappearance of endangered amphibian specimens or the possible increase in insect plagues, there is a low chance that the pathogen may have an indirect effect on tourism of the area. Furthermore, even though there is no current indication that the farm exports frogs out of Mexico, the risk of virus introduction through trade of sub-clinically infected animals cannot be entirely discarded (low risk, score 0.25, confidence score 0.5). A summary of the results of the scoring and confidence values is provided in Table 1. The original questionnaires and final risk scores provided by all four experts are available as Supplementary Tables S2–S5.

Table 1. Consensus risk classification of frog virus 3 ranavirus in Mexico. The Pandora^{+} questionnaire was solved as a consensus after all four experts had provided their individual results and discussed answers. Total number of questions: 20. Out of these, six questions comprised the initial assessment, one endemic exposure, one emerging/entry, two emerging/exposure, and two questions per each module of impact (environment, plants, animal, human and other). Color scheme and risk classification is based on cut-off values for scores by Schiphouwer et al. [28].

Module	Risk Classification	Risk Score	Certainty	Confidence
Emerging/entry	Medium	0.5	Medium	0.5
Emerging/exposure	High	1	Medium	0.75
Environmental score	High	0.875	High	1
Plant score	n/a	0	High	1
Animal score	n/a	0	High	1
Human score	n/a	0	High	1
Other score	Low	0	Medium	0.5
Consequence	High	0.875		
Entry exposure	High	0.707		
Risk score	High	0.619		

n/a inapplicable.

Comparison of the information on the threat categories for all 40 amphibian species in Sinaloa based on the IUCN list and the environmental vulnerability scores (EVS) revealed moderate differences in terms of threat classification. The IUCN list showed that 77.5% of the species were categorized as of least concern, 10% of them were not listed, 5% were considered vulnerable, 5% were classified as endangered and 2.5% as nearly threatened. On the other hand, in terms of environmental vulnerability, the EVS system categorized 45% of the species as having low vulnerability, 40% as medium, and 20% as high. All families to which the species belonged to, except for Eleutherodactylae and Microhylidae, have been reported to be affected by ranaviruses [6]. Out of the 19 amphibian species within 50 km or less of the original outbreak, the Tarahumara frog (*Lithobates tarahumarae*) and the Bell's salamander (*Isthmura bellii*) were considered vulnerable and the others as least concern by the IUCN red list. With the EVS system, the little Mexican toad (*Anaxyrus kelloggi*) was considered at high risk (>14 EVS), eight other species were considered as medium risk and the rest as low risk. Information regarding the level of threat of amphibian species in Sinaloa and the approximate distance of species distribution to the outbreak site can be found in Table 2.

Table 2. Level of threat for native amphibian species within the province of Sinaloa. The table summarizes the endemic, IUCN, and EVS, status and estimated distance of the nearest population of specific species to the affected farm.

Family	Scientific Name	Common Name	Endemic	IUCN Status	EVS Score	Distance to Farm
Ambystomatidae	*Ambystoma rosaceum*	Tarahumara salamander	Yes	LC	14	100–200 km
Bufonidae	*Anaxyrus cognatus*	Great plains toad	No	LC	9	100–200 km
Bufonidae	*Anaxyrus kelloggi*	Little mexican toad	Yes	LC	14	<50 km
Bufonidae	*Anaxyrus mexicanus*	Mexican spadefoot toad	Yes	NT	13	100–200 km
Bufonidae	*Anaxyrus punctatus*	Red-spotted toad	No	LC	5	<50 km
Bufonidae	*Incilius alvarius*	Sonoran desert toad	No	LC	11	<50 km
Bufonidae	*Incilius marmoreus*	Marbled toad	Yes	LC	11	>200 km
Bufonidae	*Incilius mazatlanensis*	Sinaloan toad	Yes	LC	12	<50 km
Bufonidae	*Incilius occidentalis*	Pine toad	Yes	LC	11	>200 km
Bufonidae	*Rhinella marina*	Cane toad	No	LC	3	<50 km
Craugastoridae	*Craugastor augusti*	Barking frog	No	LC	8	<50 km
Craugastoridae	*Craugastor horbartsmithii*	Smith's pigmy robber frog	Yes	E	15	>200 km
Craugastoridae	*Craugastor occidentalis*	Taylor's barking frog	Yes	NL	13	>200 km
Craugastoridae	*Craugastor vocalis*	Pacific stream frog	Yes	LC	13	<50 km
Eleutherodactylae	*Eleutherodactylus interorbitalis*	Spectacled chirping frog	Yes	NL	15	100–200 km
Eleutherodactylae	*Eleutherodactylus nitidus*	Shiny peeping frog	Yes	LC	12	>200 km
Eleutherodactylae	*Eleutherodactylus saxatilis*	Marbled peeping frog	Yes	E	17	>200 km
Eleutherodactylae	*Eleutherodactylus teretistes*	Whistling frog	Yes	NL	16	>200 km
Hylidae	*Agalychnis dacnicolor*	Mexican leaf frog	Yes	LC	13	<50 km
Hylidae	*Dryophytes arenicolor*	Canyon treefrog	No	LC	7	>100 km
Hylidae	*Dryophytes eximius*	Mountain treefrog	Yes	LC	10	>200 km
Hylidae	*Dryophytes wrightorum*	Wright's mountain treefrog	No	LC	9	>200 km
Hylidae	*Exerodonta smaradigna*	Emerald treefrog	Yes	LC	12	>200 km
Hylidae	*Sarcohyla bistincta*	Mexican Fringed-limbed treefrog	Yes	LC	9	>200 km
Hylidae	*Smilisca baudinii*	Common mexican treefrog	No	LC	3	<50 km
Hylidae	*Smilisca fodiens*	Lowland burrowing treefrog	No	LC	8	<50 km
Hylidae	*Tlalocohyla smithii*	Dwarf mexican treefrog	No	LC	11	<50 km
Hylidae	*Trachycephalus typhonius*	Veined treefrog	No	LC	4	>200 km
Hylidae	*Triprion spatulatus*	Mexican shovel-headed treefrog	Yes	LC	13	>200 km
Leptodactylae	*Leptodactylus melanotonus*	Black-backed frog	No	LC	6	<50 km
Mycrohylidae	*Gastrophryne mazatlanensis*	Mazatlan narrowmouth frog	No	NL	8	<50 km
Mycrohylidae	*Hypopachus ustus*	Two-spaded narrow-mouthed toad	No	LC	7	100–200 km
Mycrohylidae	*Hypopachus variolosus*	Mexican narrow-mouthed toad	No	LC	4	100–200 km
Pelobatidae	*Scaphiopus couchii*	Couch's spadefoot toad	No	LC	3	<50 km
Plethodontidae	*Isthmura bellii*	Bell's salamander	Yes	V	12	50–100 km
Ranidae	*Lithobates catesbeianus*	American bullfrog	No	LC	10	<50 km
Ranidae	*Lithobates forreri*	Forrer's leopard frog	No	LC	3	<50 km
Ranidae	*Lithobates magnaocularis*	Northwest mexican leopard frog	Yes	LC	12	<50 km
Ranidae	*Lithobates pustulosus*	Mexican cascades frog	Yes	LC	9	>200 km
Ranidae	*Lithobates tarahumarae*	Tarahumara frog	Yes	V	8	50–100 km

LC least concern, E, endangered, NL, not listed, NT, near threatened, V, vulnerable.

4. Discussion

This study investigated the first outbreak of ranavirus infection in Mexico and performed a risk analysis for the endemic wild amphibian species in the area close to the affected farm. The ranavirus outbreak took place in a farmed colony of American bullfrogs and was caused by a ranavirus clustering within the FV3 group, the dominant clade of ranaviruses present on the American continent [6].

One of the most concerning aspects of American bullfrogs is their potential to introduce high-risk pathogens such as ranavirus and *Batrachochytrium dendrobatidis* into amphibian populations [34]. Breeding facilities for these animals are considered hotspots for ranavirus outbreaks as shown in various countries such as Brazil and Japan [35,36]. Frequently, ranavirus is introduced to captive amphibian populations by means of capture of infected animals from the wild. This has taken place in Switzerland, in which a ranavirus-outbreak in a laboratory colony of water frogs (*Pelophylax* spp.) was shown to originate from a ranavirus-infected frog captured from a natural pond in Germany [17].

Despite the severe gross lesions (skin ulceration and erythema), affected animals sampled from this farm showed only scant immunohistochemical signal in the liver and kidney and no intracytoplasmic inclusion bodies in histology. The lack of inclusion bodies is a well-known feature of bullfrog ranavirus-associated outbreaks in farms [35,36]. The faint immunohistochemical signal observed in the kidneys and liver could be consistent with a chronic presentation of the disease where the most severe lesions have probably resolved in visceral organs, but skin lesions persist as severe ulcerations. This chronic presentation of the disease has also been known as "ulcerative syndrome", reported in common frogs (*Rana temporaria*) in the United Kingdom infected with FV3-like ranavirus isolates [37]. Investigation focusing on immunohistochemical analyses on animals with this condition revealed overall decrease in signal on visceral organs in comparison to animals with the per-acute or acute presentations of the disease, known as "hemorrhagic syndrome" [37].

Independent assessment of introduction and establishment risk of ranaviruses into Mexican wildlife by all four experts showed similar scoring results in the Pandora + protocol. Consensus analysis suggested that the possibilities of the pathogen being introduced into nature by bullfrogs escaping the facility was moderate (0.5). Bullfrogs are quick to colonize new habitats and capable of long-distance dispersal (>1200 meters in a week) [38]. This phenomenon has occurred precisely in the study area.

In 1956, the American bullfrog was introduced for exploitation as a harvestable food item in the vicinity of Los Mochis, Ahome (approximately 60 km north of Guasave) [39]. Only six years later, there were several populations among Ahome and Guasave counties established on agricultural irrigation channels and in ponds [40]. Such a colonization presents a threat to native amphibian species as it enables bullfrogs to become into direct contact with susceptible hosts.

Other than introduction through infected hosts, ranavirus can also gain entry into the wild through virion-contaminated water [41]. At the facility where the outbreak took place, untreated water is usually disposed through the general sewage system. It is possible that this potentially contaminated water could reach natural ponds and pools which are home to susceptible host species and are thus apt for ranavirus transmission and persistence.

The probability of the virus becoming endemic within native amphibian populations was found to be high mainly due to the amphibian species present in the area. In total, 17 species are found in the county of Guasave, and most of them live permanently or reproduce in the water, except for Barking frog (*Craugastor augusti*) that reproduces and lives in a terrestrial environment. Among these species, the little Mexican toad (*Anaxyrus kelloggi*) was found to possess a high vulnerability score [30]. Members of the family Bufonidae, to which the little Mexican toad belongs to, have been known to develop ranavirus-associated hemorrhagic syndrome when exposed to FV3 ranavirus strains in the UK [37].

The two families of anurans in which the highest amount of infected species have been reported are Ranidae and Hylidae [6]. Within the Ranidae family, one of the most prevalent amphibian species in the region is the Northwest Mexican leopard frog (*Lithobates magnaocularis*) and one of the most vulnerable ones according to the EVS is the Tarahumara frog [30].

Ranavirus has been able to establish itself among populations of other amphibian species from the genus *Lithobates*, like the Northern leopard frog (*Lithobates pipiens*) [42] and the Dusky gopher frog (*Lithobates sevosus*), in which it has caused 100% mortality under experimental settings [43]. Regarding the family Hylidae, and in contrast to all other endemic amphibian species in Guasave, the Mexican leaf frog (*A. dacnicolor*) has actually shown evidence of population declines throughout Mexico [1]. This is of particular concern, since other members of Hylidae, like the magnificent tree frog (*Litoria splendida*) and green tree frog (*Hyla cinerea*) have suffered severe outbreaks of ranavirus disease under captive settings [44]. Likewise, larval stages of the Cope's gray tree frog (*Dryophytes chrysoscelis*) have shown a high mortality rate upon experimental exposure to FV3 ranavirus highlighting the species susceptibility to develop lethal infections [45].

However, the risk of ranavirus infection is not only limited to amphibians, and the possibility of transmission from infected amphibians to susceptible fish in neighboring ponds is supported by previous evidence of ranavirus outbreaks affecting both fish and frogs and experimental interspecies transmission of ranaviruses. In the wild in the USA, an FV3-like virus caused a die-off simultaneously involving threespine stickleback fish (*Gasterosteus aculeatus)* and red-legged frogs (*Rana aurora*) [46]. Under experimental conditions, Bohle iridovirus (BIV), a ranavirus originally isolated from the ornated burrowing frog (*Limnodynastes ornatus*), induced infection and mortality of barramundi fish (*Lates calcarifer*) [47].

There is also a high probability of the virus to persist in the area due to suitable environmental conditions. The climate of the region where the outbreak took place is mostly warm and dry with an average temperature of 24 °C which can surpass 30 °C in the summer months [48]. Warm temperatures of around 20 °C, have been shown to favor ranavirus propagation and worsen signs of ranavirus disease under experimental settings [49]. Additionally, virions of Bohle iridovirus, an FV3-like ranavirus, have been known to persist in desiccated soil under a temperature of 45 °C [50]. Research has also shown that virions are able to remain infectious in waterbodies for up to one or two months in water temperatures ranging from 4 °C to 20 °C [41].

There are several measures which could be undertaken to prevent further spread into the wild. Most importantly, culling and re-stocking of animals within the farm would ensure eradication of not only animals harboring active infections but also of sub-clinically infected carriers.

Subsequent disinfection of the facilities, equipment or footgear of employees and especially waste-water treatment with cleaning agents like Virkon (at a 1% concentration) would aid in eliminating infectious virions present in the water or soil. Monitoring of amphibian populations and molecular analysis of skin swabs from a subset of animals from nearby ponds has shown to be an effective way of gaining information regarding genotype, virulence, and probability of further spread [51]. Another efficient way of monitoring and eventually controlling potential ranavirus disease outbreaks is by informing and encouraging the public to report amphibian die-offs or to keep tract of the number and life-stages of these animals in nearby ponds. Long-term science projects involving citizens have proven to be a successful approach to studying and characterizing this disease in wild amphibians in Great Britain [52].

The current study characterized the first ranavirus outbreak in captive Mexican amphibians through pathology and molecular biology. It also determined a high level of risk of the pathogen to gain entry and become established in neighboring nature areas and potentially affect endemic amphibian species of conservation concern.

Author Contributions: B.S., M.J.M., R.S.E.W.L., A.A.R.G., A.G. and C.M.E.-B. conceived and designed the experiments; B.S., M.J.M, C.M.E.-B., A.M.B. and J.M.S. performed the experiments B.S., J.M.S., M.J.M., R.S.E.W.L.,

S.J.v.B., C.M.E.-B., and A.M.B. analyzed the data; A.G., R.S.E.W.L., S.J.v.B. and J.M.S. contributed with analysis tools and reagents., B.S. wrote the paper.

Acknowledgments: We thank Alphons van Asten, Jolianne M. Rijks, Margriet Montizaan, Teresa de Jesús López, and Alinda Berends for their excellent technical and logistical support.

References

1. IUCN 2017. The IUCN Red List of Threatened Species. Version 2017-3. Available online: http://www.iucnredlist.org (accessed on 5 December 2017).
2. Bower, D.S.; Lips, K.R.; Schwarzkopf, L.; Georges, A.; Clulow, S. Amphibians on the brink. *Science* **2017**, *357*, 454–455. [CrossRef] [PubMed]
3. Blaustein, A.R.; Kiesecker, J.M. Complexity in conservation: Lessons from the global decline of amphibian populations. *Ecol. Lett.* **2002**, *5*, 597–608. [CrossRef]
4. Chinchar, V.G. Ranaviruses (family *Iridoviridae*): Emerging cold-blooded killers. *Arch. Virol.* **2002**, *147*, 447–470. [CrossRef] [PubMed]
5. Mendoza-Almeralla, C.; Burrowes, P.; Parra-Olea, G. Chytridiomycosis in amphibians from Mexico: A revision. *Rev. Mex. Biodivers.* **2015**, *86*, 238–248. [CrossRef]
6. Duffus, A.L.J.; Waltzek, T.B. Distribution and Host Range of Ranaviruses. In *Ranaviruses: Lethal Pathogenics of Ectothermic Vertebrates*, 1st ed.; Gray, M.J., Chinchar, G.V., Eds.; Springer: New York, NY, USA, 2015; Volume 1, pp. 9–57, ISBN 978-3-319-13755-1.
7. Epstein, B.; Storfer, A. Comparative genomics of an emerging amphibian virus. *G3 Genes Genom. Genet.* **2016**, *6*, 15–27. [CrossRef] [PubMed]
8. Wheelwright, N.T.; Gray, M.J.; Hill, R.D.; Miller, D.L. Sudden mass die-off of large population of wood frog (*Lithobates sylvaticus*) tadpoles in Maine, USA, likely due to ranavirus. *Herpetol. Rev.* **2014**, *45*, 240–242. [CrossRef]
9. Balseiro, A.; Dalton, K.; del Cerro, A.; Marquez, I.; Cunningham, A.A.; Parra, F.; Prieto, J.M.; Casais, R. Pathology, isolation and molecular characterisation of a ranavirus from the common midwife toad *Alytes obstetricans* on the Iberian Peninsula. *Dis. Aquat. Org.* **2009**, *84*, 95–104. [CrossRef]
10. Kik, M.; Martel, A.; van der Sluijs, A.S.; Pasmans, F.; Wohlsein, P.; Gröne, A.; Rijks, J.M. Ranavirus-associated mass mortality in wild amphibians, The Netherlands, 2010: A first report. *Vet. J.* **2011**, *190*, 284–286. [CrossRef]
11. Miaud, C.; Pozet, F.; Gaudin, N.C.G.; Martel, A.; Pasmans, F.; Labrut, S. Ranavirus causes mass die-offs of alpine amphibians in the Southwestern Alps, France. *J. Wildl. Dis.* **2016**, *52*, 242–252. [CrossRef]
12. He, J.G.; Lu, L.; Deng, M.; He, H.H.; Weng, S.P.; Wang, X.H.; Zhou, S.Y.; Long, Q.X.; Wang, X.Z.; Chan, S.M. Sequence analysis of the complete genome of an iridovirus isolated from tiger frog. *Virology* **2002**, *292*, 185–197. [CrossRef]
13. Lei, X.Y.; Ou, T.; Zhu, R.L.; Zhang, Q.J. Sequencing and analysis of the complete genome of *Rana grylio* virus (RGV). *Arch. Virol.* **2012**, *157*, 1559–1564. [CrossRef] [PubMed]
14. Geng, Y.; Wang, K.Y.; Zhou, Z.Y.; Li, C.W.; Wang, J.; He, M.; Yin, Z.Q.; Lai, W.M. First report of a ranavirus-associated with morbidity and mortality in farmed Chinese giant salamanders (*Andrias davidianus*). *J. Comp. Pathol.* **2011**, *141*, 95–102. [CrossRef] [PubMed]
15. Marschang, R.E.; Becher, P.; Posthaus, H.; Wild, P.; Thiel, H.J.; Müller-Doblies, U.; Kalet, E.F.; Bacciarini, L.N. Isolation and characterization of an iridovirus isolated from Hermann's tortoises (*Testudo hermanni*). *Arch. Virol.* **1999**, *144*, 1909–1922. [CrossRef] [PubMed]
16. Whittington, R.J.; Kearns, C.; Hyatt, A.D.; Hengstberger, S.; Rutzou, T. Spread of *Epizootic hematopoietic necrosis virus* (EHNV) in red fin perch (*Perca fluviatilis*) in southern Australia. *Aust. Vet. J.* **1996**, *73*, 112–114. [CrossRef] [PubMed]
17. Stöhr, A.C.; Hoffmann, A.; Papp, T.; Robert, N.; Pruvost, N.B.M.; Reyer, H.U.; Marschang, R.E. Long-term study of an infection with ranaviruses in a group of edible frogs (*Pelophylax* kl. *esculentus*) and partial characterization of two viruses based on four genomic regions. *Vet. J.* **2013**, *197*, 238–244. [CrossRef] [PubMed]
18. Serrano, J.M.; Berlanga-Robles, C.A.; Ruiz-Luna, A. High amphibian diversity related to unexpected environmental values in a biogeographic transitional area in north-western Mexico. *Contrib. Zool.* **2014**, *83*, 151–166.

19. Casas-Andreu, G.; Aguilar-Miguel, X.; Cruz-Aviña, R. La introducción y el cultivo de la rana toro (*Rana catesbeiana*). ¿Un atentado a la biodiversidad de México? *Revista Científica Multidisciplinaria de Prospectiva* **2001**, *8*, 62–67.
20. Cifuentes-Lemus, J.L.; Torres-García, M.P.; Frías-Mondragón, M. *El Océano y Sus Recursos XI. Acuicultura*, 2nd ed.; Serie La ciencia Para Todos; Fondo de Cultura Económica: México City, Mexico, 1997; pp. 1–97, ISBN 968-16-5242-897.
21. Luja, V.H.; Rodrìguez-Estrella, R. The invasive bullfrog *Lithobates catesbeianus* in oases of Baja California Sur Mexico: Potential effect in a fragile ecosystem. *Biol. Invasions* **2010**, *12*, 2979–2983. [CrossRef]
22. Morales-Salinas, E.; Aguilar-Arriaga, B.O.; Ramírez-Lezama, J.; Méndez-Bernal, A.; López-Garrido, S. Oral fibrosarcoma in a black iguana (*Ctenosaura pectinata*). *J. Zoo Wildl. Med.* **2013**, *44*, 513–516. [CrossRef]
23. Rijks, J.M.; Saucedo, B.; Spitzen-van der Sluijs, A.M.; Wilkie, G.S.; van Asten, A.J.A.M.; van den Broek, J.; Boonyarittichaikij, R.; Stege, M.; Van der Sterren, F.; Martel, A.; et al. Investigation of amphibian mortality events in wildlife reveals an on-going ranavirus epidemic in the North of The Netherlands. *PLoS ONE* **2016**, *11*, e0157473. [CrossRef]
24. Zhang, Y.; Uyemoto, J.K.; Kirkpatrick, B.C. A small-scale procedure for extracting nucleic acids from woody plants infected with various phytopathogens for PCR assay. *J. Virol. Methods* **1998**, *71*, 45–50. [CrossRef]
25. Mao, J.; Hendrick, R.P.; Chinchar, V.G. Molecular characterization, sequence analysis, and taxonomic position of newly isolated fish iridoviruses. *Virology* **1997**, *229*, 212–220. [CrossRef]
26. CLUSTAL Omega. Available online: https://www.ebi.ac.uk/Tools/msa/clustalo/ (accessed on 30 August 2018).
27. D'hondt, B.; Vanderhoeven, S.; Roelandt, S.; Mayer, F.; Versteirt, V.; Ducheyne, E.; San Martin, G.; Grégoire, J.C.; Stiers, I.; Quoilin, S.; et al. *Pandora: A Risk Screening Tool for Pathogens and Parasites*; Belgian Biodiversity Platform: Brussels, Belgium, 2014; pp. 1–22.
28. Schiphouwer, M.E.; Felix, R.P.W.H.; van Duinen, G.A.; de Hoop, L.; de Hullu, P.C.; Matthews, J.; van der Velde, G.; Leuven, R.S.E.W. Risk assessment of the alien smallmouth bass (*Micropterus dolomieu*). *Rep. Environ. Sci.* **2017**, *527*, 1–60.
29. Rijks, J.M.; Spitzen van-der Sluijs, A.; Leuven, R.S.E.W.; Martel, A.; Kik, M.; Pasmans, F. *Risk Analysis of the Common Midwife Toad-Like Virus (CMTV-Like Virus) in The Netherlands*; Report 60000784-2013; Netherlands Food and Consumer Product Safety Authority (NVWA), Dutch Ministry of Economic Affairs: Utrecht, The Netherlands, 2012.
30. Wilson, L.D.; Johnson, J.D.; Mata-Silva, V. A conservation reassessment of the amphibians of Mexico based on the EVS measure. *Amphib. Reptile Conserv.* **2013**, *7*, 97–127.
31. Saucedo, B.; Hughes, J.; Súarez, N.; Haenen, O.; Kik, M.J.L.; van Beurden, S.J. Complete genome sequence of frog virus 3 isolated from a Strawberry poison frog (*Oophaga pumilio*) imported from Nicaragua into The Netherlands. *Genome Announc.* **2017**, *5*, ee00863-17. [CrossRef] [PubMed]
32. Tan, W.G.; Barkman, T.J.; Chinchar, G.V.; Essani, K. Comparative genomic analyses of frog virus 3, type species of the genus *Ranavirus* (family *Iridoviridae*). *Virology* **2004**, *323*, 70–84. [CrossRef] [PubMed]
33. Morrison, E.A.; Garner, S.; Echaubard, P.; Lesbarreres, D.; Kyle, C.J.; Brunetti, C.R. Complete genome analysis of a frog virus 3 (FV3) isolate and sequence comparison with isolates of differing levels of virulence. *Virol. J.* **2014**, *12*, 11–46. [CrossRef]
34. Kolby, J.E.; Smith, K.M.; Berger, L.; Karesh, W.B.; Preston, A.; Pessier, A.P.; Skerratt, L.F. First evidence of amphibian chytrid fungus (*Batrachochytrium dendrobatidis*) and ranavirus in Hong Kong amphibian trade. *PLoS ONE* **2014**, *3*, 1–6. [CrossRef]
35. Mazzoni, R.; Mesquita, A.J.; Fleury, L.F.F.; Brito, W.; Nunes, I.A.; Robert, J.; Morales, H.; Coelho, A.S.G.; Barthasson, D.L.; Galli, L.; et al. Mass mortality associated with a Frog virus 3-like ranavirus infection in farmed tadpoles *Rana catesbeiana* from Brazil. *Dis. Aquat. Org.* **2009**, *86*, 181–191. [CrossRef]
36. Une, Y.; Sakuma, A.; Matsueda, H.; Nakai, K.; Murakami, M. Ranavirus outbreak in North American bullfrogs (*Rana catesbeiana*), Japan, 2008. *Emerg. Infect. Dis.* **2008**, *15*, 1146–1147. [CrossRef]
37. Cunningham, A.A.; Temns, C.A.; Russel, P.H. Immunohistochemical demonstration of ranavirus antigen in the tissues of infected frogs (*Rana temporaria*) with systemic hameorrhagic or cutaneous ulcerative disease. *J. Comp. Pathol.* **2008**, *138*, 3–11. [CrossRef] [PubMed]
38. Willis, Y.L.; Moyle, P.B.; Basket, T.S. Emergence, breeding, hibernation, movements and transformation of the bullfrog, *Rana catesbeiana*, in Missouri. *Copeia* **1956**, *1*, 30–41. [CrossRef]

39. Aguilar Ibarra, F. Aspectos Generales sobre las Ranas y su Cultivo. *Trab. Div. Inst. Nac. Invest. Biol. Pesq. C* **1963**, *5*, 1–21.
40. Hardy, L.M.; McDiarmid, R.W. The amphibians and reptiles of Sinaloa, Mexico. *Univ. Kansas Publ. Mus. Nat. Hist.* **1969**, *18*, 39–252. [CrossRef]
41. Nazir, J.; Spengler, M.; Marschang, R.E. Environmental persistence of amphibian and reptilian ranaviruses. *Dis. Aquat. Org.* **2012**, *98*, 177–184. [CrossRef] [PubMed]
42. Echaubard, P.; Paulli, B.D.; Trudeau, V.L.; Lesbarrères, D. Ranavirus infection in northern leopard frogs: The timing and number of exposures matter. *J. Zool.* **2015**, *298*, 30–36. [CrossRef]
43. Earl, J.E.; Chaney, J.C.; Sutton, W.B.; Lillard, C.E.; Kouba, A.J.; Langhorne, C.; Krebs, J.; Wilkis, R.P.; Hill, R.D.; Miller, D.L.; et al. Ranavirus could facilitate local extinction of rare amphibian species. *Oecologia* **2016**, *182*, 611–623. [CrossRef]
44. Jerett, I.V.; Whittington, R.J.; Weir, R.P. Pathology of a Bohle-like infection in two Australian frog species (*Litoria splendida* and *Litoria caerulea*). *J. Comp. Pathol.* **2015**, *152*, 248–259. [CrossRef]
45. Brand, M.D.; Hill, R.D.; Brenes, R.; Chaney, J.C.; Wilkes, R.P.; Grayfer, L.; Miller, D.L.; Gray, M.J. Water temperature affects susceptibility to ranavirus. *Ecohealth* **2016**, *13*, 350–359. [CrossRef]
46. Mao, J.D.E.; Green, G.; Chinchar, V.G. Molecular characterization of iridoviruses isolated from sympatric amphibians and fish. *Virus Res.* **1999**, *63*, 45–52. [CrossRef]
47. Moody, N.; Owens, L. Experimental demonstration of the pathogenicity of frog virus, Bohle iridovirus, for a fish species, barramundi, *Lates calcariger*. *Dis. Aquat. Org.* **1994**, *18*, 95–102. [CrossRef]
48. Rodríguez-Meza, D.; Rodríguez Figueroa, G.; Sapozhnikov, D.; Vargas-Ramírez, C.; Vallejo-Soto, A.; Verdugo-Quiñonez, G.; Michel-Rubio, A. Monitoreo de la calidad del agua del acuífero de Guasave, Sinaloa Mexico. Unpublished work, SIP 2008.
49. Price, S.J.; Leung, W.T.M.; Christopher, J.O.; Sergeant, C.; Cunningham, A.A.; Ballous, F.; Garner, T.W.J.; Nichols, R. A. Temperature is a key driver of a wildlife epidemic and future warming will increase impact. *bioRxiv* **2018**. [CrossRef]
50. Jancovich, J.K.; Chinchar, V.G.; Hyatt, A.; Miyazaki, T.; Zhang, Q.Y. *Virus Taxonomy: Ninth Report of the International Committee of Taxonomy of Viruses*; King, A.M.Q., Adams, M.J., Carstens, E.B., Lefkowitz, E., Eds.; Elsevier Academic Press: Cambridge, MA, USA, 2012; pp. 193–210.
51. Saucedo, B.; Hughes, J.; Spitzen van der Sluijs, A.; Kruithof, N.; Schils, M.; Rijks, J.M.; Jacinto-Maldonado, M.; Suárez, N.; Haenen, O.L.M.; Voorbergen-Laarman, M.; et al. Ranavirus genotypes in The Netherlands and their potential association with virulence in water frogs (*Pelophylax* spp.). *Emerg. Microbes Infect.* **2018**, *7*, 1–56. [CrossRef] [PubMed]
52. Cunningham, A.A.; Langton, T.E.S.; Bennet, P.W.; Lewing, J.F.; Drury, S.E.N.; Gough, R.E.; MacGregor, E.K. Pathological and microbiological findings from incidents of unusual mortality of the common frog (*Rana temporaria*). *Philos. Trans. R. Soc. Lond. B Biol. Sci.* **1996**, *351*, 1539–1557. [CrossRef] [PubMed]

13

Characterization of a Novel Megalocytivirus Isolated from European Chub (*Squalius cephalus*)

Maya A. Halaly [1], Kuttichantran Subramaniam [2], Samantha A. Koda [2], Vsevolod L. Popov [3], David Stone [4], Keith Way [4] and Thomas B. Waltzek [2,*]

1 Department of Animal Sciences, College of Agricultural and Life Sciences, University of Florida, Gainesville, FL 32611, USA; mayah123@ufl.edu
2 Department of Infectious Diseases and Immunology, College of Veterinary Medicine, University of Florida, Gainesville, FL 32611, USA; kuttichantran@ufl.edu (K.S.); samanthakoda@ufl.edu (S.A.K.)
3 Department of Pathology, University of Texas Medical Branch, Galveston, TX 77555, USA; vpopov@utmb.edu
4 Centre for Environment, Fisheries and Aquaculture Science (CEFAS), Weymouth DT4 8UB, UK; david.stone@cefas.co.uk (D.S.); keith.way@cefas.co.uk (K.W.)
* Correspondence: tbwaltzek@ufl.edu

Abstract: A novel virus from moribund European chub (*Squalius cephalus*) was isolated on *epithelioma papulosum cyprini* (EPC) cells. Transmission electron microscopic examination revealed abundant non-enveloped, hexagonal virus particles in the cytoplasm of infected EPC cells consistent with an iridovirus. Illumina MiSeq sequence data enabled the assembly and annotation of the full genome (128,216 bp encoding 108 open reading frames) of the suspected iridovirus. Maximum Likelihood phylogenetic analyses based on 25 iridovirus core genes supported the European chub iridovirus (ECIV) as being the sister species to the recently-discovered scale drop disease virus (SDDV), which together form the most basal megalocytivirus clade. Genetic analyses of the ECIV major capsid protein and ATPase genes revealed the greatest nucleotide identity to members of the genus *Megalocytivirus* including SDDV. These data support ECIV as a novel member within the genus *Megalocytivirus*. Experimental challenge studies are needed to fulfill River's postulates and determine whether ECIV induces the pathognomonic microscopic lesions (i.e., megalocytes with basophilic cytoplasmic inclusions) observed in megalocytivirus infections.

Keywords: megalocytivirus; iridovirus; European chub

---

1. Introduction

The European chub (*Squalius cephalus*) is a rheophilic cyprinid that is widely distributed throughout Eurasia [1,2]. Although they are omnivores, adults include a greater portion of fish in their diets [3,4]. They are popular among amateur anglers because they readily take a variety of live and artificial baits and reach a maximum reported standard length of 60 cm [4]. Ireland and Italy have raised concerns that the recent introductions of European chub may threaten their native biodiversity [1,2].

Cyprinid fishes are susceptible to a range of RNA and DNA viruses (reviewed in [5]). Iridoviruses have been described in a wide variety of fishes; however, few iridoviruses have been described from cyprinids [6,7]. Although an irido-like virus was isolated from the gills and kidneys of moribund common carp (*Cyprinus carpio*; i.e., common carp iridovirus; CCIV), the role of CCIV in disease could not be established (reviewed in [5,8]). In addition, two irido-like viruses were isolated from the swim bladder of healthy goldfish (*Carassius auratus*), goldfish virus-1 and goldfish virus-2, have been proposed as members of the family *Iridoviridae* based on biophysical and biochemical analyses [9].

Recent studies have reported the detection of the iridovirus, infectious spleen and kidney necrosis virus (ISKNV), in goldfish and common carp traded in Brazil [10] and in zebrafish (*Danio rerio*) from a research facility in Spain [11]. In addition, a Santee-Cooper ranavirus strain was isolated from diseased koi carp and shown experimentally to induce lethal disease [12].

Members of the family *Iridoviridae* possess large nucleocapsids that display icosahedral symmetry (120–200 nm in diameter) and encapsidate a linear, double-stranded DNA genome. The family is divided into two subfamilies, *Alphairidovirinae* and *Betairidovirinae* [13]. The former consists of three genera (*Ranavirus*, *Lymphocystivirus*, and *Megalocytivirus*) that are known to infect ectothermic vertebrates including fish, amphibians, and reptiles [13,14]. Megalocytiviruses are globally emerging viruses, causing lethal systemic infections in wild and cultured freshwater, brackish, and marine fishes [15]. Recent megalocytivirus phylogenetic analyses based on the major capsid protein (MCP) and ATPase genes revealed that the species, ISKNV, is divided into three genotypes including (1) ISKNV, which was first described from farmed mandarin fish (*Siniperca chuatsi*) reared for food in China [16,17]; (2) red sea bream iridovirus (RSIV), which was initially reported in cultured red sea bream (*Pagrus major*) from Japan [18]; and (3) turbot reddish body iridovirus (TRBIV), characterized from flatfishes (order Pleuronectiformes) reared for food in the Yellow Sea in East Asia [19]. In 2012, the threespine stickleback iridovirus (TSIV) was characterized from wild-caught threespine stickleback (*Gasterosteus aculeatus*) from Canada, and based on genetic and phylogenetic analyses, it was proposed as a novel megalocytivirus species [20]. Recently, an epizootic involving Asian seabass (*Lates calcarifer*) cultured in Singapore was found to be caused by a divergent megalocytivirus, scale drop disease virus (SDDV; [21]).

In this investigation, we described the *in vitro* growth characteristics, ultrastructural pathology, and phylogenomic characterization of the first iridovirus isolated from European chub. The cytopathic effect (i.e., enlarged, rounded, refractile cells) induced by this virus and its ultrastructural features are consistent with that of a megalocytivirus. The genetic and phylogenetic analyses further supported this virus as a divergent megalocytivirus, referred to as European chub iridovirus.

2. Materials and Methods

2.1. Case History

A hatchery located in the English Midlands of the UK, rearing cyprinids for wild stock enhancement, experienced increased mortality in several species, including European chub. Live European chub juveniles were transported to the Center for Environmental Fisheries and Aquaculture Science (CEFAS), in Weymouth, England and were subsequently euthanized after they appeared moribund upon arrival. Internal tissue homogenates were inoculated onto the following cell lines using standard virological methods: bluegill fry (BF-2), *epithelioma papulosum cyprini* (EPC), chinook salmon embryo (CHSE-214), koi fin (KF-1), and common carp brain (CCB). Cytopathic effects (CPE) were observed as early as 48 h post-inoculation, resulting in the appearance of enlarged refractile cells in all cell lines at 20 °C.

2.2. Cell Culture

The European chub iridovirus (ECIV) isolate grown on EPC cells was sent from CEFAS to the Wildlife and Aquatic Veterinary Disease Laboratory in Gainesville, Florida, USA. The virus isolate was then inoculated onto confluent monolayers of EPC cells maintained in MEM media with 10% fetal bovine serum and 1% HEPES (4-(2-hydroxyethyl)-1-piperazineethanesulfonic acid) at 25 °C and monitored daily for CPE. Flasks of EPC cells displaying CPE were used to characterize the ultrastructural and genetic properties of ECIV.

2.3. Transmission Electron Microscopy

The ECIV isolate was propagated in a 75 cm^2 flask of EPC cells until CPE was observed. The supernatant from the infected flask was discarded, and the monolayer was fixed in 15 mL of modified Karnovsky's fixative (2P + 2G, 2% formaldehyde prepared from paraformaldehyde and 2% glutaraldehyde in 0.1 M cacodylate buffer pH 7.4) at room temperature for 1 h. The monolayer was then washed in cacodylate buffer, scraped off the flask, and pelleted. The pellet was shipped via PBS overnight on ice packs to the University of Texas Medical Branch Department of Pathology Electron Microscopy Laboratory (UTMB-EML). At UTMB-EML, the cell pellet was washed in cacodylate buffer and left in 2P + 2G fixative overnight at 4 °C. The next day, the cell pellet was washed twice in cacodylate buffer, post-fixed in 1% OsO_4 in 0.1 M cacodylate buffer pH 7.4, en bloc stained with 2% aqueous uranyl acetate, dehydrated in ascending concentrations of ethanol, processed through propylene oxide, and embedded in Poly/Bed 812 epoxy plastic (Polysciences, Warrington, PA, USA). Ultrathin sections were cut on a Leica ULTRACUT EM UC7 ultramicrotome (Leica Microsystems, Buffalo Grove, IL, USA), stained with 0.4% lead citrate, and examined in a JEM-1400 electron microscope (JEOL USA) at 80 kV.

2.4. DNA Extraction, Whole Genome Sequencing, and Assembly

Inoculation of the ECIV isolate onto EPC cells in four 175 cm^2 flasks at a high multiplicity of infection provided third-passage material harvested after 21 days post-infection when CPE was extensive. Cell culture supernatant was clarified at 5520× *g* for 20 min at 4 °C. The pelleted virus was obtained by centrifugation of the clarified supernatant at 100,000× *g* for 1.25 h at 4 °C. The viral pellet was resuspended in 360 μL of animal tissue lysis (ATL) buffer prior to extraction of viral genomic DNA using a DNeasy Blood and Tissue Kit (Qiagen, Germantown, MD, USA) according to the manufacturer's instructions. A DNA library was generated using a Nextera XT DNA Kit, and sequencing was performed using a V3 chemistry 600 cycle Kit on a MiSeq sequencer (Illumina, Germantown, MD, USA). *De novo* assembly of the paired-end reads was performed in SPAdes 3.5.0 genome assembly algorithm [22]. The quality of the genome assembly was verified by mapping the reads back to the consensus sequence in Bowtie 2 2.1.0 [23] and visually inspecting the alignment in Tablet 1.14.10.20 [24].

2.5. Genome Annotation, Genetic, and Phylogenetic Analysis

The viral genome was annotated using GenemarkS [25], and the functions were predicted based on BLASTP searches against the National Center for Biotechnology Information (NCBI) GenBank non-redundant (nr) protein sequence database and conserved domain database. A total of 25 iridovirus core genes were used to conduct the Maximum Likelihood (ML) analysis for 47 iridoviruses, including ECIV (Table 1). The amino acid (AA) sequence alignments were performed for each gene in MAFFT 5.8 using default parameters [26] and concatenated using Geneious R10 [27]. The final dataset contained 19,340 AA characters, and the phylogenetic tree was constructed using IQ-Tree [28] with default parameters. In addition, genetic analyses were performed using the Sequence Demarcation Tool v1.2 with the MAFFT alignment option implemented [29] to compare the nucleotide sequence identity of megalocytiviruses based on the major capsid protein and ATPase gene alignments.

Table 1. GenBank accession numbers for the full genome sequences of iridoviruses used in the 25 iridovirus core gene phylogenetic analysis.

Species Name (Virus Abbreviation)	Genus	GenBank Acc. No.
Anopheles minimus iridovirus (AMIV)	*Chloriridovirus*	KF938901
Invertebrate iridovirus 22 (IIV-22)	*Chloriridovirus*	HF920633
Invertebrate iridovirus 22a (IIV-22a)	*Chloriridovirus*	HF920634
Invertebrate iridescent virus 3 (IIV-3)	*Chloriridovirus*	DQ643392
Invertebrate iridescent virus 30 (IIV-30)	*Chloriridovirus*	HF920636
Invertebrate iridescent virus 9 (IIV-9)	*Chloriridovirus*	GQ918152
Invertebrate iridescent virus 25 (IIV-25)	*Chloriridovirus*	HF920635

Table 1. *Cont.*

Species Name (Virus Abbreviation)	Genus	GenBank Acc. No.
Invertebrate iridescent virus 31 (IIV-31)	*Iridovirus*	HF920637
Invertebrate iridescent virus 6 (IIV-6)	*Iridovirus*	AF303741
Lymphocystis disease virus 1 (LCDV-1)	*Lymphocystivirus*	L63545
Lymphocystis disease virus 2 (LCDV-C)	*Lymphocystivirus*	AY380826
Lymphocystis disease virus 3 (LCDV-Sa)	*Lymphocystivirus*	PRJEB12506
European chub iridovirus (ECIV)	*Megalocytivirus*	MK637631
Giant seaperch iridovirus (GSIV-K1)	*Megalocytivirus*	KT804738
Infectious spleen and kidney necrosis virus (ISKNV)	*Megalocytivirus*	AF371960
Infectious spleen and kidney necrosis virus (RSIV-Ku)	*Megalocytivirus*	KT781098
Orange-spotted grouper iridovirus (OSGIV)	*Megalocytivirus*	AY894343
Red seabream iridovirus (RSIV)	*Megalocytivirus*	AB104413
Red seabream iridovirus (RSIV RIE12–1)	*Megalocytivirus*	AP017456
Rock bream iridovirus (RBIV-KOR-TY1)	*Megalocytivirus*	AY532606
Rock bream iridovirus (RBIV-C1)	*Megalocytivirus*	KC244182
Scale drop disease virus (SDDV)	*Megalocytivirus*	KR139659
South American cichlid iridovirus (SACIV)	*Megalocytivirus*	MG570131
Turbot reddish body iridovirus (TRBIV)	*Megalocytivirus*	GQ273492
Three spot gourami iridovirus (TSGIV)	*Megalocytivirus*	MG570132
Ambystoma tigrinum virus (ATV)	*Ranavirus*	AY150217
Andrias davidianus ranavirus (ADRV)	*Ranavirus*	KC865735
Bohle iridovirus (BIV)	*Ranavirus*	KX185156
Cod iridovirus (CoIV)	*Ranavirus*	KX574342
Common midwife toad virus (CMTVM)	*Ranavirus*	JQ231222
Common midwife toad virus (CMTVVB)	*Ranavirus*	KP056312
Epizootic haematopoietic necrosis virus (EHNV)	*Ranavirus*	FJ433873
European catfish virus (ECV)	*Ranavirus*	KT989885
European sheatfish virus (ESV)	*Ranavirus*	JQ724856
Frog virus 3 (FV3)	*Ranavirus*	AY548484
Frog virus 3 isolate SSME (SSME)	*Ranavirus*	KF175144
German gecko ranavirus (GGRV)	*Ranavirus*	KP266742
Grouper iridovirus (GIV)	*Ranavirus*	AY666015
Pike perch iridovirus (PPIV)	*Ranavirus*	KX574341
Rana grylio iridovirus (RGV)	*Ranavirus*	JQ654586
Ranavirus maximus (Rmax)	*Ranavirus*	KX574434
Short-finned eel ranavirus (SERV)	*Ranavirus*	KX353311
Singapore grouper iridovirus (SGIV)	*Ranavirus*	AY521625
Soft-shelled turtle iridovirus (STIV)	*Ranavirus*	EU627010
Testudo hermanni ranavirus (CH8/96)	*Ranavirus*	KP266741
Tiger frog virus (TFV)	*Ranavirus*	AF389451
Tortoise ranavirus isolate (ToRV1)	*Ranavirus*	KP266743

3. Results

3.1. Cell Culture

The CPE in the EPC cell line consisted of enlarged refractile cells observed within 48 h post-inoculation (hpi). Extensive CPE of the monolayer was observed by 96 hpi, at which point, clumps of affected cells were observed and began to detach from the monolayer (Figure 1).

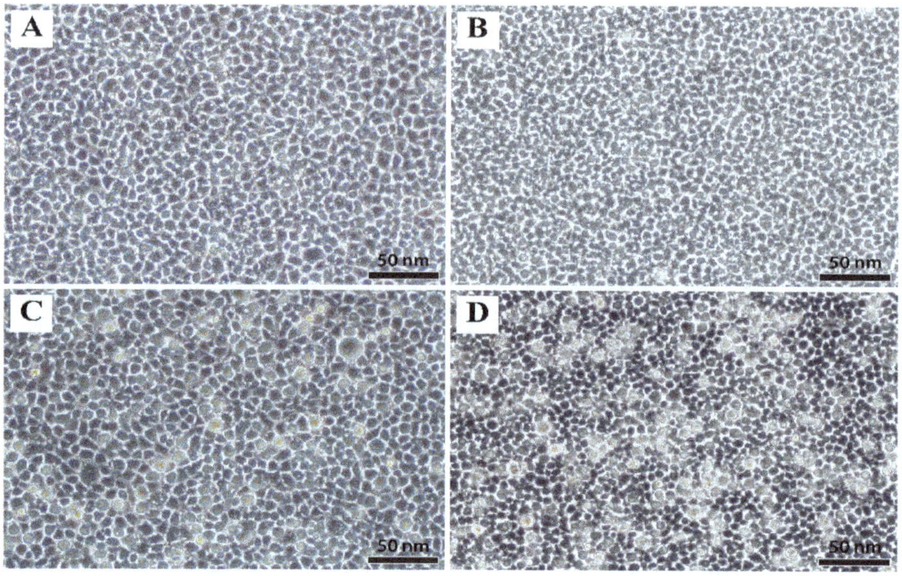

Figure 1. Microscopic examination of *epithelioma papulosum cyprini* cells infected with European chub iridovirus. (**A**) Control flask at 48 h post-inoculation (hpi); (**B**) control flask 96 hpi; (**C**) infected flask showing enlarged and refractile cells at 48 hpi; (**D**) infected flask showing enlarged and refractile cells at 96 hpi. Scale bars are 50 μm.

3.2. *Transmission Electron Microscopy*

Non-enveloped, hexagonal virus particles with electron-lucent or electron-dense cores were observed within viral assembly sites in the cytoplasm of infected EPC cells (Figure 2). The mean diameter of individual virus particles was 127 nm from opposite sides (n = 20, standard deviation = 9) and 147 nm from apex to apex (n = 20, standard deviation = 10).

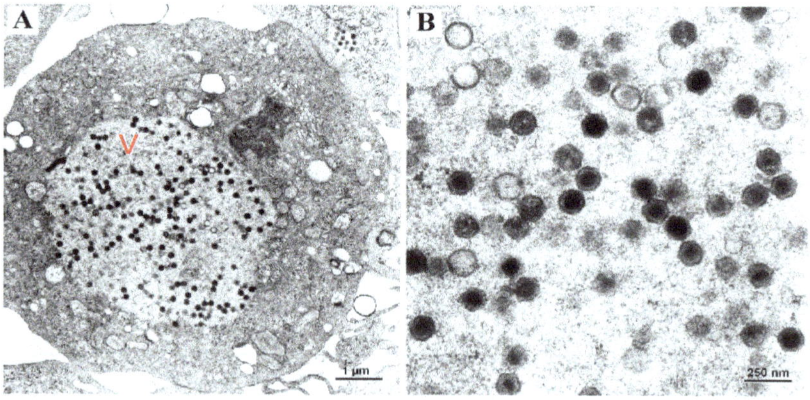

Figure 2. (**A**) Transmission electron photomicrograph of an *epithelioma papulosum cyprini* cell infected with European chub iridovirus, displaying numerous non-enveloped, hexagonal viral particles within the viral assembly site (labeled as V) in the cytoplasm. Scale bar is 1 μm. (**B**) Higher magnification of the virus particles. Scale bar is 250 nm.

3.3. *Genome Annotation, Genetic and Phylogenetic Analyses*

The *de novo* assembly of the 14,420,600 paired-end reads recovered a contiguous sequence of 128,216 bp with an overall coverage of 2948 reads/nucleotide. The %GC of the genome was 38.83, and a total of 108 open reading frames (ORFs) were predicted (Table S1). Comparative genomic analysis revealed the absence of one iridovirus core gene (ISKNV ORF 32R encoding putative thymidine kinase, GenBank accession number AF371960; SDDV ORF 125L, GenBank accession number KR139659) in Europoean chub iridovirus (ECIV). Eighty-seven genes showed the highest amino acid (AA) sequence identity to SDDV, seven genes to various members of the family *Iridoviridae*, and six genes to other organisms (e.g., eukaryotes including fish and fungi). Eight genes were found to be unique to the ECIV genome and did not display similarity to existing genes within the NCBI GenBank nr protein sequence database. Of these, seven genes (ORFs 2, 5, 18, 58, 65, 82, and 89) were predicted as hypothetical proteins, and ORF 66 predicted to be a chromosome segregation protein (Table S1). As with cherax quadricarinatus iridovirus strain CQIV-CN01 (GenBank acc. MF197913), shrimp hemocyte iridescent virus isolate 20141215 (GenBank accession number MF599468), grouper iridovirus (GenBank accession number AY666015), and Singapore grouper iridovirus (GenBank acc. no. AY521625), the ECIV ORF 84 is predicted to encode an ubiquitin family protein. The ECIV ORFs 31 and 57 were predicted as members of the serpin superfamily and showed the highest AA to SDDV ORFs 97L and 45R, respectively. The ECIV encoded a HIRAN domain containing protein (ORF 49) and a family of ankyrin (ANK) proteins (ORFs 3, 4, 44, 46, 47, 63, 92, 99, and 102) that ranged in size from 143 to 478 AA residues. Each ECIV ANK protein possesses between 1 and 7 ANK motifs. ECIV ORF 97 was predicted to encode an US22 protein and displayed the highest AA sequence identity (39.2%) to an US22 protein from Asian swamp eel (*Monopterus albus*). The complete genome sequence of ECIV has been deposited in NCBI GenBank under the accession number MK637631.

The ML analysis of the concatenated 25 iridovirus core gene sequences produced a well-resolved and supported tree (Figure 3). The ECIV was found to be the sister species to the SDDV, which together form the basal branch of the megalocytivirus tree. Genetic comparisons of the ECIV ATPase and MCP nucleotide sequences to other megalocytiviruses ranged from 66.4 to 76.9% and 62.8 to 73.1%, respectively (Tables S2 and S3). The highest identities were observed between ECIV and SDDV.

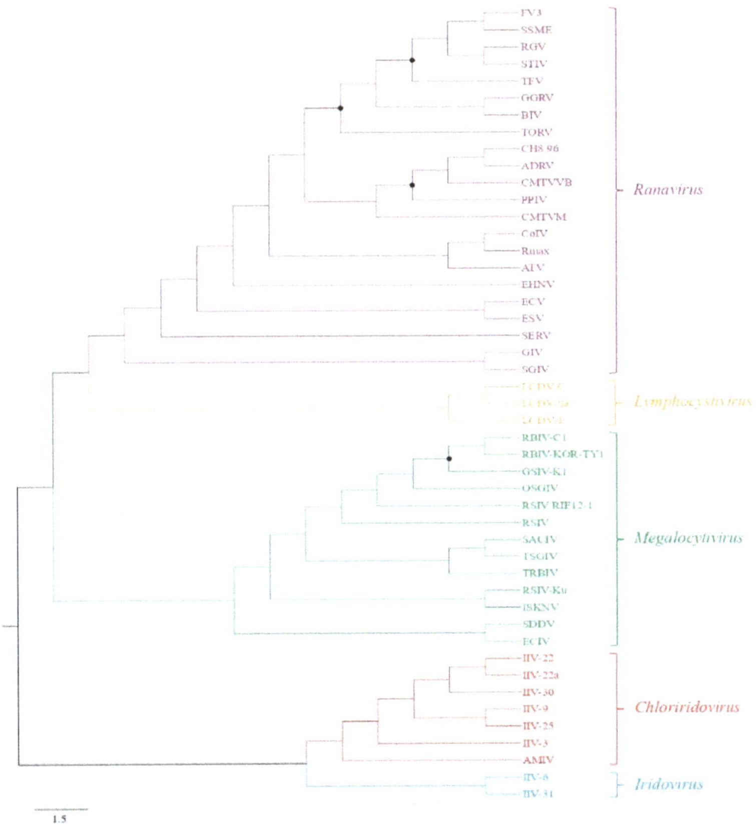

Figure 3. Cladogram depicting the relationship of European chub iridovirus to 47 other members of the family *Iridoviridae* based on 25 core genes. The Maximum Likelihood tree was generated using 1000 bootstraps and the branch lengths are based on the number of inferred substitutions, as indicated by the scale. All nodes were supported by bootstrap values >80% except those labeled with black circles. See Table 1 for virus abbreviations.

4. Discussion

In this investigation, we report the complete genome sequence of a novel iridovirus isolated from moribund European chub (*Squalius cephalus*) in England. The *in vitro* characteristics (i.e., enlarged, rounded, and refractile cells), virion ultrastructure and morphogenesis (i.e., non-enveloped hexagonal virus particles within the cytoplasm), and genetic/phylogenetic analyses supported the identification of European chub iridovirus (ECIV) as a divergent megalocytivirus most closely related to the recently described SDDV [21]. To our knowledge, this study represents the first isolation and genomic characterization of a megalocytivirus in a cyprinid fish. The results of these studies add to a growing body of literature on the host range and threat megalocytiviruses pose to wild and cultured fishes, including their potential impacts in cyprinids [7].

The type species of the genus *Megalocytivirus* (*Infectious spleen and kidney necrosis virus*; ISKNV) exhibits low host specificity, with strains infecting >150 species of freshwater, brackish, and marine fishes [7,30]. The recent characterization of the TSIV from Canadian threespine stickleback (*Gasterosteus aculeatus*) [20] and SDDV from Asian seabass (*Lates calcarifer*) [21] have revealed the existence of genetically-divergent megalocytiviruses that have been argued to represent new species. The discovery of iridoviruses distantly related to ISKNV has stimulated discussion by members of the International Committee on Taxonomy of Viruses study group on iridoviruses to begin re-evaluating the criteria used in defining megalocytivirus species [13]. The genome annotation of ECIV revealed that it possesses 108 predicted genes (including eight unique genes), and compared to other fully-sequenced megalocytivirus genomes, ECIV has the longest genome and a low %GC content similar to SDDV. These data, taken together with the genetic and phylogenetic analyses, suggest ECIV represents yet another novel megalocytivirus, and we propose the formal species designation of European chub iridovirus to be considered for approval by the International Committee on Taxonomy of Viruses.

The HIRAN domain-containing protein (ECIV ORF 49) is not observed in other viruses, except in some bacteriophages [31], and displayed the highest amino acid (AA) sequence identity (39.1%) to the protein of a zygomycete fungus (*Basidiobolus meristosporus*). The HIRAN domain has been found as a standalone protein in a wide range of bacteria or fused to other catalytic domains in eukaryotes [31]. The HIRAN domain is predicted to function as a DNA-binding domain that recognizes damaged DNA or stalled replication forks and recruits repair and remodeling enzymes to these sites [31]. Although a variety of DNA viruses encode serpin proteins, lymphocystis disease virus Sa isolate SA9 (ORF 50R; GenBank accession number KX643370), SDDV (ORFs 45R and 97L; GenBank accession number KR139659), and ECIV (ORFs 31 and 57) are the only iridoviruses that possess these genes [32]. Recent studies have demonstrated that poxvirus-encoded serpins subvert host immune responses by inhibiting the inflammatory response and apoptosis [33]. Megalocytiviruses are the only member of the family *Iridoviridae* to encode ANK repeat proteins, and ECIV encodes the greatest number of copies (ORFs 2, 5, 18, 58, 65, 82, and 89) among members of the genus. ANK repeat proteins have also been described in poxviruses, mimiviruses, and phycodnaviruses. The ISKNV ANK repeat protein (ISKNV ORF 124L; GenBank accession number AF371960) has been shown to interfere with TNF-α-induced NF-κB activation, an important immune regulatory pathway [34]. Poxvirus-encoded ANK repeat proteins are suggested to be involved with host cell tropism [35] and manipulation of the host cell ubiquitin-proteasome machinery [36]. The US22 proteins are present in all megalocytiviruses, except in SDDV and ISKNV, and these proteins are believed to counter diverse host immune responses by interacting with specific host proteins [37,38]. The highest AA sequence identity of the ECIV US22 protein (ORF 97) to Asian swamp eel suggests it was acquired from a fish host.

The *in vitro* cultivation of megalocytiviruses is challenging, with propagation reported in a handful of cell lines including the grunt fin cell line for RSIV and three spot gourami iridovirus [17,30], the mandarin fish fry cell line for ISKNV [39], and the turbot fin cell line for TRBIV [40]. Commonly-used cell lines failed in the propagation of the Banggai cardinalfish iridovirus, a strain of the ISKNV genotype, including the *epithelioma papulosum cyprini* (EPC), bluegill fry (BF-2), chinook salmon embryo (CHSE-214), and fathead minnow (FHM) cell lines [41]. Similarly, TSIV was refractory to culture on

EPC, BF-2, and CHSE-214 cell lines [20]. In contrast, ECIV is less fastidious than other megalocytiviruses growing on EPC, BF-2, CHSE-214, KF-1, and CCB cell lines. Whether the related SDDV shares similar *in vitro* growth characteristics with ECIV remains to be determined as SDDV has only been tested and cultivated in the seabass kidney cell line [21].

Future challenge studies will be needed to determine whether ECIV causes disease in European chub and related cyprinids. These experiments will also help determine whether ECIV induces the pathognomonic microscopic lesions (i.e., megalocytes with basophilic cytoplasmic inclusions) observed in all other megalocytivirus infections to date [17,20,21,30]. Finally, the genomic sequence presented here will facilitate the development of molecular diagnostic assays that could be used to determine the prevalence of ECIV among European chub populations across Eurasia.

Author Contributions: Conceptualization, T.B.W.; methodology, M.A.H., K.S., T.B.W.; validation, K.S., S.A.K., T.B.W.; formal analysis, M.A.H., K.S., S.A.K.; investigation, M.A.H., V.L.P.; resources, D.S., K.W., T.B.W.; data curation, M.A.H., K.S., S.A.K.; writing—original draft preparation, M.A.H.; writing—review and editing, K.S., S.A.K., V.L.P., D.S., K.W., T.B.W.; visualization, M.A.H., K.S., S.A.K.; supervision, T.B.W.; project administration, T.B.W.

Acknowledgments: We thank Patrick M. Thompson for his technical assistance throughout the study, and Jeffrey Go for his critical review of the manuscript.

References

1. Caffrey, J.M.; Acevedo, S.M.; Gallagher, K.; Britton, R. Chub (*Leuciscus cephalus*): A new potentially invasive fish species in Ireland. *Aquatic Invasions* **2008**, *3*, 201–209. [CrossRef]
2. Kottelat, M.; Freyhof, J. *Handbook of European Freshwater Fishes*; Publications Kottelat: Cornol, Switzerland, 2007.
3. Mann, R.H.K. Observations on the age, growth, reproduction and food of the pike *Esox lucius* (L.) in two rivers in southern England. *J. Fish Biol.* **1976**, *8*, 179–197. [CrossRef]
4. Vitali, R.; Braghieri, L. Population dynamics of *Barbus plebejus* (Valenciennes) and *Leuciscus cephalus cabeda* (Risso) in the middle River Po (Italy). *Hydrobiologia* **1984**, *109*, 105–124. [CrossRef]
5. Dixon, P.F. Virus diseases of cyprinids. In *Fish diseases*; Eiras, J.C., Segner, H., Kapoor, B.G., Eds.; Science Publishers: Enfield, NH, USA, 2008; Volume 1, pp. 87–184.
6. Yanong, R.P. Lymphocystis Disease in Fish (FA181). University of Florida Institute of Food and Agricultural Sciences, 2010. Available online: http://edis.ifas.ufl.edu/fa181 (accessed on 11 December 2018).
7. Yanong, R.P.; Waltzek, T.B. Megalocytivirus infections in fish, with emphasis on ornamental species (FA182). University of Florida Institute of Food and Agricultural Sciences, 2010. Available online: http://edis.ifas.ufl.edu/fa182 (accessed on 3 December 2018).
8. Shchelkunov, I.S.; Shchelkunova, T.I. Infectivity experiments with *Cyprinus carpio* iridovirus (CCIV), a virus unassociated with carp gill necrosis. *J. Fish Dis.* **1990**, *13*, 475–484. [CrossRef]
9. Berry, E.S.; Shea, T.B.; Gabliks, J. Two iridovirus isolates from *Carassius auratus* (L.). *J. Fish Dis.* **1983**, *6*, 501–510. [CrossRef]
10. Maganha, S.R.; Cardoso, P.H.; Balian, S.; Almeida-Queiroz, S.R.; Fernandes, A.M.; de Sousa, R.L. Molecular detection and phylogenetic analysis of megalocytivirus in Brazilian ornamental fish. *Arch. Of Vir.* **2018**, *163*, 2225–2231. [CrossRef]
11. Bermúdez, R.; Losada, A.P.; de Azevedo, A.M.; Guerra-Varela, J.; Pérez-Fernández, D.; Sánchez, L.; Padrós, F.; Nowak, B.; Quiroga, M.I. First description of a natural infection with spleen and kidney necrosis virus in zebrafish. *J. Fish Dis.* **2018**, *41*, 1283–1294. [CrossRef]
12. George, M.R.; John, K.R.; Mansoor, M.M.; Saravanakumar, R.; Sundar, P.; Pradeep, V. Isolation and characterization of a ranavirus from koi, *Cyprinus carpio* L., experiencing mass mortalities in India. *J. Fish Dis.* **2015**, *38*, 389–403. [CrossRef]
13. Chinchar, V.R.; Hick, P.; Ince, I.A.; Jancovich, J.K.; Marschang, R.; Qin, Q.; Subramaniam, K.; Waltzek, T.B.; Whittington, R.; Williams, T.; Zhang, Q.; ICTV Report Consortium. ICTV Virus Taxonomy Profile: *Iridoviridae*. *J. Gen. Virol.* **2017**, *98*, 890–891.
14. Jancovich, J.K.; Chinchar, V.G.; Hyatt, A.; Miyazaki, T.; Williams, T.; Zhang, Q.Y. Family Iridoviridae. In *Virus*

Taxonomy: Ninth Report of the International Committee on Taxonomy of Viruses; King, A.M.Q., Adams, M.J., Carstens, E.B., Lefkowitz, E.J., Eds.; Elsevier Academic Press: San Diego, CA, USA, 2012; pp. 193–210.
15. Chinchar, V.G.; Hyatt, A.; Miyazaki, T.; Williams, T. Family *Iridoviridae*: poor viral relations no longer. *Curr. Top. Microbiol. Immunol* **2009**, *328*, 123–170.
16. He, J.G.; Wang, S.P.; Zeng, K.; Huang, Z.J.; Chan, S. Systemic disease caused by an iridovirus-like agent in cultured mandarin fish, *Siniperca chuatsi* (Basilewsky), in China. *J. Fish Dis.* **2000**, *23*, 219–222. [CrossRef]
17. He, J.G.; Deng, M.; Weng, S.P.; Li, Z.; Zhou, S.Y.; Long, Q.X.; Wang, X.Z.; Chan, S. Complete genome analysis of the mandarin fish infectious spleen and kidney necrosis iridovirus. *Virol. J.* **2001**, *291*, 126–139. [CrossRef] [PubMed]
18. Inouye, K.; Yamano, K.; Maeno, Y.; Nakajima, K.; Matsuoka, M.; Wada, Y.; Sorimachi, M. Iridovirus infection of cultured red sea bream, *Pagrus major*. *Fish. Pathol.* **1992**, *27*, 19–27. [CrossRef]
19. Kawato, Y.; Subramaniam, K.; Nakajima, K.; Waltzek, T.; Whittington, R. Iridoviral Diseases: Red Sea Bream Iridovirus and White Sturgeon Iridovirus. In *Fish Viruses and Bacteria: Pathobiology and Protection*; Woo, P.T.K., Cipriano, R.C., Eds.; CABI Publishing: Wallingford, UK, 2017; pp. 147–159.
20. Waltzek, T.B.; Marty, G.D.; Alfaro, M.E.; Bennett, W.R.; Garver, K.A.; Haulena, M.; Weber, E.S., 3rd.; Hedrick, R.P. Systemic iridovirus from threespine stickleback Gasterosteus aculeatus represents a new megalocytivirus species (family *Iridoviridae*). *Dis. Aquat. Org.* **2012**, *98*, 41–56. [CrossRef]
21. De Groof, A.; Guelen, L.; Dejis, M.; van der Wal, Y.; Miyata, M.; Ng, K.S.; van Grinsven, L.; Simmelink, B.; Viermann, Y.; Grisez, L.; et al. A novel virus causes scale drop disease in *Lates calcarifer*. *PLoS Pathog.* **2015**, *11*, e1005074. [CrossRef] [PubMed]
22. Bankevich, A.; Nurk, S.; Antipov, D.; Gurevich, A.A.; Dvorkin, M.; Kulikov, A.S.; Lesin, V.M.; Nikolenko, S.I.; Pham, S.; Prjibelski, A.D.; et al. SPAdes: A new genome assembly algorithm and its applications to single-cell sequencing. *J. Comput. Biol.* **2012**, *19*, 455–477. [CrossRef] [PubMed]
23. Langmead, B.; Salzberg, S.L. Fast gapped-read alignment with Bowtie 2. *Nat. Methods* **2012**, *9*, 357–359. [CrossRef]
24. Milne, I.; Bayer, M.; Cardle, L.; Shaw, P.; Stephen, G.; Wright, F.; Marshall, D. Tablet—next generation sequence assembly visualization. *Bioinformatics* **2010**, *26*, 401–402. [CrossRef]
25. Besemer, J.; Lomsadze, A.; Borodovsky, M. GeneMarkS: a self-training method for prediction of gene starts in microbial genomes. Implications for finding sequence motifs in regulatory regions. *Nucleic Acids Res.* **2001**, *29*, 2607–2618. [CrossRef] [PubMed]
26. Katoh, K.; Toh, H. Recent developments in the MAFFT multiple sequence alignment program. *Brief Bioinform.* **2008**, *9*, 286–298. [CrossRef]
27. Kearse, M.; Moir, R.; Wilson, A.; Stones-Havas, S.; Cheung, M.; Sturrock, S.; Buxton, S.; Cooper, A.; Markowitz, S.; Duran, C.; et al. Geneious Basic: An integrated and extendable desktop software platform for the organization and analysis of sequence data. *Bioinformatics* **2012**, *28*, 1647–1649. [CrossRef] [PubMed]
28. Trifinopoulos, J.; Nguyen, L.T.; von Haeseler, A.; Minh, B.Q. W-IQ-TREE: A fast online phylogenetic tool for maximum likelihood analysis. *Nucleic Acids Res.* **2016**, *44*, W232–W235. [CrossRef] [PubMed]
29. Muhire, B.M.; Varsani, A.; Martin, D.P. SDT: A virus classification tool based on pairwise sequence alignment and identity calculation. *PLoS ONE* **2014**, *9*, e108277. [CrossRef]
30. Koda, S.A.; Subramaniam, K.; Francis-Floyd, R.; Yanong, R.P.; Frasca, S., Jr.; Groff, J.M.; Popov, V.L.; Fraser, W.A.; Yan, A.; Mohan, S.; Waltzek, T.B. Phylogenomic characterization of two novel members of the genus *Megalocytivirus* from archived ornamental fish samples. *Dis. Aquat. Org.* **2018**, *130*, 11–24. [CrossRef]
31. Iyer, L.M.; Babu, M.; Aravind, B. The HIRAN domain and recruitment of chromatin remodeling and repair activities to damaged DNA. *Cell Cycle* **2006**, *5*, 775–782. [CrossRef] [PubMed]
32. Bao, J.; Pan, G.; Poncz, M.; Wei, J.; Ran, M.; Zhou, Z. Serpin functions in host-pathogen interactions. *Peer J.* **2018**. [CrossRef] [PubMed]
33. Johnston, J.B.; McFadden, G. Poxvirus immunomodulatory strategies: Current perspectives. *J. Virol.* **2003**, *77*, 6093–6100. [CrossRef]
34. Guo, C.J.; Chen, W.J.; Yuan, L.Q.; Yang, L.S.; Weng, S.P.; Yu, X.Q.; He, J.G. The viral ankyrin repeat protein (ORF124L) from infectious spleen and kidney necrosis virus attenuates nuclear factor-κB activation and interacts with IκB kinase β. *J. Gen. Virol.* **2011**, *92*, 1561–1570. [CrossRef] [PubMed]
35. Werden, S.J.; McFadden, G. The role of cell signaling in poxvirus tropism: the case of the M-T5 host range protein of myxoma virus. *Biochim. Biophys. Acta* **2008**, *1784*, 228–237. [CrossRef] [PubMed]

36. Sonnberg, S.; Seet, B.T.; Pawson, T.; Fleming, S.B.; Mercer, A.A. Poxvirus ankyrin repeat proteins are a unique class of F-box proteins that associate with cellular SCF1 ubiquitin ligase complexes. *Proc. Natl. Acad. Sci. USA* **2008**, *105*, 10955–10960. [CrossRef] [PubMed]
37. Zhang, D.; Iyer, L.M.; Aravind, L. A novel immunity system for bacterial nucleic acid degrading toxins and its recruitment in various eukaryotic and DNA viral systems. *Nucleic Acids Res.* **2011**, *39*, 4532–4552. [CrossRef] [PubMed]
38. Claytor, S.C.; Subramaniam, K.; Landrau-Giovannetti, N.; Chinchar, V.G.; Gray, M.J.; Miller, D.L.; Mavian, C.; Salemi, M.; Wisely, S.; Waltzek, T.B. Ranavirus phylogenomics: Signatures of recombination and inversions among bullfrog ranaculture isolates. *Virology* **2017**, *511*, 330–343. [CrossRef]
39. Dong, C.; Weng, S.; Shi, X.; Xu, X.; Shi, N.; He, J. Development of a mandarin fish *Siniperca chuatsi* fry cell line suitable for the study of infectious spleen and kidney necrosis virus (ISKNV). *Virus Res.* **2008**, *135*, 273–281. [CrossRef] [PubMed]
40. Fan, T.J.; Ren, B.X.; Geng, X.F.; Yu, Q.T.; Wang, L.Y. Establishment of a turbot fin cell line and its susceptibility to turbot reddish body iridovirus. *Cytotechnology* **2010**, *62*, 217–223. [CrossRef] [PubMed]
41. Weber, E.S.; Waltzek, T.B.; Young, D.A.; Twitchell, E.L.; Gates, A.E.; Vagelli, A.; Risatti, G.R.; Hedrick, R.P.; Frasca, S., Jr. Systemic iridovirus infection in the Banggai cardinalfish (*Pterapogon kauderni* Koumans 1933). *J. Vet. Diagn. Investig.* **2009**, *21*, 306–320. [CrossRef]

14

Modelling Ranavirus Transmission in Populations of Common Frogs (*Rana temporaria*) in the United Kingdom

Amanda L.J. Duffus [1,*], Trenton W.J. Garner [2], Richard A. Nichols [3], Joshua P. Standridge [1] and Julia E. Earl [4]

1 Department of Mathematics and Natural Sciences, Gordon State College, Barnesville, GA 30204, USA; js197212@gordonstate.edu
2 Institute of Zoology, Zoological Society of London, London NW1 4RY, UK; trent.garner@ioz.ac.uk
3 School of Biological and Chemical Sciences, Queen Mary, University of London, London E1 4NS, UK; r.a.nichols@qmul.ac.uk
4 School of Biological Sciences, Louisiana Tech University, Ruston, LA 71272, USA; jearl@latech.edu
* Correspondence: aduffus@gordonstate.edu

Abstract: Ranaviruses began emerging in common frogs (*Rana temporaria*) in the United Kingdom in the late 1980s and early 1990s, causing severe disease and declines in the populations of these animals. Herein, we explored the transmission dynamics of the ranavirus(es) present in common frog populations, in the context of a simple susceptible-infected (SI) model, using parameters derived from the literature. We explored the effects of disease-induced population decline on the dynamics of the ranavirus. We then extended the model to consider the infection dynamics in populations exposed to both ulcerative and hemorrhagic forms of the ranaviral disease. The preliminary investigation indicated the important interactions between the forms. When the ulcerative form was present in a population and the hemorrhagic form was later introduced, the hemorrhagic form of the disease needed to be highly contagious, to persist. We highlighted the areas where further research and experimental evidence is needed and hope that these models would act as a guide for further research into the amphibian disease dynamics.

Keywords: transmission modelling; susceptible-infected (SI) models; emerging infection; ranavirosis; *Iridoviridae*; disease dynamics

1. Introduction

Ranaviruses are double-stranded DNA viruses in the *Iridoviridae* family that can infect ectothermic vertebrates [1] and are found worldwide [2]. Ranaviruses cause systemic hemorrhaging and edema in amphibian, reptile, and fish hosts [3]. In amphibians, mortality can occur in only three days in very susceptible host species and life history stages [4], resulting in die-offs of both adults and larvae, in Europe [5,6], and tadpoles and metamorphs, in North America [7,8]. Ranaviruses have been shown to alter amphibian population dynamics, with declines of the common frog (*Rana temporaria*) in the United Kingdom [9] and whole amphibian communities in Spain [10]. Additionally, population simulation models have shown that ranaviruses could potentially cause the local extinction of populations of three, highly susceptible species of the United States anurans [11,12]. However, these models do not include the transmission dynamics, which are not well-understood and could potentially alter host population dynamics. Farrell et al. [13] modeled host population declines in the case of the highly pathogenic *Ambystoma tigrinum virus* (ATV), but they found that extinction of host populations would not occur even in the cases where the host population was severely reduced. These diverse conclusions showed that there were many different possible outcomes of ranavirus infections that influenced disease dynamics, and they appeared to be system-specific.

Mathematical models are helpful tools for understanding transmission dynamics and the persistence of pathogen and host populations [14,15]. Models might help determine which transmission conditions are most likely, given the information about disease dynamics in wild populations. In the case of the ranavirus, Brunner and Yarber [16] developed a model which suggested that ranavirus transmission from water and scavengers are likely minimal in most circumstances. Previous attempts have also been made to formalize the transmission dynamics at, both, the species level [17] and at the community level [18]. Most models have been based on North American amphibian populations or communities. The dynamics of amphibian–ranavirus systems in the UK, the ranavirus–common frog (*Rana temporaria*) system, are distinctly different from those in North America and thus require specific investigation.

Here, we developed susceptible-infection (SI) models for ranavirus infection in common frogs, in the UK. The ranavirus–common frog system in the UK differs from the infection/disease dynamics seen in North American anuran species, because infections in eggs or tadpoles of wild populations are absent, for the most part [19,20], despite experimental evidence that tadpoles can be infected and have died with signs of ranavirosis. Given this information, the route of transmission in common frogs is most likely among adults. With the benefit of long-term data sets on the persistence of ranavirus infections in common frogs, we know that the ranavirus can persist in adult frog populations for many years [9] and involves at least two distinct disease syndromes (hemorrhagic and ulcerative) with variations in the prevalence of disease, across sites and years [21]. We, therefore, developed SI models to test the hypothesis that ranaviruses have the potential to remain in the UK common frog populations, under various conditions, with only horizontal transmission between adults. We examined scenarios with host population declines that included both ranavirus syndromes.

2. Basic Model Formulation

In the simplest case, we used an SI model with no recovery, based on the high mortality rate associated with ranavirus infection. We followed the method for developing mathematical models described by Otto and Day [22]. The population consisted of recruits (A_R), susceptible individuals (A_S), and infected individuals (A_I) and mortality occurred at two different rates—natural mortality (M_N) and mortality due to disease (M_D). We assumed that the adult population size remained constant (i.e., mortality was fully compensated irrespective of the cause of death), all recruits to the population were susceptible to the ranavirus, and all individuals were equally susceptible to infection. The contact rate (Ψ) was defined as the number of different individuals that one animal physically contacts, the likelihood of transmission (σ) was the probability of transmission given the physical contact, and R_0 was the basic reproductive rate of the virus. We used a discrete time model, because common frogs aggregate annually for breeding and we assumed that transmission primarily occurred from contacts during breeding. We also assumed that mortality due to disease occurred primarily during summer; ranavirus-associated mortality peaks between mid-July and mid-August, after breeding, has previously been concluded [20]. (See Figure 1 for a schematic view of the important life history events and timing.) Based on the above assumptions, we developed the following equations:

$$A_s(t+1) = A_s(t) - \sigma\Psi \cdot A_s(t) \cdot A_I(t) - M_N(t) + A_R(t) \tag{1}$$

$$A_I(t+1) = A_I(t) + \sigma\Psi \cdot A_s(t) \cdot A_I(t) - [M_N(t) + M_D(t)] \tag{2}$$

$$R_0 = \sigma\Psi \cdot A_s(t)/[M_N(t) + M_D(t)] \tag{3}$$

To determine under what conditions the ranavirus would remain or spread in a population, we modified Equation (3) by removing M_D (t) to assume a successful introduction:

$$R_o = \sigma\Psi \cdot A_s/M_N(t) \tag{4}$$

The important interactions in this equation are between σ and Ψ, and this relationship needed to be explored, graphically, to determine when the conditions for $R_o \geq 1$ exist. If we assume a population size of 99 (A_S), with an initial introduction of one infected individual (A_I) and an M_N of 20% [23] (although other estimates do exist [24]), values of σ and Ψ under which $R_o \geq 1$ are illustrated in Figure 2. To test if similar assumptions are valid at a smaller population size, we repeated the process using alternative values for $A_S = 49$ and $M_N = 10\%$ (i.e., 5 individuals/annum) and introduced one infected individual (total population size 50; Figure 2).

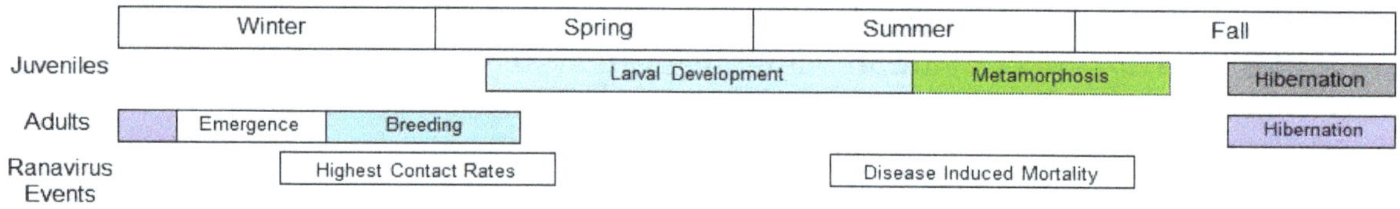

Figure 1. Annual cycle of important life history events for common frogs (*Rana temporaria*) and important events for ranavirus infections and diseases, for these animals. Boxes shaded in blue are those that occur in the aquatic environment, green boxes are those that straddle the land and water, grey boxes occur at an unknown location, and mauve boxes are life history events that are known to happen on both the land and in the water.

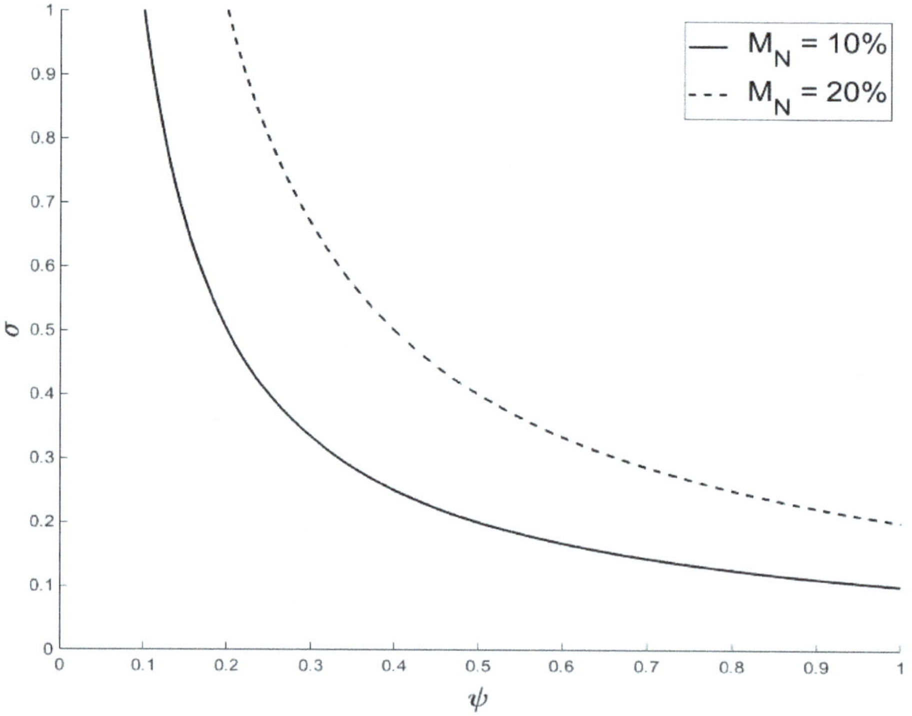

Figure 2. The interaction between σ and Ψ, under which $R_o \geq 1$, when $A_S = 99$ and an $M_N = 20\%$ (upper dashed-curve) and when the initial conditions of $A_S = 49$ and an $M_N = 10\%$ (lower curve), where Ψ is the contact rate and σ is the likelihood of transmission for the model.

After establishing the conditions which permit ranavirus persistence, we next examined the behavior of A_S and A_I under potential biologically relevant conditions. We used experimental data from ranavirus exposures in the literature, to estimate σ (Table 1). It is important to note that the data used to estimate σ was extremely variable; the viral titers used in the experiments range from $TCID_{50}$ of 10^1 to 10^2 mL for virus obtained from crude organ homogenates to $TCID_{50}$ of $10^{4.2}$ to $10^{6.2}$ mL, for virus produced via tissue culture [25]. This variability made the biological relevance of our estimates less than ideal; however, these are the best estimates that can be made with the available data.

Table 1. Estimates for σ derived from the literature. Note: Experiments where the exposure was via inoculation have not been included in these estimates. No distinction has been made between the types of ranavirus-associated disease that the virus was derived from. Development of disease data, $TCID_{50}$ and type of experiment information are summarized from Cunningham et al. [25]. (HS = Hemorrhagic and US = Ulcerative forms of ranavirosis.).

Development of Disease		Disease Prevalence	Type of Experiment/Exposure Type	Estimate of σ	$TCID_{50}$
No. with Disease	Total No. Exposed				
3	20	15%	Immersion with virus from naturally disease tissue, with and without bacteria	0.15	10^1/mL
9	20	45%	Immersion with virus from naturally disease tissue homogenate to animals with skin wounds, with and without bacteria	0.45	10^2/mL HS $10^{1.5}$/mL US
9	10	90%	Immersion in virus from culture	0.90	10^2/mL HS $10^{1.5}$/mL US
5	5	100%	Immersion in virus from virus culture to animals with wounded skin	1	$10^{5.6}$ to $10^{6.2}$/mL
2	5	40%	Immersion in virus from tissue homogenate from naturally diseased animals to animals with wounded skin	0.40	10^3/mL

Estimates of Ψ were unavailable, so we arbitrarily chose Ψ = 0.3→0.6 (Figure 3). We ignored sex specific contact rates, as male–male contact rates are extremely high and polyandry is common [26]. Additionally, Teacher et al. [9] found that the median population size of common frogs was 31 for those places where ranaviruses have emerged in the UK, so we used an A_{Total} of 30 (Figure 4).

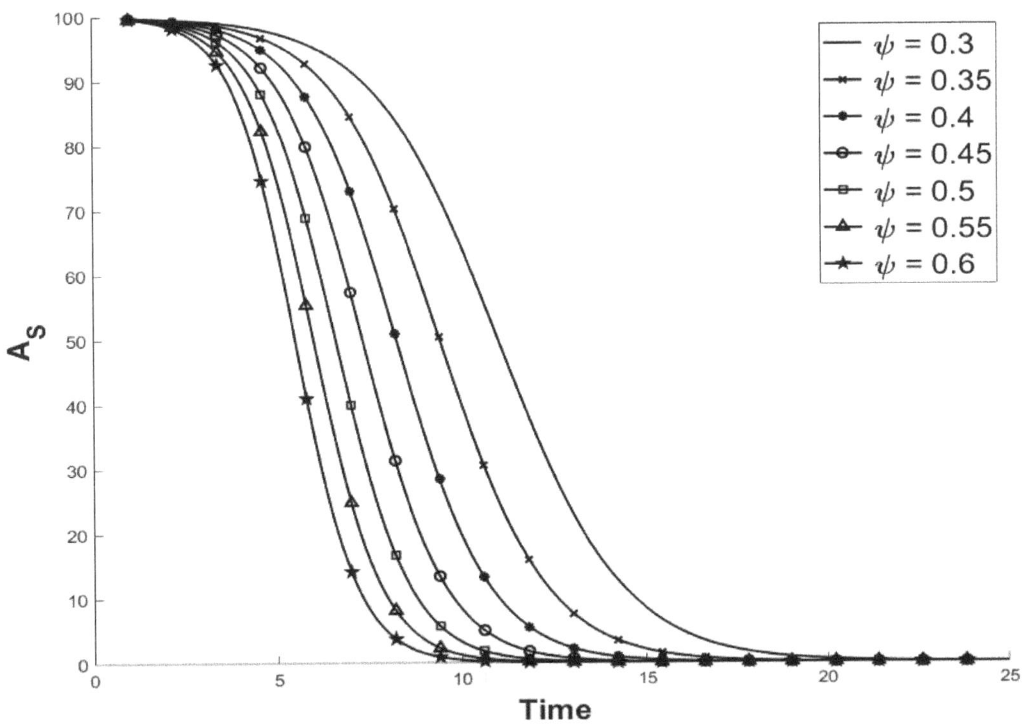

Figure 3. Predicted values of A_S with varying values of Ψ while other values remained constant at: σ = 0.3; M_N = 0.2; and M_D = 0.75. The starting population composition is A_I = 1 and A_S = 99 and time is in years. A_S is the number of susceptible individuals, Ψ is the contact rate, σ is the likelihood of transmission, M_N is the natural mortality rate, and M_D is the mortality rate associated with ranavirosis.

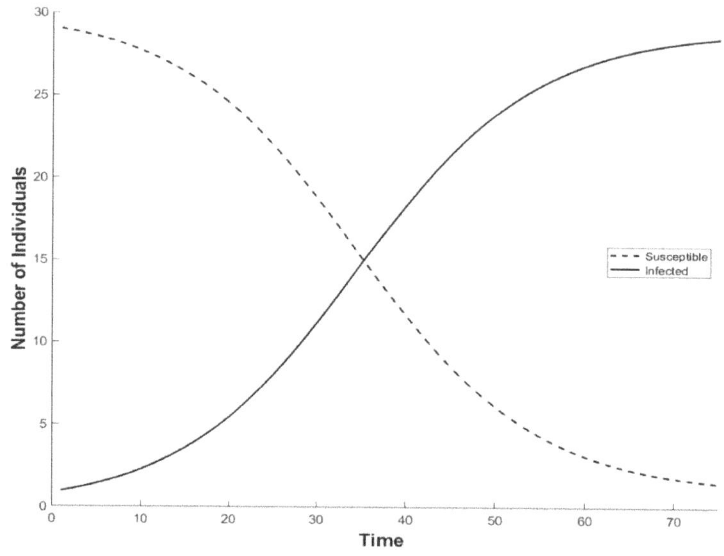

Figure 4. The average expectation of the ranavirus dynamics in a population of adult common frogs (*Rana temporaria*) through time (years). ($\Psi = 0.45$; $\sigma = 0.3$; $M_N = 0.2$; $M_D = 0.775$; starting population comprised of $A_I = 1$ and $A_S = 29$.) A_I is the number of infected individuals, A_S is the number of susceptible individuals, Ψ is the contact rate, σ is the likelihood of transmission, M_N is the natural mortality rate, and M_D is the mortality rate associated with ranavirosis.

Since ranaviruses have emerged recently in the UK, and common frogs and the severity of the disease is affected by climate change, it is unlikely that the disease dynamics have reached an equilibrium ([27,28]; Figure 4). Using an estimated σ of 0.3 (from data from Table 1), an average M_D of 0.775, an average Ψ value of 0.45, and an A_{total} of 30 [9] with the initial conditions of an A_s of 29 and an A_I of 1, we found that the interaction between A_S and A_I requires greater than 60 years to stabilize to post-epidemic dynamics (Figure 5). These results suggest that, under the present conditions, ranavirus in populations of common frogs can be sustained entirely through adult–adult transmission.

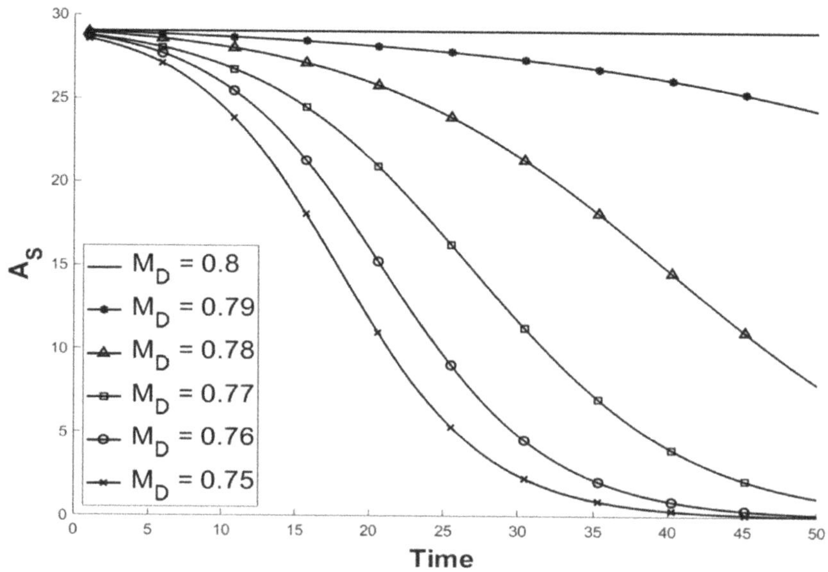

Figure 5. Illustration of the predicted values for A_S with different disease-induced mortality rates, while other values remained constant at: $\Psi = 0.45$; $\sigma = 0.3$; $M_N = 0.2$; the starting population comprised of $A_I = 1$ and $As = 29$. A_s is the number of susceptible individuals, Ψ is the contact rate, σ is the likelihood of transmission, M_N is the natural mortality rate and M_D is the mortality rate associated with ranavirosis.

2.1. Factoring in Population Decline

In UK common frogs, ranavirosis has caused an 81% decline in some affected populations, over a 10 year period [9]; this appears to be a clear violation of the assumption in the model that population size remains constant. Teacher et al. [9] found that the declines in these populations were proportional to their size (i.e., the larger the population, the larger the decline experienced). This is possibly consistent with the assumption that all adults are equally susceptible, when a ranavirus first invades a population. Contrarily, larger declines in larger populations might be due to density-dependent transmission of the virus. This scenario can also account for the differences among individuals, in their susceptibility to the ranavirus.

However, Teacher et al. [29] report that populations of common frogs maintain allelic diversity through the immigration of adults from nearby ponds. This immigration has two potentially important consequences for the emergence of ranavirosis: (1) the population of adult frogs in the affected population will remain susceptible to infection because of the homogenizing effect of immigration; thus, the immigration of susceptible individuals might dilute the prevalence of the resistance genes, as long as the immigrants come from susceptible populations; and (2) immigration might bolster population numbers and reduce the observed decline. Consequently, the estimates of population decline due to ranavirosis that were made by Teacher et al. [9] might, in fact, be underestimates. In addition, Campbell et al. [30] have found that the population structure shifts from older adult frogs to proportionally more juveniles, with a much smaller total population size in the affected populations. These smaller populations have also been shown to be more susceptible to stochastic events, via population modelling [30]. For simplicity, we assumed that declines of 81% over a 10 year period, as seen by Teacher et al. [9], correspond to a steady annual decline of 8.1%. Under such conditions, it takes the model approximately 65 years to stabilize, with all adults in the population suffering from ranavirus infections (Figure 6).

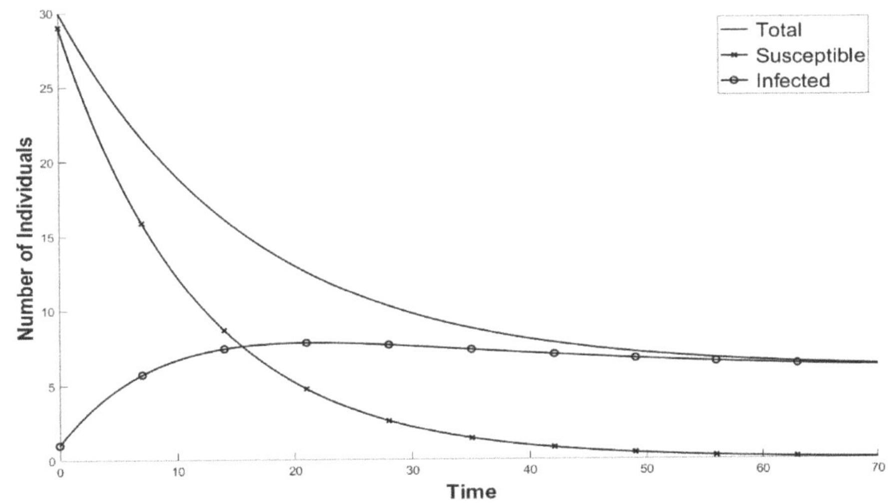

Figure 6. Illustration of the predicted dynamics of a common frog population, with the ranavirus factoring in an annual population decline of 8.1% for adult common frogs. ($\Psi = 0.45$; $\sigma = 0.3$; $M_N = 0.2$; $M_D = 0.775$; starting population comprised of $A_I = 1$ and $A_S = 29$; time is in years.) A_s is the number of susceptible individuals, Ψ is the contact rate, σ is the likelihood of transmission, M_N is the natural mortality rate, and M_D is the mortality rate associated with the ranavirosis.

2.2. Accounting for Different Disease Syndromes

Ranavirosis in the UK common frogs presents as two syndromes which are not mutually exclusive. The ulcerative form of the disease is characterized by ulcers of the skin and the skeletal muscle, and sometimes necrosis of the digits, while the hemorrhagic form of the disease is characterized by internal hemorrhages, most commonly involving the gastrointestinal and reproductive tracts ([21];

personal observation). Adult common frogs exposed to a tissue homogenate derived from skin ulcers only developed the ulcerative form of the disease (with a prevalence of ~30%; See Table 2; [25]). Conversely, adult frogs exposed to a virus isolate obtained from skin ulcers generated both ulcerative and hemorrhagic signs of the disease, while virus isolated from a hemorrhage caused the hemorrhagic form of the disease in exposed adults [25]. These, and other data, indicate that ranaviruses associated with different pathologies in the UK might have different transmission rates [25]. Our estimates of σ for both the ulcerative and hemorrhagic forms can be found in Table 2. If we take the view that infection using viral isolates derived from the cell culture does not mimic the natural process, the other estimates from Table 2 would be preferred, in which case, we estimated σ for the ulcerative and hemorrhagic syndromes as 0.33 and 0.20, respectively.

Table 2. Estimates for σ derived from the literature, taking into account the different disease syndromes and type of syndrome that the virus was obtained from. U indicates the ulcerative form of the disease; H is the hemorrhagic form. The estimate of σ is simply the prevalence of the disease based on the presence of the signs of disease when the experiment terminated. The average estimate of σ is simply the mean of the estimates for each type of virus used for exposure. Development of disease data and type of experiment information are summarized from Tables 3–5 of Cunningham et al. [25].

Development of Disease				Type of Experiment/ Exposure Type	Estimate of σ	Average Estimate of σ
No. with U	No. with H	No. with U & H	Total Exp.			
2	0	0	5	Immersion with virus from naturally disease tissue with bacteria (Ulcerative)	0.4	0.36
1	0	0	5	Immersion with virus from naturally disease tissue without bacteria (Ulcerative)	0.2	
2	0	0	5	Immersion with virus from naturally disease tissue to animals with skin wounds with bacteria (Ulcerative)	0.4	
0	0	0	5	Immersion with virus from naturally disease tissue to animals with skin wounds without bacteria (Ulcerative)	0	
2	2	0	5	Immersion in virus isolated from naturally diseased animals from virus culture (RUK 13, Ulcerative)	0.8	
0	0	0	5	Immersion with virus from naturally disease tissue without bacteria (Hemorrhagic)	0	0.44
0	0	0	5	Immersion with virus from naturally disease tissue with bacteria (Hemorrhagic)	0	
1	1	1	5	Immersion with virus from naturally disease tissue to animals with skin wounds with bacteria (Hemorrhagic)	0.6	
0	3	1	5	Immersion with virus from naturally disease tissue to animals with skin wounds without bacteria (Hemorrhagic)	0.8	
1	2	1	5	Immersion in virus isolated from naturally diseased animals from virus culture (RUK 11, Hemorrhagic)	0.8	

We investigated three scenarios applying different values of σ, where the first two versions explored the two syndromes in isolation (ulcerative form, A_U, and hemorrhagic form, A_H, present, respectively, in Figure 7A,B), represented by Equations (5)–(9) for the initial frequencies and Equations (10)–(14) at time (t). Then we developed a new series of equations for when both disease syndromes are present and including individuals that show signs of both forms of the disease, A_{U+H}. For simplicity, we assumed that all animals are equally susceptible to both forms of the disease (Figure 8) and that there was no difference in the disease-induced mortality rate between syndromes:

$$A_S(t) = A_S(i) - [\sigma_1\Psi\cdot A_S(i)\cdot A_U(i) + \sigma_2\Psi\cdot A_S(i)\cdot A_{(U+H)}(i) + \sigma_3\Psi\cdot A_S(i)\cdot A_H(i)] - M_N(i) + A_R(i) \quad (5)$$

$$A_U(t) = A_U(i) + [\sigma_1\Psi\cdot A_S(i)\cdot A_U(i)] - [\sigma_1\Psi\cdot A_U(i)\cdot A_H(i) + \sigma_3\Psi\cdot A_H(i)\cdot A_U(i)] - [M_N(i) + M_{D(U)}(i)] \quad (6)$$

$$A_{(U+H)}(t) = A_{(U+H)}(i) + [\sigma_2\Psi\cdot A_S(i)\cdot A_{(U+H)}] + [\sigma_1\Psi\cdot A_U(i)\cdot A_H(i)] + [\sigma_3\Psi\cdot A_H(i)\cdot A_U(i)] - [M_N(i) + M_{D(U+H)}(i)] \quad (7)$$

$$A_H(t) = A_H(i) + [\sigma_3\beta\cdot A_S(i)\cdot A_H(i)] - [\sigma_1\beta\cdot A_U(i)\cdot A_H(i) + \sigma_3\beta\cdot A_H(i)\cdot A_U(i)] - [M_N(i) + M_{D(H)}(i)] \quad (8)$$

$$A_s(t+1) = A_S(t) - [\sigma_1\Psi\cdot A_S(t)\cdot A_U(t) + \sigma_2\Psi\cdot A_S(t)\cdot A_{(U+H)}(t) + \sigma_3\Psi\cdot A_S(t)\cdot A_H(t)] - M_N(t) + A_R(t) \quad (9)$$

$$A_U(t+1) = A_U(t) + [\sigma_1\Psi\cdot A_S(t)\cdot A_U(t)] - [\sigma_1\Psi\cdot A_U(t)\cdot A_H(t) + \sigma_3\Psi\cdot A_H(t)\cdot A_U(t)] - [M_N(t) + M_{D(U)}(t)] \quad (10)$$

$$A_{(U+H)}(t+1) = A_{(U+H)}(i) + [\sigma_2\Psi\cdot A_S(i)\cdot A_{(U+H)}] + [\sigma_1\Psi\cdot A_U(i)\cdot A_H(i)] + [\sigma_3\Psi\cdot A_H(i)\cdot A_U(i)] - [M_N(i) + M_{D(U+H)}(i)] \quad (11)$$

$$A_H(t+1) = A_H(t) + [\sigma_3\Psi\cdot A_S(t)\cdot A_H(t)] - [\sigma_1\Psi\cdot A_U(t)\cdot A_H(t) + \sigma_3\Psi\cdot A_H(t)\cdot A_U(t)] - [M_N(t) + M_{D(H)}(t)] \quad (12)$$

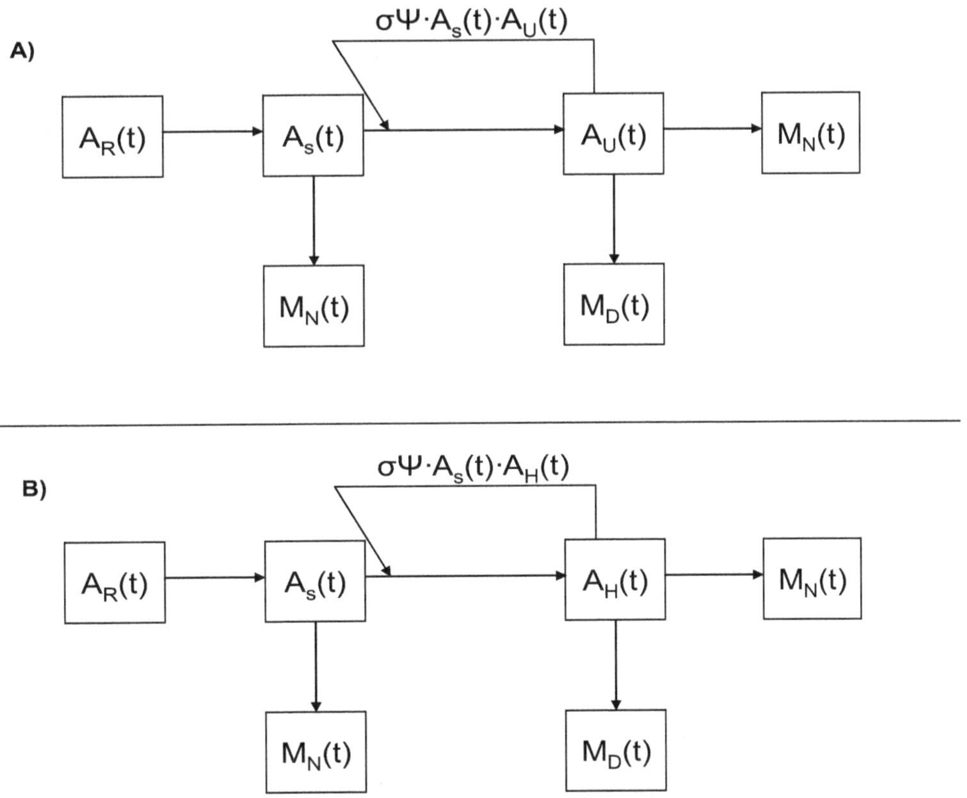

Figure 7. Diagrammatic representations of the transmission dynamics of the ranavirus when only the A_S or A_H causing isolate of the ranavirus is present. (**A**) When only the ulcerative form of the ranavirus is present within the population. (**B**) When only the hemorrhagic form of the disease is present in the population. All of the variables present are the same as described above and all have a time component associated with them. Where A_s is the number of susceptible individuals, Ψ is the contact rate, σ is the likelihood of transmission, M_N is the natural mortality rate, and M_D is the mortality rate associated with ranavirosis.

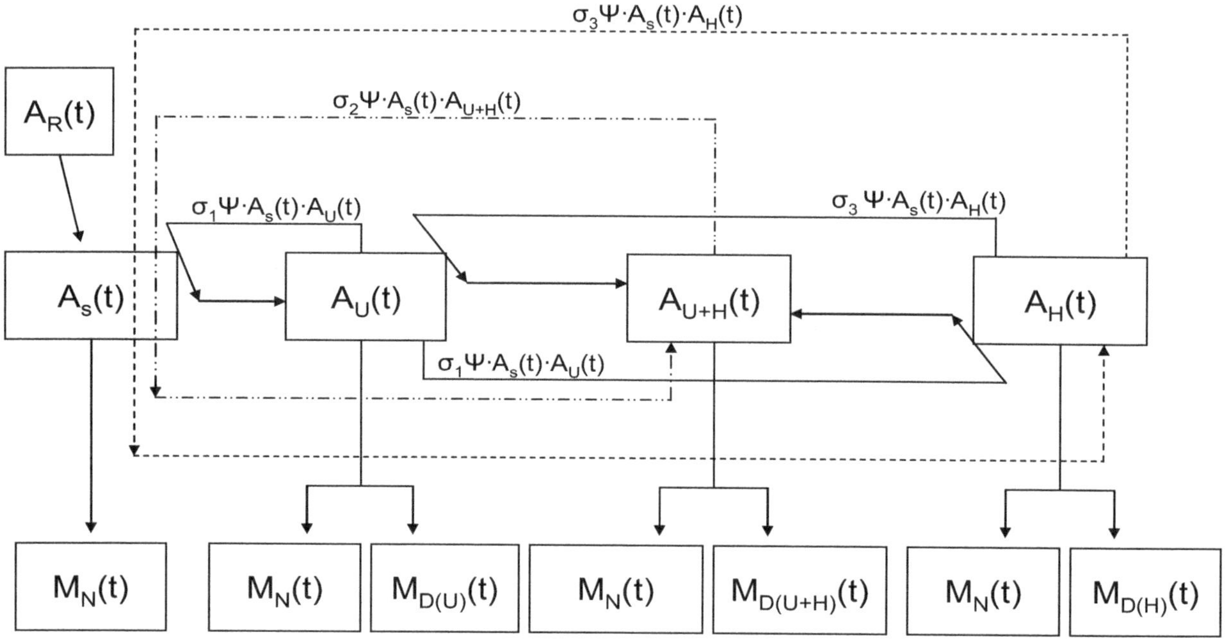

Figure 8. Illustration of the complex transmission dynamics of the ranavirus, when both of the observed disease syndromes are present in the population. Dashed lines are used to make the disease syndrome-specific vectors of the transmission easier to follow. The box sizes are not representative of the number of individuals in each category. The order of the boxes does not indicate when the given disease syndrome was introduced. All parameters have time components associated with them. A_s is the number of susceptible individuals, Ψ is the contact rate, σ is the likelihood of transmission, M_N is the natural mortality rate, and M_D is the mortality rate associated with ranavirosis.

Ulcerative Syndrome

$$R_{oU} = \sigma_1\Psi\ [A_S(t) + A_H(t)]/M_N(t)+M_{D(U)}(t) + M_{D(H)}(t) \quad (13)$$

Hemorrhagic Syndrome

$$R_{oH} = \sigma_3\Psi\ [A_S(t) + A_U(t)]/M_N(t) + M_{D(H)}(t) + M_{D(U)}(t) \quad (14)$$

Under the assumptions that $A_S = 28 \rightarrow 14$, $A_H = 1 \rightarrow 14$, with the introduction of $1 \rightarrow 5$ A_U, $M_{D(U)} = M_{D(H)} = 0.775$, $\sigma_1 = 0.3$, $\sigma_3 = 0.25$, $\Psi = 0.45$, and $M_N = 0.2$, based on the R_O values, A_U and A_H should not coexist over the long term. It does not matter if A_U is introduced into a population with A_H or vice-versa (Figure 9). Both of these disease syndromes co-occur in nature [21] and in experimental infections [25].

Increasing the number of A_U individuals in a population requires a higher transmission rate at each contact, for A_H to become established (i.e., $R_o \geq 1$). However, establishment does not guarantee coexistence over the long term. While at first this might seem counter intuitive, it can be explained as follows—when there are more A_U individuals in the population there is a greater overall mortality rate because $M_{D(U)} >> M_N$. Hence, there are actually fewer individuals to infect. Even when all of the population ($n = 29$) is composed of A_U individuals, A_H can become established ($\sigma_3 \approx 0.85$; see Figure 10). Although this is a high transmission rate, it is not unlikely in the situation where animals have broken skin (see [25]; adult common frogs with broken skin are more likely to become infected), which is characteristic of the ulcerative form of the ranaviral disease. The above analysis demonstrates that the two disease syndromes could persist in the short term, in the same population, by adult-to-adult transmission, albeit under conditions which have been estimated from data that are incomplete and not necessarily biologically relevant, based on the differing R_o values.

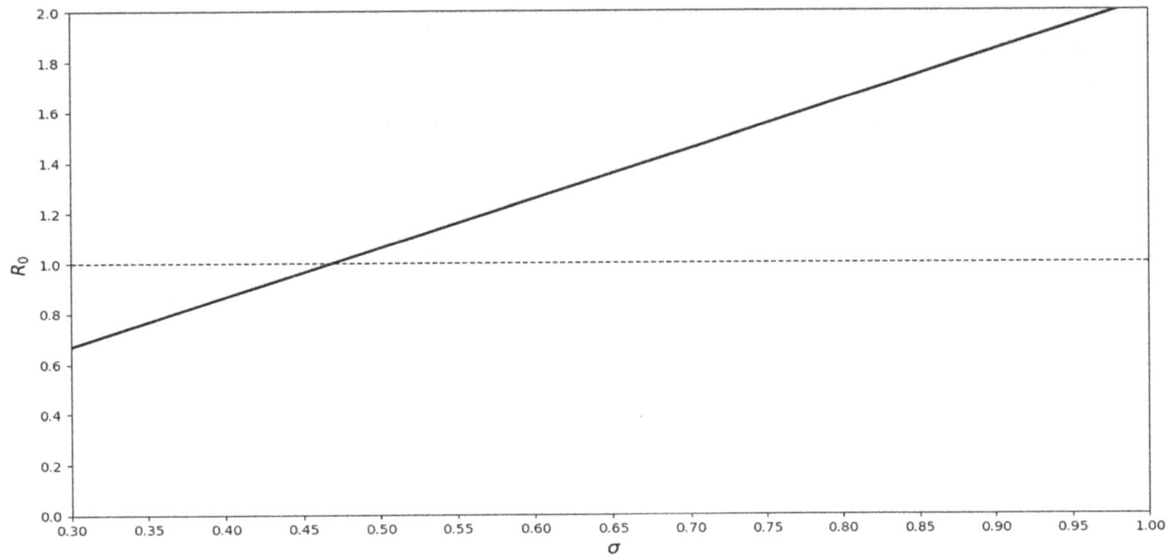

Figure 9. R_o values for the introduction of one A_H individual into a population of $A_S = 28$ and $A_U = 1$ ($M_{D(U)} = M_{D(H)} = 0.775$, $\Psi = 0.45$, $M_N = 0.2$). A_S is the number of susceptible individuals, Ψ is the contact rate, σ is the likelihood of transmission, M_N is the natural mortality rate, and M_D is the mortality rate associated with ranavirosis.

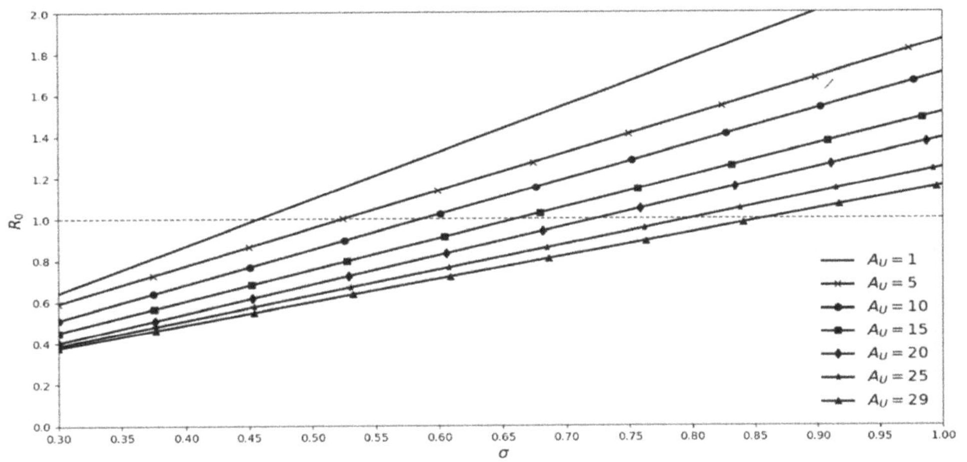

Figure 10. Ro values for the introduction of one A_H individual to populations with differing numbers of A_U, while the total population size remains constant at 30. The number associated with each line indicates the number of A_U individuals present in the population. ($M_{D(U)} = 0.775$, $\Psi = 0.45$, $M_N = 0.2$) A_s is the number of susceptible individuals, Ψ is the contact rate, σ is the likelihood of transmission, M_N is the natural mortality rate, and M_D is the mortality rate associated with ranavirosis.

3. Discussion

Transmission dynamics are key to understanding host and pathogen persistence. In amphibian ranavirus systems, transmission can occur through direct contact, from scavenging, from virus particles persisting in the environment, and even between vertebrate classes, as seen in laboratory experiments. However, there is little understanding of the transmission routes that are most important in natural communities. In North America, ranavirus transmission appears to be primarily through direct contact with minimal transmission from water and scavenging, in most circumstances [16]. Our model is consistent with these results, demonstrating that the ranavirus(es) present in the UK common frog populations might persist in the short-term through horizontal, adult-to-adult transmissions alone.

When the declines of 8.1% per annum—observed by Teacher et al. [9] in the common frog populations where ranavirosis have emerged—are factored into the model, it took approximately

65 years for all adults in the population to become infected. This model did not take into account immigration, which has been noted by Teacher et al. [29] or the change in population structure that is caused by the emergence of ranavirosis, as described by Campbell et al. [30]. These additional factors, along with the effects of climate change [28], might change the outcomes of the models, and perhaps even require adjustments to the model parameters.

Strict adult-to-adult transmission of ranaviruses appears to be relatively rare. It is likely to have occurred in a mass mortality event of over 1000 adult and metamorphic water frogs (*Pelophylax* spp.) in the Netherlands, caused by a strain of *Common Midwife Toad Virus* (CMTV) [31]. In this case, there were also common newts (*Lissotriton vulgaris*) involved in the mortality event [31], but they made up only approximately 1% of the animals killed, and ranavirus transmission to the newts might be best attributed to pathogen spillover. In North America, reports of adult anurans infected with ranavirus are rare (e.g., only one adult wood frog [18]); morbidity and mortality events tend to occur in the tadpole stage. However, in the UK, adult-only mortality and morbidity events are typical, therefore, adult-to-adult transmission must play a major role in the transmission, which is consistent with our model results. This is an important finding since, there often is a lack of tadpoles in the ponds at the time of the adult mortality events [20]. However, other species might act as a reservoir of hosts that can infect common frogs [32], such as common toads, common newts (*Lissotriton vulgaris*) and the introduced common midwife toad (*Alytes obstetricans*). Future models and studies should explore interspecies transmission and subsequent population dynamics, especially since the presence of other species, namely common toads (*Bufo bufo*), can reduce disease risk in common frogs [5,27].

Our models showed that both disease syndromes can co-exist in the short term, despite competition between the ranavirus(es) associated with the different disease syndromes. This is not surprising if there are multiple strains of ranavirus present in the population. Exposure to different ranaviruses has been shown to result in enhanced viral infectivity in larval amphibians in the USA [33], and it is possible that this same pattern is occurring in the adult UK common frogs. The susceptibility of the host might depend on which ranavirus strain (ulcerative or hemorrhagic) is first introduced into the population. A similar effect has been previously observed in tadpoles exposed to *Frog virus 3* (FV3) and *Ambystoma tigrinum virus* [33]. This might be the case in the UK, because there are at least two different types of ranaviruses present [27]. CMTV-like and FV3-like ranaviruses have been previously identified in the UK populations of common frogs [27]; however, no association with the distinct disease syndromes were found in this study. In a previous study, molecular differences between two of the isolates (RUK 11 and RUK 13) were found, both of which were responsible for different disease syndromes. Duffus et al. [34] found that while the major capsid protein sequences for these two isolates were similar to FV3, the partial sequence of open reading frame 57r (an eIF-2α homologue) was similar to that found in an isolate of Chinese giant salamander virus (a common midwife toad-like virus). These differences could be indicative of larger scale molecular differences between the RUK isolates that result in the different ranaviral syndromes seen in common frogs in the UK.

Our models were greatly limited by a lack of robust parameter estimates. The contact rates were unknown for common frogs and the transmission coefficients were based on experiments with small sample sizes and unrealistic viral titres. Better parameter estimation (e.g., contact rates, individual susceptibility to ranavirosis, and disease-induced mortality in adults) would be key to improving the predictive values of all presented models.

Author Contributions: A.L.J.D. conceived the models. T.W.J.G. and R.A.N. provided valuable guidance to A.L.J.D. on model development. J.P.S. ran and modified the original models, as necessary. A.L.J.D., J.E.E., and T.W.J.G. wrote the manuscript. All authors edited the manuscript and approved it.

Acknowledgments: We would like to thank Robert Knell for the helpful discussions about model development and two anonymous reviewers who's comments and suggestions greatly improved the quality of the manuscript.

References

1. Gray, M.J.; Chinchar, V.G. Introduction: History and future of ranaviruses. In *Ranaviruses: Lethal Pathogens of Ectothermic Vertebrates*; Gray, M.J., Chinchar, V.G., Eds.; Springer: New York, NY, USA, 2015; pp. 1–7.
2. Duffus, A.L.J.; Waltzek, T.B.; Stöhr, A.C.; Allender, M.C.; Gotesman, M.; Whittington, R.J.; Hick, P.; Hines, M.K.; Marschang, R.E. Distribution and host range of ranviruses. In *Ranaviruses: Lethal Pathogens of Ectothermic Vertebrates*; Gray, M.J., Chinchar, V.G., Eds.; Springer: Secaucus, NJ, USA, 2015.
3. Miller, D.L.; Pessier, A.P.; Hick, P.; Whittington, R.J. Comparative pathology of ranaviruses and diagnostic techniques. In *Ranaviruses: Lethal Pathogens of Ectothermic Vertebrates*; Gray, M.J., Chinchar, V.G., Eds.; Springer: Secaucus, NJ, USA, 2015.
4. Hoverman, J.T.; Gray, M.J.; Haislip, N.A.; Miller, D.L. Phylogeny, life history, and ecology contribute to differences in amphibian susceptibility to ranaviruses. *EcoHealth* **2011**, *8*, 301–319. [CrossRef] [PubMed]
5. North, A.C.; Hodgson, D.J.; Price, S.J.; Griffiths, A.G.F. Anthropogenic and ecological drivers of amphibian disease (ranavirosis). *PLoS ONE* **2015**, *10*, e0127037. [CrossRef] [PubMed]
6. Balseiro, A.; Dalton, K.P.; del Cerro, A.; Marquez, I.; Cunningham, A.A.; Parra, F.; Prieto, J.M.; Casais, R. Pathology, isolation and molecular characterisation of a ranavirus from the common midwife toad *Alytes obstetricans* on the iberian peninsula. *Dis. Aquat. Org.* **2009**, *84*, 95–104. [CrossRef] [PubMed]
7. Green, D.E.; Converse, K.A.; Schrader, A.K. Epizootiology of sixty-four amphibian morbidity and mortality events in the USA, 1996–2001. *Ann. N. Y. Acad. Sci.* **2002**, *969*, 323–339. [CrossRef] [PubMed]
8. Muths, E.; Gallant, A.L.; Grant, E.H.C.; Battaglin, W.A.; Green, D.E.; Staiger, J.S.; Walls, S.C.; Gunzburger, M.S.; Kearney, R.F. *The Amphibian Research and Monitoring Iniative (ARMI): 5-Year Report*; US Geological Survey Scientific Investigations Report 2006-5224; USDOI, USGS, Eds.; US Geological Survey: Reston, VA, USA, 2006; p. 77.
9. Teacher, A.G.F.; Cunningham, A.A.; Garner, T.W.J. Assessing the long-term impact of *ranavirus* infection in wild common frog populations. *Anim. Conserv.* **2010**, *13*, 514–522. [CrossRef]
10. Price, S.J.; Garner, T.W.J.; Nichols, R.A.; Balloux, F.; Ayres, C.; Mora-Cabello de Alba, A.; Bosch, J. Collapse of amphibian communities due to an introduced *ranavirus*. *Curr. Biol.* **2014**, *24*, 1–6. [CrossRef] [PubMed]
11. Earl, J.E.; Gray, M.J. Introduction of ranavirus to isolated wood frog populations could cause local extinction. *EcoHealth* **2014**, *11*, 581–592. [CrossRef]
12. Earl, J.E.; Chaney, J.C.; Sutton, W.B.; Lillard, C.E.; Kouba, A.J.; Langhorne, C.; Krebs, J.; Wilkes, R.P.; Hill, R.D.; Miller, D.L.; et al. Ranavirus could facilitate local extinction of rare amphibian species. *Oecologia* **2016**, *182*, 611–623. [CrossRef]
13. Farrell, A.P.; Collins, J.P.; Greer, A.L.; Thieme, H.R. Times from infection to disease-induced death and their influence on final population sizes after epidemic outbreaks. *Bull. Math. Biol.* **2018**, *80*, 1937–1961. [CrossRef]
14. Anderson, R.M.; May, R.M. Population biology of infectious disease. *Nature* **1979**, *280*, 361–461. [CrossRef]
15. May, R.M.; Anderson, R.M. Population biology of infectious diseases: Part II. *Nature* **1979**, *280*, 455–461. [CrossRef] [PubMed]
16. Brunner, J.L.; Yarber, C.M. Evaluating the importance of environmental persistence for *Ranavirus* transmission and epidemiology. *Adv. Virus Res.* **2018**, *101*, 129–148. [PubMed]
17. Brunner, J.L.; Schock, D.M.; Davidson, E.W.; Collins, J.P. Intraspecific reservoirs: Complex life history and the persistence of a lethal ranavirus. *Ecology* **2004**, *85*, 560–566. [CrossRef]
18. Duffus, A.L.J.; Pauli, B.D.; Wozney, K.; Brunetti, C.R.; Berrill, M. Frog virus 3-like infections in aquatic amphibian communities. *J. Wildl. Dis.* **2008**, *44*, 109–120. [CrossRef] [PubMed]
19. Duffus, A.L.J.; Nichols, R.A.; Garner, T.W.J. Experimental evidence in support of single host maintenance of a multihost pathogen. *Ecosphere* **2014**, *5*, 1–11. [CrossRef]
20. Duffus, A.L.J.; Nichols, R.A.; Garner, T.W.J. Investigations into the life history stages of the common frog (*Rana temporaria*) affected by an amphibian ranavirus in the United Kingdom. *Herpetol. Rev.* **2013**, *44*, 260–263.
21. Cunningham, A.A.; Langton, T.E.S.; Bennett, P.M.; Lewin, J.F.; Drury, S.E.N.; Gough, R.E.; MacGregor, S.K. Pathological and microbiological findings from incidents of unusual mortality of the common frog (*Rana temporaria*). *Philos. Trans. R. Soc. Lond. B Biol. Sci.* **1996**, *351*, 1539–1557.
22. Otto, S.P.; Day, T. *A Biologist's Guide to Mathematical Modeling in Ecology and Evolution*; Princeton University Press: Princeton, NJ, USA, 2007; p. 732.
23. Miaud, C.; Guyetant, R.; Elmberg, J. Variations in life-history traits in the common frog *Rana temporaria* (Amphibia: Anura): A literature review and new data from the French Alps. *J. Zool.* **1999**, *249*, 61–73. [CrossRef]

24. Gibbons, M.M.; McCarthy, T.K. Growth, maturation and survival of frogs *Rana temporaria* L. *Holarct. Ecol.* **1984**, *7*, 419–427. [CrossRef]
25. Cunningham, A.A.; Hyatt, A.D.; Russell, P.; Bennett, P.M. Emerging epidemic diseases of frogs in britain are dependant on the source of ranavirus agent and the route of exposure. *Epidemiol. Infect.* **2007**, *135*, 1200–1212. [CrossRef] [PubMed]
26. Vieites, D.R.; Nieto-Román, S.; Barluenga, M.; Palanca, A.; Vences, M.; Meyer, A. Post-mating clutch piracy in an amphibian. *Nature* **2004**, *431*, 305–308. [CrossRef] [PubMed]
27. Price, S.J.; Wadia, A.; Wright, O.N.; Leung, W.T.; Cunningham, A.A.; Lawson, B. Screening of a long-term sample set reveals two ranavirus lineages in british herpetofauna. *PLoS ONE* **2017**, *12*, e0184768. [CrossRef] [PubMed]
28. Price, S.J.; Leung, W.T.M.; Owen, C.; Puschendorf, R.; Sergeant, C.; Cunningham, A.A.; Balloux, F.; Garner, T.W.J.; Nichols, R.A. Effects of historic and projected climate change on the range and impacts of an emerging wildlife disease. *Glob. Chang. Biol.* **2019**, in press. [CrossRef] [PubMed]
29. Teacher, A.G.F.; Garner, T.W.J.; Nichols, R.A. Evidence for directional selection at a novel major histocompatability class i marker in wild common frogs (*Rana temporaria*) exposed to a viral pathogen (*Ranavirus*). *PLoS ONE* **2009**, *4*, e4616. [CrossRef]
30. Campbell, L.J.; Garner, T.W.J.; Tessa, G.; Scheele, B.C.; Griffiths, A.G.F.; Harrison, X.A. An emerging viral pathogen truncates population age structure in a european amphibian and may reduce population viability. *PeerJ* **2018**, *6*, e5949. [CrossRef] [PubMed]
31. Kik, M.; Martel, A.; Spitzen-van der Sluijs, A.; Pasmans, F.; Wohlsein, P.; Gröne, A.; Rijks, J.M. Ranavirus-associated mass mortality in wild amphibians, the netherlands, 2010: A first report. *Vet. J.* **2011**, *190*, 284–286. [CrossRef]
32. Duffus, A.L.J.; Nichols, R.A.; Garner, T.W.J. Detection of a frog virus 3-like ranavirus in native and introduced amphibians in the United Kingdom in 2007 and 2008. *Herpetol. Rev.* **2014**, *45*, 608–610.
33. Mihaljevic, J.R.; Hoverman, J.T.; Johnson, P.T.J. Co-exposure to multiple ranavirus types enhances viral infectivity and replication in a larval amphibian system. *Dis. Aquat. Org.* **2018**, *132*, 23–35. [CrossRef]
34. Duffus, A.L.J.; Garner, T.W.J.; Davis, A.R.; Dean, A.W.; Nichols, R.A. Phylogentic analysis of 24 ranavirus isolates from English amphibians using 2 partial loci. *J. Emerg. Dis. Virol.* **2017**, *3*. [CrossRef]

15

Evaluating the Within-Host Dynamics of *Ranavirus* Infection with Mechanistic Disease Models and Experimental Data

Joseph R. Mihaljevic [1,*], **Amy L. Greer** [2] **and Jesse L. Brunner** [3]

1 School of Informatics, Computing and Cyber Systems, Northern Arizona University, Flagstaff, AZ 86011, USA
2 Department of Population Medicine, University of Guelph, Guelph, ON N1G 2W1, Canada; agreer@uoguelph.ca
3 School of Biological Sciences, Washington State University, Pullman, WA 99163, USA; jesse.brunner@wsu.edu
* Correspondence: joseph.mihaljevic@nau.edu

Abstract: Mechanistic models are critical for our understanding of both within-host dynamics (i.e., pathogen replication and immune system processes) and among-host dynamics (i.e., transmission). Within-host models, however, are not often fit to experimental data, which can serve as a robust method of hypothesis testing and hypothesis generation. In this study, we use mechanistic models and empirical, time-series data of viral titer to better understand the replication of ranaviruses within their amphibian hosts and the immune dynamics that limit viral replication. Specifically, we fit a suite of potential models to our data, where each model represents a hypothesis about the interactions between viral replication and immune defense. Through formal model comparison, we find a parsimonious model that captures key features of our time-series data: The viral titer rises and falls through time, likely due to an immune system response, and that the initial viral dosage affects both the peak viral titer and the timing of the peak. Importantly, our model makes several predictions, including the existence of long-term viral infections, which can be validated in future studies.

Keywords: amphibian; *Ranavirus*; frog virus 3; mathematical models; Bayesian inference

---

1. Introduction

Understanding the specific interactions between a host's immune system and an infectious agent can be critical for understanding population-level patterns of infection prevalence and the dynamics of prevalence through time [1,2]. Mechanistic models of within-host dynamics use mathematical expressions to represent the processes of pathogen population growth and immune defenses that fight against pathogen growth. These types of models have revealed key insights, especially regarding the effects of various immune strategies on the evolution of pathogen virulence [3,4]. Within-host models also help us to understand the effects of chronic and acute infections on disease progression, pathogen evolution, and the links between within-host and among-host processes [5]. For example, Kennedy et al. [6] showed that models of within-host viral replication and immune processes accurately recapitulate the effects of viral exposure dose on the probability of death in an insect host. Here, we develop within-host models to understand the replication of ranaviruses in their amphibian hosts, and we compare our models' expectations to laboratory data.

Viruses in the genus *Ranavirus* can be impressively lethal, causing rapid host mortality and mass mortality events, especially in larval amphibians. However, the outcome of infection varies tremendously across host taxa [7,8] and life stage [9–11], as well as dose of exposure and numerous environmental factors (reviewed in Reference [12]). Much of this variation is presumably due to variation in the host-virus interaction, where viral replication is countered by immune responses. Indeed, some qualitatively distinct outcomes appear to have their basis in different immune responses. Adult *Xenopus laevis*, for instance, are refractory to ranavirus infections, while larvae often succumb, and this is due to their distinct immune responses [13,14]. The much more effective adult response to infection involves rapid, substantial CD8+ cellular responses, whereas tadpoles' responses are dramatically less effective and slower. Larval *X. laevis* produce type I and type III interferon responses, which are effective at reducing viral replication, but are eventually overwhelmed by the virus [15,16]. However, while there has been great progress in elucidating the components of the immune system that interact with and help control ranavirus, many questions remain.

In particular, the dynamics of viral population growth are largely unknown outside of cell culture experiments. What are the dynamics of the virus population within the host? Does the host's immune system effectively control the virus population, outside of the *Xenopus* model system? Does the immune response lead to viral clearance or can persistent infections remain? Does this outcome depend on the initial exposure dose? We propose that mathematical models of the within-host dynamics of ranaviruses, which provide a quantitative understanding of the host-virus interaction, will help answer these questions and will provide a means for integrating the numerous factors that appear to be important for the outcome of ranavirus infections. There are several additional advantages of using a mechanistic, model-based approach. First, parameterized disease models yield quantitative insights that can be validated by collecting new data. Second, we can simulate parameterized models under novel scenarios to predict how the system might behave. Third, we can use models to identify which novel sources of data would best inform our understanding under different scenarios. Thus, our approach offers a rigorous method to generate new hypotheses and guide future studies to understand the interactions between ranaviruses and host immune defenses.

To this end we set out to develop several versions of a mechanistic, within-host model that embodied hypotheses about the replication of frog virus 3 (FV3) and the host's immune response to the infection. We then fit these models to a time series of FV3 titers within bullfrog (*Lithobates catesbeianus*) tadpoles that were experimentally infected with FV3. We used a model-comparison approach to determine the most parsimonious model(s) that could describe the observed viral titer through time and the effects of viral exposure dose. The salient features of these dynamics are: (1) a clear rise and then slower decline in viral titers over time, suggesting viral clearance by the host's immune response and motivating our model-fitting exercise, (2) an earlier and higher peak in viral titers with higher exposure dose, and (3) low-level infections remaining detectable for seven weeks. We then use these models to generate testable hypotheses and guide future studies to understand the interactions between ranaviruses and host immune defenses.

2. Materials and Methods

Our goal is to test a set of hypotheses about how ranaviral populations replicate within their larval amphibian hosts and how the virus interacts with the larval host's immune system. Our strategy is to develop mechanistic within-host growth models with various assumptions that correspond to our hypotheses. We then attempt to fit these models to serially-collected data on within-host titers, and we use a model-comparison approach to determine the most parsimonious model structure, and therefore the most parsimonious support for our hypotheses.

2.1. Experimental Assessment of Within-Host Ranaviral Dynamics

Details of the experiment can be found in Brunner et al. [17], but the essential details are that tadpoles (Gosner [18] stages 25–28) were exposed in 400 mL water baths for 24 h to one of three doses of the ranavirus FV3: a high dose (10^5 plaque-forming units [pfu] mL^{-1}; $n = 150$), medium-dose (10^3 pfu mL^{-1}; $n = 150$), or a low, but unknown dose ($n = 90$ in "mock" exposure to inadvertently contaminated cell culture media). The tadpoles were assigned to be euthanized and sampled on days

2, 4, 6, 8, 14, 21, 28, 35, 42, and 49 post exposure (n = 15 high-dose, n = 15 low-dose, and n = 9 "mock" tadpoles per time point), although some died and were sampled before their assigned date. Viral titers in liver and kidney samples were measured with a quantitative real-time PCR assay [19].

We do not include individuals that died before their pre-determined sampling date, because we focus on how the immune system is functioning against the virus, and we assume that individuals that died due to virus had immune systems that were overwhelmed. However, in the supplement (Table S1, Figures S1–S4), we demonstrate that our results are robust, and that including the dead individuals does not change the rank-order of our models or our general conclusions about the system.

2.2. Mechanistic Models of Within-Host Dynamics

We fit models to these time-series data that embodied two assumptions about viral replication and two about the host's immune response. First, we fit models with exponential viral growth ($V' = \phi V$) or with logistic growth [$V' = \phi V\left(1 - \frac{V}{K}\right)$]. In these differential equations, V' represents $\frac{dV(t)}{dt}$, where V is the density of virus within the host, ϕ is the per capita replication rate of the virus, and K is the viral carrying capacity. In some models, K can be viewed as the viral titer above which the virus kills the host [5]; however, because we are modeling mean-field dynamics, we do not explicitly account for virus-induced host mortality.

We then layer on possible immune system dynamics. From previous experimental work, we know that larval amphibians (at least *Xenopus*) show interferon type I and type III responses to FV3, although the latter appears more effective [15,16]. While we lack detailed knowledge of the immune response to FV3, especially in non-model amphibian species, we begin with the reasonable assumption that the production of the immune components, Z, responds to viral infection following the Michaelis-Menten form of enzymatic activity [20,21]:

$$Z' = \psi Z\left(\frac{V}{V+\gamma}\right) \tag{1}$$

In this case, the engagement of the immune component ramps up to a maximum, ψZ, as the virus population increases. The rate of immune component production is mediated by γ, which is the half-saturation constant. In this formulation the density of immune components never returns to pre-infection levels. We therefore also considered a second formulation

$$Z' = (N_Z - \delta Z) + \psi Z\left(\frac{V}{V+\gamma}\right) \tag{2}$$

Here, N_Z is the constant rate of production of the immune components, and δ is the per-capita background loss of the immune components. Moreover, we define $N_Z = \delta Z(0)$, such that, when no virus is present, the immune system stabilizes to a homeostatic level of immune component density equal to $Z(0)$. In other words, we assume a balance of the immune components' production and loss at equilibrium, when no virus is present. This means that there is a constant background level of immune components ($Z(0)$) in a host, and when a virus infects the host, the production of immune components ramps up until the virus is cleared.

Finally, we assume a mass-action attack rate of the immune system against the virus, which is a Type I Hollings' functional response, such that our most complex model becomes:

$$V' = \phi V\left(1 - \frac{V}{K}\right) - \beta VZ \tag{3.1}$$

$$Z' = (N_Z - \delta Z) + \psi Z\left(\frac{V}{V+\gamma}\right). \tag{3.2}$$

We therefore model immune component production in response to virus infection as a fundamentally different process compared to the function of the immune component that limits viral population growth. We further note that all of our models are nested, and Table 1 shows all model formulations that we fit to our data set. We divide the models into two classes, A and B, which are differentiated by the assumptions about the immune system's production dynamics. Class A models are mostly distinguished by the fact that, at equilibrium, if the immune system response is strong enough, the virus is driven extinct (i.e., full viral clearance). For Class B models, the virus is not driven extinct, such that the virus persists at low levels within the hosts (Appendix A).

On a less technical note, in some ways our models are not specific about how the immune system of the larval amphibian functions. For example, our model could equally represent a population of immune cells or a population of molecules generated by the immune system (e.g., interferon molecules) that act to control viral replication more or less indirectly. In other words, although we model a mass-action attack rate of the immune system against the virus, we do not necessarily mean that the virus and immune component must come into direct contact. This immune action could be indirect, for instance, by controlling apoptosis of virus-infected cells.

2.3. Fitting the Models to the Experimental Data

We employ a Bayesian approach to inference to fit the suite of dynamical models to the time-series of within-host viral titers. Then we use Bayesian information criteria to compare the within-sample predictive accuracy of our models, while penalizing model complexity. In this way, we seek to understand the most parsimonious model structure that captures the main features of the time series. All model-fitting code is available on our open-source Bitbucket repository (https://bitbucket.org/jrmihalj/ranavirus-within-host-dynamics/src/master/).

We use the open-source statistical programming language, Stan [22], to fit our differential equation models to the time-series data. Stan employs a Hamiltonian Monte Carlo (HMC) algorithm to sample from the model's posterior. For each model, we sampled the posterior using three Markov chains, with a 2000 iteration warm-up, followed by 2000 iterations, for a total of 2000 samples recorded from each chain. We conducted various graphical and quantitative diagnoses of the chain behavior, including inspections for temporal auto-correlation, some of which are shown in the supplement (Figures S5 and S6). We assessed convergence with the Gelman-Rubin statistic, $\hat{\boldsymbol{R}}$ [23]. All models converged after 4000 iterations.

Model dynamics are sensitive to parameter magnitudes. Therefore, in constructing our prior probability distributions for the parameters, we often restricted parameters to vague, but realistic ranges (see open-source model statements). All parameters were restricted to positive values, and for parameters that could take very large values (e.g., $\boldsymbol{K}$) we set the parameter on a natural-logarithmic scale. To allow for potential correlations in our model parameters, we assumed that the main model parameters (i.e., $\boldsymbol{\phi}, \boldsymbol{\beta}, \boldsymbol{K}, \boldsymbol{\delta}, \boldsymbol{\psi}$) had prior probabilities that were zero-truncated, multivariate normal, with estimated means and an estimated covariance matrix (based on underlying correlations). We used Cholesky Factorization to estimate the correlation structure, assuming vague priors for the correlations. Specifically, we used an LKJ prior on the correlations with a shape parameter equal to 2, which creates a correlation matrix closer to zero correlation (i.e., a conservative assumption) [22]. Additional details on prior structures and model specification can be seen in our open-source code repository.

Besides estimating the model parameters, we also had to estimate the initial conditions of the system: The initial viral titer, $\boldsymbol{V}(0)$, and the initial immune component density, $\boldsymbol{Z}(0)$. As we have three viral dosage treatments, we estimated three independent $V(0)$. We used a hierarchical prior structure, such that each treatment's initial titer was drawn from a distribution with an estimated mean and variance (see model code). We further assumed that the initial immune component density, $Z(0)$, did not vary among treatments. We therefore assumed that all larval amphibians start with the same average immune system component density that then responds to the invasion of the virus. As the magnitude of $V(0)$ and $Z(0)$ could be large, we again set these parameters on a natural-logarithmic scale.

The viral titer values spanned approximately six orders of magnitude. To improve numerical integration performance, by reducing the error in the integration algorithms due to such large viral titer values, we fit the differential equation models using natural log-transformed titer. Assuming that $v = log(V)$ and $z = log(Z)$, we can reformulate Equation (3) as follows:

$$v' = \phi\left(1 - \frac{e^v}{K}\right) - \beta e^z \quad (4.1)$$

$$z' = \left(\frac{N_Z}{e^z} - \delta\right) + \psi\left(\frac{e^v}{e^v + \gamma}\right). \quad (4.2)$$

Thus, the scaling of the parameters does not change, just the scaling of the state variables. This also allowed us to more reliably assume a Gaussian likelihood for log-transformed viral titer. We further allowed the residual variance to be different for each dosage treatment.

We also note that we only included non-zero titers in our data set for the model-fitting routine. Our reasoning is that a zero titer could represent an individual that never became infected, even upon exposure, or an individual that cleared infection later on. Because we could not distinguish between these two scenarios, we excluded these values. We did re-fit our models to the full data set, including zeros, to understand the effects on our inference. Importantly, this did not change the rank-order of our models using model comparison, but it generally made the models fit more poorly to the data, bringing down the average predicted titer and altering the marginal posterior estimates of the model parameters. We therefore do not consider these model fits further.

2.4. Model Comparisons

Our goal for model comparison was to determine the most parsimonious model(s) with the best goodness-of-fit. In practice, this means assessing the within-sample predictive performance of each model and correcting for model complexity. Given that we used a Bayesian approach to fitting our models to the data, we chose to use a Bayesian information theoretic approach to model comparison. Therefore, we calculated the leave-one-out information criterion (LOO-IC) for each model. Analogous to the Akaike information criterion (AIC), we can compare models based on the LOO-IC value, because they follow the deviance scale [24]. Thus, following the rule-of-thumb that ΔLOO-IC > 3 designates models with significantly different model performance, and where lower LOO-IC values indicate better-fitting, more parsimonious models.

Table 1. Model structures and model comparisons. The bolded model (B2) is the most parsimonious based on LOO-IC selection and the lower number of parameters compared to B3.

Class	ID	Structure	Notes	Penalty (pLOO)	LOO-IC	ΔLOO-IC
A	A1	$V' = \phi V - \beta VZ$ $Z' = \psi Z \frac{V}{(V+\gamma)}$	Drives virus extinct. Z goes to equilibrium $Z_{(\infty)}$, which is above $Z_{(0)}$.	8.5	782.9	15.5
	A2	$V' = \phi V\left(1 - \frac{V}{K}\right) - \beta VZ$ $Z' = \psi Z \frac{V}{(V+\gamma)}$	Conditions under which virus goes to carrying capacity. Or virus goes extinct.	7.1	811.7	44.3
B	B1	$V' = \phi V - \beta VZ$ $Z' = (N_Z - \delta Z) + \psi ZV$	Damped oscillations to a stable point equilibrium, where virus is persistent in host. The model fit shows several oscillations before equilibrium.	12.7	826.6	59.2
	B2	$V' = \phi V - \beta VZ$ $Z' = (N_Z - \delta Z) + \psi Z \frac{V}{(V+\gamma)}$	Spike in viral load, then decline to stable point equilibrium, where virus is persistent in host.	**10.1**	**768.6**	**1.2**
	B3	$V' = \phi V\left(1 - \frac{V}{K}\right) - \beta VZ$ $Z' = (N_Z - \delta Z) + \psi Z \frac{V}{(V+\gamma)}$	Over-fitting. Extra parameter (carrying capacity, K) unnecessary.	9.1	767.4	0

3. Results

A full qualitative and statistical analysis of the experimental data can be found in [17]. For our purposes, it is important to note the clear rise and fall of viral titers over time in the medium and high dosage treatments, which suggested a possible immune response that was leading towards viral clearance, and which motivated our modeling-fitting and model comparison routine.

Model B2, our best-fitting and most parsimonious model, fits to the experimental data very well (Figure 1). This model includes exponential growth of the virus population, and an assumption of homeostatic levels of immune components within the host, and the model's parameter estimates are shown in Table 2. Importantly, model B2 captures key features of the data that the other models failed to capture. First, the model correctly predicts that higher initial viral doses lead to higher peaks of within-host titer. Second, the model also shows that, with lower initial doses, the timing of the peak titer is delayed (Figure S7). In other words, it takes longer for the virus population to build up within the host. This is due, in part, to the exponential growth of the virus population and, in part, to the action of the immune system that inhibits viral replication.

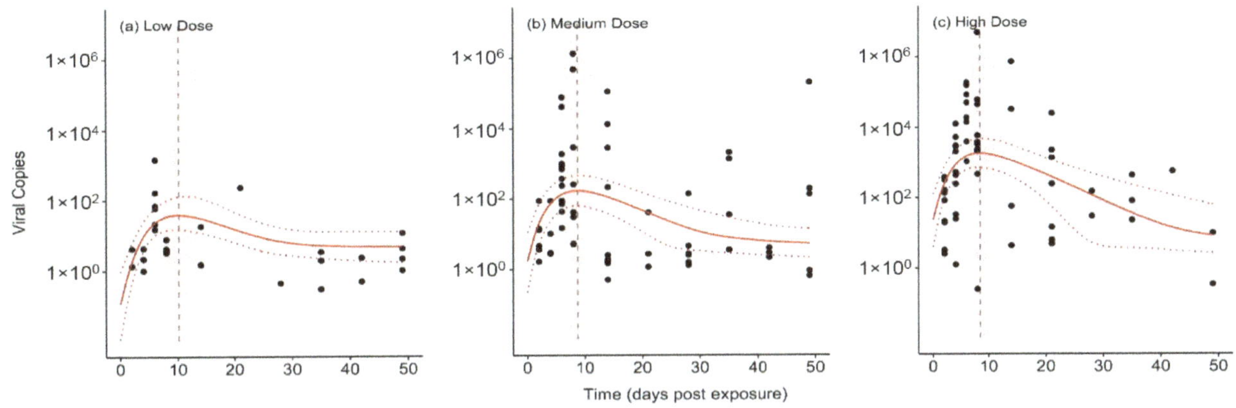

Figure 1. The fit of model B2 to the experimental data. Circles are data points representing the viral DNA copies from individual bullfrog tadpoles that were sampled on a given day. The median model fit (solid red line) and 95% Bayesian credible interval (CI) of the fit (dashed red lines) are shown. Additionally, the median (dashed vertical line) and 95% CI (light red polygon) are shown for the time of the maximum viral titer predicted by the model.

Our model-fitting and model comparison approach suggests strong differences between the appropriateness of our different model structures. Models B2 and B3 had indistinguishable model fits (Table 1, Figure S8), but model B3 included an additional parameter, assuming logistic growth of the virus population (i.e., a carrying capacity within the host). This is a classic case of over-fitting, however, because the within-sample prediction for the model with more parameters was indistinguishable from the simpler model. Therefore, we conclude that model B2 is more parsimonious.

We had trouble fitting model A1 to the data, because it fits so poorly to the low-dose treatment. Therefore, the model ends up predicting a more or less linear decline in virus load across time, with high variance (Figure S9). This linear effect allowed the model to fit very well to the data from the low dose treatment, increasing its overall likelihood and improving its LOO-IC value. However, this parameterized version of the model fails to capture basic features of the data set. In contrast, model A2 fits to the data reasonably well (Figure S10). However, this model predicts the same peak titer across dosage treatments, leading to a very poor fit to the data from the low-dose treatment.

Table 2. Parameter estimates (median and 95% credible intervals) from the most parsimonious model, B2.

Parameter	Description	Units	Estimate
$V(0)$ low			0.12 (0.01–0.89)
$V(0)$ medium	Initial viral densities (per dosage)	Viral DNA copy (VC)	1.47 (0.24–11.55)
$V(0)$ high			24.10 (5.31–146.85)
$Z(0)$	Initial immune component denisty	Immune component (IC)	0.35 (0.04–4.43)
ϕ	Viral replication rate	day^{-1}	2.39 (1.07–4.63)
β	Mass-action attack rate	(IC)$^{-1}$ day^{-1}	1.75 (0.15–6.28)
N_Z	Rate of production that ensures return of immune system to homeostasis	(IC) day^{-1}	$N_Z = \delta Z(0)$
δ	Rate of decline that ensures return of immune system to homeostasis	day^{-1}	1.29 (0.41–3.92)
ψ	Immune component growth rate in response to virus	day^{-1}	0.99 (0.19–3.56)
γ	Half saturation constant	VC	0.13 (0.02–1.02)

Model B1 also has a decent fit to the data (Figure S11). However, with this model's estimated parameters, the dynamics show damped oscillations towards the endemic equilibrium. This pattern does not seem biologically realistic, as the data do not show any clear oscillations.

We do note that we had difficulty identifying the attack rate parameter β using model B2. This is likely because of the strong effects of viral replication rate ϕ and the growth rate of the immune components in response to viral density ψ on the overall model dynamics. Future experiments should be used to better estimate the attack rate parameter.

4. Discussion

Our study represents an initial step towards making model-based, quantitative predictions about ranavirus growth and immune system functioning within larval amphibian hosts, which complements the history of empirical work in these research areas. The experimental data with bullfrog (*Lithobates catesbeianus*) tadpoles suggests a non-monotonic relationship between within-host viral titer and time post-exposure. That is, an immune response appears to be reducing the growth of the virus over time. This is broadly consistent with the immunological studies with *Xenopus laevis* tadpoles, showing a robust interferon response [15,16]. Through model comparisons, we identified a model of exponential viral growth and a mass-action immune system response that captures key features of our data and generally fits our data quite well. Importantly, this model-based approach allows us to make predictions and to develop testable hypotheses for follow-up experiments. However, we did not measure immune responses directly in our experiment, and therefore cannot conclude with certainty that the patterns we observe in viral titer are driven by immune system processes. To clarify these mechanisms, below we suggest specific ways to collect and analyze data in future studies.

The best-fitting model predicts that individuals that survive infection should exhibit a viral titer pattern that rises and falls over time. This prediction contrasts with the empirical work that uses *Xenopus*, because larval *Xenopus* almost always die from frog virus 3 (FV3) exposure. Even a pre-treatment of type III interferon, which should boost the immune response, only delays mortality in the *Xenopus* system [16]. Unfortunately, we do not know of any studies that have tested whether similar host immune responses exist in our study organism, the American bullfrog. However, it is well known that the American bullfrog is uncommon in its high resistance and tolerance to multiple diseases, including FV3 [7,25] and the amphibian chytrid fungus (*Batrachochytrium dendrobatidis*) [26,27]. Therefore, it is not surprising that we could infer a more robust immune response from this species compared to *X. laevis*. Still, our current data cannot validate that, in surviving individuals, virus loads rise and then fall, because we do not have individual host-level time series of viral load. Future studies could use various sources of eDNA (e.g., swabbing, skin scrapes, or water filtration), which are correlated with internal titers [17], to validate whether there is a rise and fall of viral load in individuals.

Alternatively, it may be fruitful to experiment with larger-bodied animals that would permit sampling blood repeatedly through time.

Our model also predicts that, if a host does not die from the initial exponential growth of the virus, the immune system will decrease the virus to an endemic equilibrium (Appendix A). In other words, we should see persistent infections with very low levels of virus in some infected bullfrog tadpoles. It is worth noting that this model prediction is not relevant to quiescent virus persisting in particular tissues, such as that found in peritoneal leukocytes in adult *Xenopus* [28]. Rather, the persistence in our model is due to a dynamic balance between viral replication and host immune responses. In any case, our model predicts low-level infections that would be detectable beyond 60 days from initial exposure time and should be actively replicating. That said, there are also several reasons that an endemic equilibrium may not hold. First, demographic stochasticity in the virus population could cause low-level infections to fade out [6,29]. Second, variability in the immune response over time (e.g., waning immunity) could cause fluctuations in viral titer. Third, the costs of continued immune responses could lead to host death even with low levels of infection [30].

Several of the tested model structures predict full viral clearance at equilibrium, but none of these models fit as well to the observed dynamics in bullfrog tadpoles. Future studies that collect time-series data and that have some long-term exposures (e.g., greater than 50 days) will improve our ability to distinguish between viral clearance and viral persistence. Furthermore, because we did not have simultaneous empirical data from the larval immune system, our model fitting algorithm was agnostic to the model's prediction of immune system dynamics. If future studies simultaneously capture time-series data on viral titer and measures of larval immune system, we would have much more power to distinguish between different models of within-host dynamics.

There are key features of the data set that our deterministic modeling scheme does not fully capture, and these are areas for future model development. First, the main process that our model ignores is host mortality due to viral growth. For instance, our current model predicts that immunocompromised individuals would show exponential viral growth, and in reality, this likely lead to death. Future versions of our model could include stochasticity, following Kennedy and Dwyer [6], such that demographic stochasticity in the viral population could lead to the virus exceeding a threshold growth rate or titer that causes host death. Then, we could analyze the probability of this occurring over many realizations of the stochastic model. This would allow us to explicitly evaluate how initial exposure dosage affects the probability of host mortality, from a within-host, mechanistic framework.

Viral titers within a host can be important determinants of the infection dynamics among hosts [3], affecting both the duration of infections and their propensity to be transmitted. Many dose-response experiments, for instance, show that hosts exposed to higher doses of ranaviruses, and so presumably harbor larger viral populations, are more likely to die and die more rapidly [31–33]. More intensely infected individuals may also shed more virus (e.g., Reference [17]) and thus be more infectious. Similar effects of the intensity of infection are common in other host-virus systems [34,35], as well as for other amphibian pathogens (e.g., *B. dendrobatidis*; [36]). Pathogen titers or intensity of infection can be a natural way in which to link within and between-host dynamics [3,37,38]. Within-host models of host-virus interactions may therefore be a fruitful way of integrating the myriad effects that are known to influence the outcome of ranavirus infections.

Author Contributions: Conceptualization, J.R.M., A.L.G., J.L.B.; methodology, J.R.M., A.L.G., J.L.B.; software, J.R.M.; validation, J.R.M.; formal analysis, J.R.M.; investigation, J.R.M., A.L.G., J.L.B.; data curation, J.L.B.; writing—original draft preparation, J.R.M., J.L.B.; writing—review and editing, J.R.M., A.L.G., J.L.B.; visualization, J.R.M.; supervision, J.R.M., A.L.G., J.L.B.; project administration, J.R.M., A.L.G., J.L.B.

Appendix A

We can express the coexistence equilibrium of model B2 analytically, which defines the amount of virus and host immune components at the stable equilibrium with a persistent infection. Equilibrium values of the state variables, defined by the $*$, are as follows:

$$Z^* = \frac{\phi}{\beta} \quad \text{(A1.1)}$$

$$V^* = \frac{\frac{\gamma}{\psi}\left(\delta - \frac{N_Z}{Z^*}\right)}{\left(1 - \frac{\delta}{\psi} + \frac{N_Z}{\psi Z^*}\right)} \quad \text{(A1.2)}$$

References

1. Hawley, D.M.; Altizer, S.M. Disease ecology meets ecological immunology: Understanding the links between organismal immunity and infection dynamics in natural populations. *Funct. Ecol.* **2011**, *25*, 48–60. [CrossRef]
2. Ezenwa, V.O.; Jolles, A.E. From host immunity to pathogen invasion: the effects of helminth coinfection on the dynamics of microparasites. *Integr. Comp. Biol.* **2011**, *51*, 540–551. [CrossRef]
3. Mideo, N.; Alizon, S.; Day, T. Linking within- and between-host dynamics in the evolutionary epidemiology of infectious diseases. *Trends Ecol. Evol.* **2008**, *23*, 511–517. [CrossRef]
4. Day, T.; Alizon, S.; Mideo, N. Bridging scales in the evolution of infectious disease life histories: Theory. *Evolution (N. Y).* **2011**, *65*, 3448–3461. [CrossRef]
5. Alizon, S.; van Baalen, M. Acute or Chronic? Within-Host Models with Immune Dynamics, Infection Outcome, and Parasite Evolution. *Am. Nat.* **2008**, *172*, E244–E256. [CrossRef] [PubMed]
6. Kennedy, D.A.; Dukic, V.; Dwyer, G. Pathogen Growth in Insect Hosts: Inferring the Importance of Different Mechanisms Using Stochastic Models and Response-Time Data. *Am. Nat.* **2014**, *184*, 407–423. [CrossRef] [PubMed]
7. Hoverman, J.; Gray, M.; Miller, D. Anuran susceptibilities to ranaviruses: role of species identity, exposure route, and a novel virus isolate. *Dis. Aquat. Organ.* **2010**, *89*, 97–107. [CrossRef]
8. Duffus, A.L.J.; Waltzek, T.B.; Stöhr, A.C.; Allender, M.C.; Gotesman, M.; Whittington, R.J.; Hick, P.; Hines, M.K.; Marschang, R.E. Distribution and Host Range of Ranaviruses. In *Ranaviruses*; Gray, M.J., Chinchar, V.G., Eds.; Springer: Cham, Switzerland, 2015; pp. 9–57, ISBN 978-3-319-13754-4.
9. Hoverman, J.T.; Gray, M.J.; Haislip, N.A.; Miller, D.L. Phylogeny, life history, and ecology contribute to differences in amphibian susceptibility to ranaviruses. *Ecohealth* **2011**, *8*, 301–319. [CrossRef]
10. Haislip, N.A.; Gray, M.J.; Hoverman, J.T.; Miller, D.L. Development and disease: how susceptibility to an emerging pathogen changes through anuran development. *PLoS ONE* **2011**, *6*, e22307. [CrossRef] [PubMed]
11. Warne, R.W.; Crespi, E.J.; Brunner, J.L. Escape from the pond: Stress and developmental responses to ranavirus infection in wood frog tadpoles. *Funct. Ecol.* **2011**, *25*, 139–146. [CrossRef]
12. Brunner, J.L.; Storfer, A.; Gray, M.J.; Hoverman, J.T. Ranavirus Ecology and Evolution: From Epidemiology to Extinction. In *Ranaviruses*; Gray, M.J., Chinchar, V.G., Eds.; Springer: Cham, Switzerland, 2015; pp. 71–104.
13. De Jesús Andino, F.; Chen, G.; Li, Z.; Grayfer, L.; Robert, J. Susceptibility of Xenopus laevis tadpoles to infection by the ranavirus Frog-Virus 3 correlates with a reduced and delayed innate immune response in comparison with adult frogs. *Virology* **2012**, *432*, 435–443. [CrossRef] [PubMed]
14. Wendel, E.S.; Yaparla, A.; Koubourli, D.V.; Grayfer, L. Amphibian (Xenopus laevis) tadpoles and adult frogs mount distinct interferon responses to the Frog Virus 3 ranavirus. *Virology* **2017**, *503*, 12–20. [CrossRef] [PubMed]
15. Grayfer, L.; De Jesus Andino, F.; Robert, J. The Amphibian (Xenopus laevis) Type I Interferon Response to Frog Virus 3: New Insight into Ranavirus Pathogenicity. *J. Virol.* **2014**, *88*, 5766–5777. [CrossRef] [PubMed]
16. Grayfer, L.; de Jesús Andino, F.; Robert, J. Prominent Amphibian (Xenopus laevis) Tadpole Type III Interferon Response to the Frog Virus 3 Ranavirus. *J. Virol.* **2015** *89*, 5072–5082. [CrossRef] [PubMed]

17. Brunner, J.L.; Olson, A.D.; Rice, J.G.; Meiners, S.E.; Le Sage, M.J.; Cundiff, J.A.; Goldberg, C.S.; Pessier, A.P. Ranavirus infection dynamics and shedding in American bullfrogs: Consequences for spread and detection in trade. Submitted.
18. Gosner, K.L. A Simplified Table for Staging Anuran Embryos Larvae with Notes on Identification. *Herpetologica* **1960**, *16*, 183–190.
19. Stilwell, N.; Whittington, R.J.; Hick, P.; Becker, J.A.; Ariel, E.; van Beurden, S.; Vendramin, N.; Olesen, N. Waltz Partial validation of a TaqMan real-time quantitative PCR for the detection of ranaviruses. *Dis. Aquat. Organ.* **2018**, *128*, 105–116. [CrossRef] [PubMed]
20. Menten, L.; Michaelis, M.I. Die kinetik der invertinwirkung. *Biochem. Z* **1913**, *49*, 5.
21. Rosenbaum, B.; Rall, B.C. Fitting functional responses: Direct parameter estimation by simulating differential equations. *Methods Ecol. Evol.* **2018**, *9*, 2076–2090. [CrossRef]
22. Carpenter, B.; Gelman, A.; Hoffman, M.D.; Lee, D.; Goodrich, B.; Betancourt, M.; Brubaker, M.; Guo, J.; Li, P.; Riddell, A. Stan: A Probabilistic Programming Language. *J. Stat. Softw.* **2017**, *76*. [CrossRef]
23. Gelman, A. Inference and monitoring convergence. In *Markov Chain Monte Carlo in Practice*; Gilks, W., Richardson, S., Spiegelhalter, D., Eds.; CRC Press: Boca Raton, FL, USA, 1996; pp. 131–143.
24. Vehtari, A.; Gelman, A.; Gabry, J. Practical Bayesian model evaluation using leave-one-out cross-validation and WAIC. *Stat. Comput.* **2016**, *27*, 1–20.
25. Majji, S.; Lapatra, S.; Long, S.M.; Sample, R.; Bryan, L.; Sinning, A.; Chinchar, V.G. Rana catesbeiana virus Z (RCV-Z): A novel pathogenic ranavirus. *Dis. Aquat. Organ.* **2006**, *73*, 1–11. [CrossRef] [PubMed]
26. Daszak, P.; Strieby, A.; Cunningham, A.A.; Longcore, J.E.; Brown, C.C.; Porter, A.D. Experimental Evidence That the Bullfrog (Rana Catesbeiana) Is a Potential Carrier of Chytridiomycosis, an Emerging Fungal Disease of Amphibians. *Herpetol. J.* **2004**, *14*, 201–207.
27. Peterson, A.C.; McKenzie, V.J. Investigating differences across host species and scales to explain the distribution of the amphibian pathogen Batrachochytrium dendrobatidis. *PLoS ONE* **2014**, *9*, e107441. [CrossRef] [PubMed]
28. Morales, H.D.; Abramowitz, L.; Gertz, J.; Sowa, J.; Vogel, A.; Robert, J. Innate immune responses and permissiveness to ranavirus infection of peritoneal leukocytes in the frog Xenopus laevis. *J. Virol.* **2010**, *84*, 4912–4922. [CrossRef] [PubMed]
29. Kennedy, D.A.; Dwyer, G. Effects of multiple sources of genetic drift on pathogen variation within hosts. *PLoS Biol.* **2018**, *16*, 1–17. [CrossRef] [PubMed]
30. Hanssen, S.A.; Hasselquist, D.; Folstad, I.; Erikstad, K.E. Costs of immunity: Immune responsiveness reduces survival in a vertebrate. *Proc. R. Soc. B Biol. Sci.* **2004**, *271*, 925–930. [CrossRef] [PubMed]
31. Plumb, J.A.; Zilberg, D. The lethal dose of largemouth bass virus in juvenile largemouth bass and the comparative susceptibility of striped bass. *J. Aquat. Anim. Health* **1999**, *11*, 246–252. [CrossRef]
32. Brunner, J.L.; Richards, K.; Collins, J.P. Dose and host characteristics influence virulence of ranavirus infections. *Oecologia* **2005**, *144*, 399–406. [CrossRef] [PubMed]
33. Duffus, A.L.J.; Nichols, R.A.; Garner, T.W.J. Experimental evidence in support of single host maintenance of a multihost pathogen. *Ecosphere* **2014**, *5*, 1–11. [CrossRef]
34. Cattadori, I.M.; Albert, R.; Boag, B. Variation in host susceptibility and infectiousness generated by co-infection: the myxoma-Trichostrongylus retortaeformis case in wild rabbits. *J. R. Soc. Interface* **2007**, *4*, 831–840. [CrossRef] [PubMed]
35. Lowen, A.C.; Mubarka, S.; Steel, J.; Palese, P. Influenza Virus Transmission Is Dependent on Relative Humidity and Temperature. *PLoS Pathog.* **2007**, *3*, e151. [CrossRef] [PubMed]
36. Wilber, M.Q.; Weinstein, S.B.; Briggs, C.J. Detecting and quantifying parasite-induced host mortality from intensity data: method comparisons and limitations. *Int. J. Parasitol.* **2016**, *46*, 59–66. [CrossRef] [PubMed]
37. Gog, J.R.; Pellis, L.; Wood, J.L.N.; McLean, A.R.; Arinaminpathy, N.; Lloyd-Smith, J.O. Seven challenges in modeling pathogen dynamics within-host and across scales. *Epidemics* **2014**, *10*, 45–48. [CrossRef]
38. Metcalf, C.J.; Graham, A.L.; Martinez-Bakker, M.; Childs, D.Z. Opportunities and challenges of Integral Projection Models for modelling host-parasite dynamics. *J. Anim. Ecol.* **2016**, *85*, 343–355. [CrossRef] [PubMed]

16

Critical Role of an MHC Class I-Like/Innate-Like T Cell Immune Surveillance System in Host Defense against Ranavirus (Frog Virus 3) Infection

Eva-Stina Isabella Edholm [1,2,†], Francisco De Jesús Andino [1,†], Jinyeong Yim [1], Katherine Woo [1] and Jacques Robert [1,*]

1 Department of Microbiology and Immunology, University of Rochester Medical Center, Rochester, NY 14642, USA; eva-stina.i.edholm@uit.no (E.-S.I.E.); Francisco_Dejesus@URMC.Rochester.edu (F.D.J.A.); jyim3@u.rochester.edu (J.Y.); kwoo6@u.rochester.edu (K.W.)

2 The Norwegian College of Fishery Science, University of Tromsø, the Arctic university of Norway, 9037, Tromsø, Norway

* Correspondence: Jacques_Robert@urmc.rochester.edu

† These authors contributed equally to this work.

Abstract: Besides the central role of classical Major Histocompatibility Complex (MHC) class Ia-restricted conventional Cluster of Differentiation 8 (CD8) T cells in antiviral host immune response, the amphibian *Xenopus laevis* critically rely on MHC class I-like (mhc1b10.1.L or XNC10)-restricted innate-like (i)T cells (iVα6 T cells) to control infection by the ranavirus Frog virus 3 (FV3). To complement and extend our previous reverse genetic studies showing that iVα6 T cells are required for tadpole survival, as well as for timely and effective adult viral clearance, we examined the conditions and kinetics of iVα6 T cell response against FV3. Using a FV3 knock-out (KO) growth-defective mutant, we found that upregulation of the XNC10 restricting class I-like gene and the rapid recruitment of iVα6 T cells depend on detectable viral replication and productive FV3 infection. In addition, by in vivo depletion with XNC10 tetramers, we demonstrated the direct antiviral effector function of iVα6 T cells. Notably, the transitory iVα6 T cell defect delayed innate interferon and cytokine gene response, resulting in long-lasting negative inability to control FV3 infection. These findings suggest that in *Xenopus* and likely other amphibians, an immune surveillance system based on the early activation of iT cells by non-polymorphic MHC class-I like molecules is important for efficient antiviral immune response.

Keywords: Unconventional T cell; nonclassical MHC; antiviral immunity; interferon

---

1. Introduction

In mammals, potent adaptive immune response against viral pathogens largely depends on conventional Cluster of Differentiation (CD8) T cell effectors that express a broad T cell receptor (TCR) repertoire capable of recognizing a large array of antigens presented by polymorphic classical Major Histocompatibility Complex (MHC) class I molecules (class Ia). However, it is increasingly appreciated that other unconventional or innate-like T cell effectors are critically involved in antiviral immunity [1]. These iT cells interact with non-polymorphic MHC class I-like molecules, have a limited or invariant T Cell Receptor (TCR) repertoire, and are thought to serve as important early responders and immune regulators. One of the best characterized iT cell populations are invariant Natural Killer T (iNKT) cells, which are restricted by the MHC class I-like molecule Cluster of Differentiation CD1d that presents lipids [2,3]. The antiviral role of iNKT cells has been shown for murine cytomegalovirus

(MCMV; [4]), lymphocytic choriomeningitis virus (LCMV; [5]), herpes simplex virus 1 (HSV1; [6]), HIV and influenza [7], and even for vaccinia virus (VAC; [8]).

In the amphibian *Xenopus laevis*, while adult frogs exhibit an immune system whose conventional class Ia-restricted CD8 T cell compartment is dominant, as in mammals, tadpoles are immunocompetent but are naturally class Ia-deficient and primarily rely on MHC class I-like and iT cells to mediate immunity [9]. Deep-sequencing repertoire analysis has revealed the over-representation of 6 invariant TCRα rearrangements, implying the predominance of 6 larval iT cell subsets [10]. One of the 6 putative iT cell subsets, the iV6 T cell subset, expresses the rearranged Vα6-J1.43 TCRα chain and requires the MHC class I-like molecule XNC10 (encoded by mhc1b10.1.L) for its thymic development and function [10]. Notably, host immune responses against the ranavirus pathogen Frog Virus 3 (FV3) are significantly impaired by lack of iVα6 T cells resulting from XNC10 loss-of-function. FV3 is a major pathogen of amphibians, fish, and even reptiles [11]. The dramatic worldwide increases in host ranges (i.e., populations and species infected) and amphibian die-off caused by RVs (see eBook [12]) raise alarming concerns for biodiversity and aquaculture, and pose fundamental issues related to evolution of host/pathogen interactions. As such, the involvement of iT cells in addition to conventional CD8 T cells in antiviral host resistance is of high relevance.

In adult *X. laevis*, XNC10 deficiency considerably delays viral clearance, resulting in increased tissue damage [13]. As expected from iT cell prevalence in tadpoles, XNC10 deficiency markedly increases the mortality rate, especially at early stages of FV3 infection [10]. The generation of XNC10 tetramers by expressing XNC10 fused to *X. laevis* beta 2-microglobulin (b2m) has permitted a better characterization of the iVα6 T cell response kinetics and shows a rapid recruitment (within less than 24 h.) at the site of infection. This fast targeted response suggests a sensitive detection mechanism of FV3 infection. However, many facets of this response remain to be investigated, including whether iVα6 T cells require stimulation by XNC10 bound to viral or other ligands, or can be activated by XNC10-independent co-stimulation signals (e.g., danger signals, interferon [IFN] response). To explore some of these issues, we have taken advantage of our XNC10 tetramer, as well as a FV3 recombinant virus deficient for a putative immune evasion gene, to determine the role of productive versus ablated viral infection, as well as the effect of transiently impairing iVα6 T cell function at early stages of FV3 infection.

2. Material and methods:

2.1. Animals

All outbred *X. laevis* were from the *X. laevis* research resource for immunology at the University of Rochester (https://www.urmc.rochester.edu/microbiology-immunology/xenopus-laevis.aspx). All animals were handled in accordance with stringent laboratory and University Committee on Animal Research regulations, minimizing suffering (Approval number 100577/2003-151). For all experiments, three-week old tadpoles (stage 55, 1.5 cm long; [14]) and 1 year-old young adult frogs were used.

2.2. Frog Virus 3 Stocks and Infection

Baby hamster kidney cells (BHK-21, ATCC No. CCL-10) were maintained in Dulbecco's modified Eagle's medium DMEM (Invitrogen, Thermo Fisher Scientific, Inc. Waltham, MA, USA) containing 10% fetal bovine serum (Invitrogen), streptomycin (100μg/mL), and penicillin (100 U/mL) with 5% CO_2 at 37 °C, then 30°C for infection. The generation and characterization of the FV3 knock-out (KO) mutant ΔvCARD (or Δ64R) has been detailed elsewhere [15–17]. Wild type (WT) and KO FV3 were grown using a single passage through BHK-21 cells and were subsequently purified by ultracentrifugation on a 30% sucrose cushion. Adult frogs were infected by intraperitoneal (i.p.) injection of 1×10^6 PFUs in 100 μL of Amphibian Phosphate buffered saline (APBS) and tadpoles were infected by i.p. injection of 10,000 PFUs in 5 μL APBS. Uninfected control animals were mock-infected with an equivalent volume of APBS. At different days post-infection (dpi), animals were euthanized using 1 μg/L tricaine methanesulfonate (TMS) buffered with bicarbonate prior to dissection and extraction of nucleic acids from tissues.

2.3. *Quantitative Gene Expression Analyses*

Total RNA was extracted from peritoneal leukocytes and kidneys using Trizol reagent, following the manufacturer's protocol (Invitrogen). The cDNA was synthetized with 0.5 μg of RNA in 20 μL using the iScript cDNA synthesis kit (Bio-Rad, Hercules, CA, USA), and 1 μL of cDNA template was used in all RT-PCRs and 150 ng DNA for PCR. Minus RT controls were included for every primer pair. A water-only control was included in each reaction. The qPCR analysis was performed using the ABI 7300 real-time PCR system with PerfeCT SYBR Green FastMix, ROX (Quanta, Beverly, MA, USA), and ABI sequence detection system (SDS) software. Glyceraldehyde-3-phosphate dehydrogenase (GAPDH) controls were used in conjunction with the ΔΔ CT method to analyze cDNA for gene expression. All primer sequences are listed in Table S1.

2.4. *Viral Load Quantification by qPCR and Plaque Assay*

FV3 viral loads were assessed by absolute qPCR by analysis of isolated DNA in comparison to a serially diluted standard curve. Briefly, an FV3 DNA Pol II PCR fragment was cloned into the pGEM-T Easy vector (Promega, Madison, WI, USA). This construct was amplified in bacteria, quantified, and serially diluted to yield 1010-101 plasmid copies of the FV3 DNA polymerase II. These dilutions were employed as a standard curve in subsequent absolute qPCR experiments to derive the viral genome transcript copy numbers relative to this standard curve. Virus quantification by plaque assay was performed on BHK-21 monolayers in 6-well plates under an overlay of 1% methylcellulose [18]. Infected cells were cultured for 7 days at 30 °C in 5% CO_2. Overlay media was aspirated and cells were stained for 10 min with 1% crystal violet in 20% ethanol.

2.5. *XNC10 Tetramer Production*

XNC10 tetramers were generated as previously described [10]. Briefly, beta 2 microglobulin (b2m) was linked via a 23-aa Glycine rich C-terminal flexible linker to the α1-α3 domains of XNC10 containing a BirA site-specific biotinylation site at the end of the α3 domain and cloned into the pMIBV5-HisA expression vector (Invitrogen). The b2m-linker-XNC10 construct was expressed in Sf9 insect cells and monomeric b2m-linker-XNC10 was purified by Ni-NTA-Agarose Chromatography (Qiagen, Hilden, Germany) and concentrated to 1 μg/μL using Amicon Ultra Centrifugal Filter (Millipore, Burlington, MA, USA). BirA enzymatic biotinylation was performed for 18 h. at 30 °C according to the manufacturer's protocol (Avidity, Aurora, CO, USA), and the purified biotinylated proteins were extensively dialyzed against APBS, pH 7.5, to remove any unbound biotin. XNC10 tetramers were generated by incubating b2m-linker-XNC10 with fluorochrome-labeled streptavidin at a 5:1 ratio at room temperature for 4 h. before use. Purified XNC10 tetramers (1 μg) were injected intra i.p. at a volume of 5 μL.

2.6. *Statistical Analysis*

One way ANOVA followed by Dunn's or Tukey's multiple comparison tests were used for statistical analysis of expression and viral load data. Analyses were performed using a Vassar Stat online resource (http://vassarstats.net/utest.html). Statistical analysis of survival data was performed using a Log-Rank Test (GraphPad Prism 6, San Diego, CA, USA). A probability value of $p < 0.05$ was used in all analyses to indicate significance. Error bars on all graphs represent the standard error of the mean (SEM).

3. Results

3.1. Relationships Between FV3 Infection Magnitude and iVα6 T cell Response in Adult X. Laevis

We have previously demonstrated that FV3 infection in adult X. *laevis* elicits a transitory influx of the innate (i)T cell subset iVα6 T into the peritoneal cavity [13]. To delineate the factors governing iVα6 T cell recruitment during a FV3 infection, we took advantage of the FV3 KO mutant ΔvCARD (64R)-FV3 that has previously been shown to have attenuated virulence and growth in vivo, eliciting different host responses [14,17]. Accordingly, we i.p infected adult X. *laevis* with wild type (WT) FV3 or Δ64-FV3 for 24 h. before collecting peritoneal leukocytes (PLs) and quantified transcript levels of the specific invariant Vα6-Jα1.43 rearrangement. As a control we also challenged frogs with heat-killed *E. coli*. This type of bacterial stimulation has been shown to induce a strong nonspecific inflammatory response but does not stimulate the recruitment or accumulation of iVα6 T cells in the peritoneal cavity [13]. Because iVα6 T cells interact with the MHC class I-like molecule XNC10, we also examined its gene expression profile. To evaluate viral loads, we determined the FV3 genome copy number using absolute qPCR. Consistent with previous findings, Δ64-FV3 infection resulted in significantly ($p = 0.005$) lower viral load compared to WT-FV3 already at 1 dpi [17]. All treatments resulted in a significantly increased XNC10 expression compared to PLs collected from APBS-injected control frogs at the same time point, with no difference among the treatment groups (Figure 1B). However, only WT-FV3 infection resulted in significantly elevated iVα6-Jα1.43 transcript levels compared to uninfected controls ($p = 0.018$), whereas in frogs injected with heat-killed bacteria, iVα6-Jα1.43 expression was lower compared to controls, presumably due to the large influx of immune cells into the peritoneal cavity. Although the high individual variation prevented statistical significance among the two FV3 infected groups, iVα6-Jα1.43 transcript levels observed with Δ64-FV3 mutants were elevated above sham-infect APBS controls in only 20% of frogs (3 out of 15 frogs) compared to 70% (9 out of 13) for WT-FV3 infected frogs, suggesting a correlation between iVα6 recruitment and the level of viral replication.

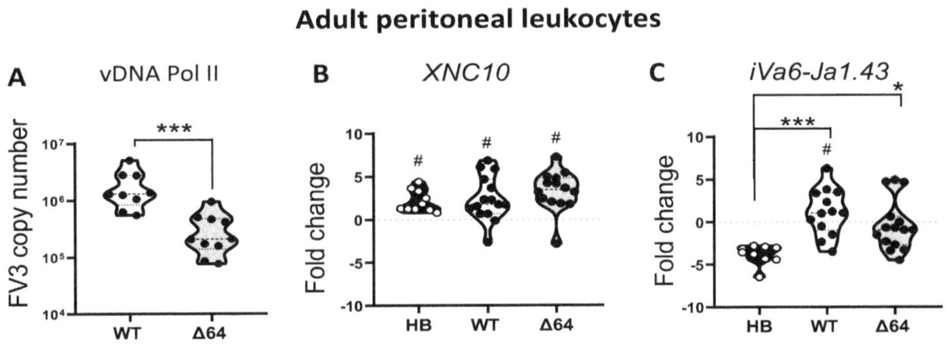

Figure 1. Effects of infection with attenuated knockout (KO) FV3 recombinant or bacterial stimulation on iVα6 T cell response. Peritoneal leukocytes (PLs) were collected at 1 dpi from adult frogs infected with 1×10^6 PFUs of WT-FV3 or Δ64-FV3, or 100 μl heat-killed (HB) *E. coli*. (**A**) Genome copy number using absolute qPCR with primers against FV3 DNA polymerase II. (**B**) XNC10 relative gene expression and (**C**) iVα6-Jα1.43 relative gene expression. Gene expression was determined relative to an endogenous control (GAPDH) and fold changes were calculated using the unstimulated sample (injected with equivalent volume of APBS) collected at the same time point. Data are pooled from three independent experiments with n = 4–5 animals in each experiment and each dot represents an individual animal. The line intersecting the y-axis at 0 represents the unstimulated control that the fold changes of the treatments are in relation to; (#) $p < 0.05$ significant differences compared to unchallenged (APBS) injected controls; (*) $p < 0.05$ and (***) $p < 0.001$ statistically significant differences between the indicated groups (one way ANOVA and Dunns's multiple comparison test).

To obtain additional evidence of the correlation between active viral replication and iVα6 T cell recruitment, we i.p. infected adult X. *laevis* with WT- and Δ64-FV3, and then monitored the transcript levels of iVα6-Jα1.43 and XNC10 in the peritoneal cavity (the site of infection) and kidney (main site for viral replication) at 1, 3, and 6 dpi (Figure 2). To evaluate viral replication, we determined the genome copy number by absolute qPCR (Figure 2A,B) and to assess productive infection we performed plaque assays (Figure 2C and Figure S1). Again Δ64-FV3 exhibited a severe replication defect preventing the production of infectious particles (Figure 2C and Figure S1). Using iVα6-Jα1.43 expression as a proxy for monitoring the kinetics of iVα6 T cell recruitment, we detected an increase in iVα6-Jα1.43 transcript levels in the peritoneal cavity as early as 1 dpi with WT-FV3 compared to uninfected controls, which became significant by 3 dpi, and then returned to low but detectable levels by 6 dpi (Figure 2D). In contrast, Δ64-FV3 infection did not trigger a significant increase in iVα6-Jα1.43 transcript levels. In kidneys, iVα6-Jα1.43 expression was significantly elevated at 3 dpi and then returned to baseline at 6 dpi following infection with WT-FV3 (Figure 2E). In contrast, no significant changes in iVα6-Jα1.43 expression were detected with Δ64-FV3 infection.

The examination of the expression response of the iVα6 T cell restricting the MHC class I-like XNC10 gene revealed some interesting correlations. In PLs, WT-FV3 infection induced a rapid (from 1 dpi) and sustained (until 6 dpi) XNC10 increase (Figure 2F), which was delayed (from 3 to 6 dpi) in kidneys (Figure 2G). Consistent with the poorly induced iVα6-Jα1.43 expression, infection with Δ64-FV3 also resulted in impaired induction of XNC10 gene expression. Indeed, Δ64-FV3 infection did not induce any significant change in XNC10 gene expression at 3 and 6 dpi in both PLs and kidneys.

Collectively, these data indicate that XNC10-restricted iVα6 T cell recruitment is relative to viral burden and requires active viral replication.

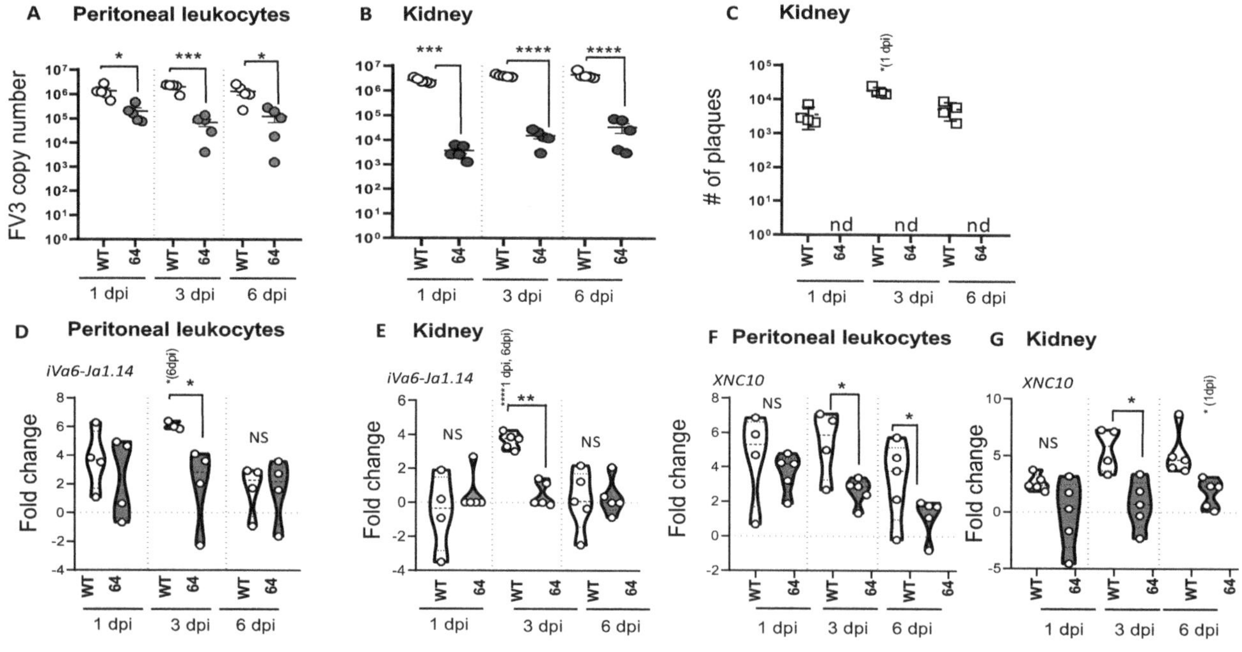

Figure 2. Magnitude of iVα6 T cell response in adult frogs is associated to the level of viral replication and production of infectious particles. PLs and kidneys were collected at 1, 3, and 6 dpi from adult frogs infected with 1×10^6 PFUs of WT-or Δ64-FV3. FV3 genome copy number by absolute qPCR in PLs (**A**) and kidneys (**B**) were determined. The total number of infectious particles (nd: not detected) in the kidney was determined by plaque assay (**C**).Gene expression of iVα6-Jα1.43 and XNC10 in PLs (**D**,**F**) and kidneys (**E**,**G**) were determined relative to an endogenous control (GAPDH) and fold changes were calculated using mock-infected frogs as a control. Each dot represents an individual animal (n = 4–5). The line intersecting the y-axis at 0 represents the APBS control that the fold changes of the treatments are in relation to. Note: * $p < 0{,}05$; ** $p < 0.01$; *** $p < 0.001$, and **** $p < 0.0001$ above the line denotes statistically significant differences between the different treatment groups; significant differences between time points within each treatment group are indicated within parentheses; NS indicates no significant differences (one way ANOVA and Dunns's multiple comparison test).

3.2. Relationships Between FV3 Infection Magnitude and iVα6 T Cell Response in Tadpoles

In contrast to adult frogs that exhibit a prominent conventional antiviral CD8 T cell response restricted by classical MHC class Ia, tadpoles have barely detectable class Ia surface protein expression and rely heavily on XNC10-restricted iVα6 T cells [10]. Indeed, XNC10-deficient transgenic tadpoles lacking iVα6 T cells succumb to FV3 infection much faster than controls. This suggests that a rapid iVα6 T cell response is required during the early stage of FV3 infection. Similar to adult frogs, we postulated that the efficient recruitment of iVα6 T cells at the site of infection in tadpoles would depend on a productive FV3 infection. Accordingly, we i.p. infected tadpoles with either WT- or Δ64-FV3. Consistent with previous reports, Δ64-FV3 exhibited a severe growth defect in tadpoles, as shown by the very low number of plaques detected, even at 6 dpi (Figure 3A and Figure S2). As expected, WT-FV3 infection elicited an iVα6 T cell response that was overall comparable to adults, including an early (by 1 dpi) increase in iVα6-Jα1.43, as well as XNC10 transcript levels that were maintained by 3 dpi and then declined close to background levels by 6 dpi. In contrast, there were no significant changes in XNC10 and iVα6-Jα1.43 transcript levels following Δ64-FV3 infection (Figure 3B,C).

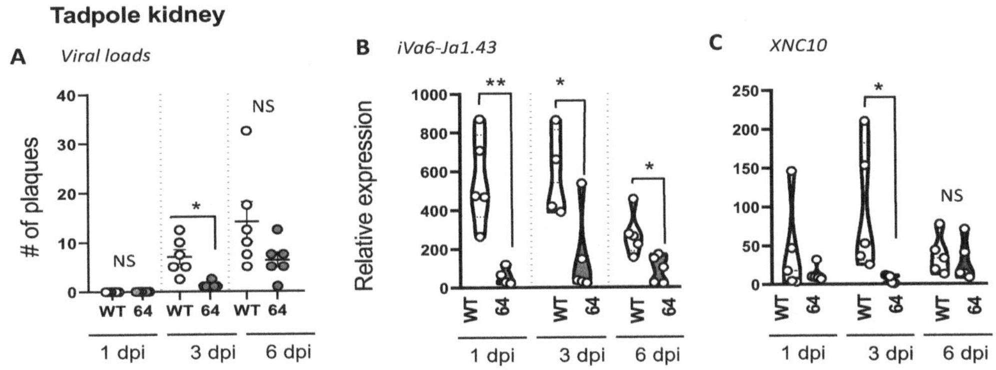

Figure 3. The iVα6 T cell response in tadpoles depends on active viral replication and productive FV3 infection. Three week-old (stage 55) tadpoles were infected with 10,000 PFUs of WT-or Δ64-FV3. At 1, 3, and 6 dpi, kidneys were collected and the total number of infectious particles was determined by plaque assay, respectively (**A**). Gene expression of iVα6-Jα1.43 (**B**) and XNC10 (**C**) was determined relative to an endogenous control (GAPDH) and relative expression was calculated against the lowest observed expression according to the ΔΔCt method ($n = 5$). No iVα6-Jα1.43 transcripts were detected in the kidney uninfected tadpoles. * $p < 0.05$ and ** $p < 0.01$ above the line denotes statistically significant differences between treatment groups; NS indicates no significant differences (one way ANOVA and Dunn's multiple comparison test).

3.3. Targeted iVα6T cell depletion in vivo

As mentioned, intraperitoneal (i.p.) infection of *X. laevis* tadpoles with FV3 results in a rapid and transitory increase in Vα6-Jα1.43 transcript levels, both at the site of infection and in the kidney (main FV3 target organ), consistent with a direct and active involvement of this specific iT cell population in the early anti-FV3 response [13]. In support of this possibility, the developmental disruption of iVα6 T cells in MHC class I-like XNC10-deficient transgenic tadpoles results in increased susceptibility to FV3 [10]. However, silencing of XNC10 from early embryogenesis is likely to result in biological effects other than those inferred via the reciprocal loss of a unique iT cell population. Therefore, to more directly determine the role of iVα6 T cells in anti-viral defense against FV3 in the presence of a functional XNC10 gene, we targeted iVα6 T cells using XNC10 tetramers. XNC10 tetramers have been shown to selectively induce iVα6T cell death ex vivo and to temporarily ablate Vα6-Jα1.43 transcript levels in vivo (Banach et al., submitted). Accordingly, we injected *X. laevis* tadpoles with 1 μg of XNC10 tetramers 1 day prior and 1 day following i.p. injection with FV3 and quantified transcript levels of the invariant Vα6-Jα1.43 rearrangement in PLs and kidneys at different times following infection (Figure 4). Control groups received injections with an equivalent volume of amphibian PBS at the same time points. To evaluate viral loads and dissemination in the different groups, we determined the FV3 genome copy number using absolute qPCR (Figure 4D,E). In addition, we examined the gene expression of XNC10 to evaluate any impact of the injection regime on the expression of this MHC class I-like gene (Figure S3).

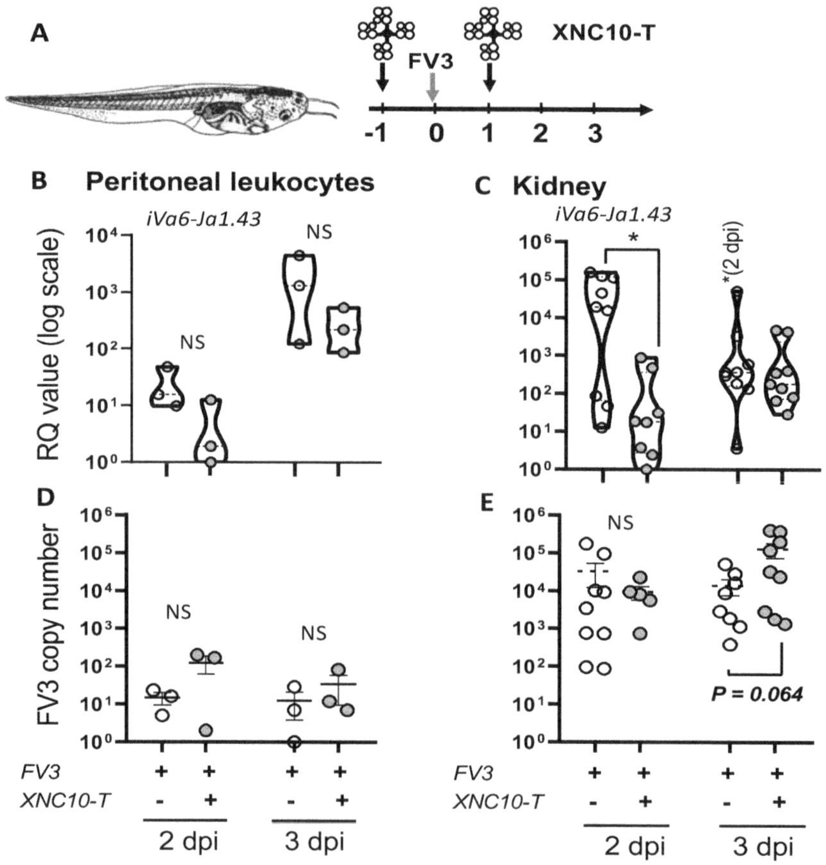

Figure 4. XNC10 tetramer-mediated iVα6T cell depletion in tadpoles affects iVα6-Jα1.43 transcript levels and viral replication. PLs and kidneys were collected from three week-old (stage 55) tadpoles that had been injected with 1 µg XNC10 tetramers (XNC10-T) or vehicle control 1 day pre, and 1 day post i.p. injected with 10,000 PFUs of FV3, at the indicated time points ($n = 9$). A schematic of the injection regime is shown in (**A**). Gene expression of iVα6-Jα1.43 in PLs (**B**) and kidneys (**C**) is shown. Results are normalized to an endogenous control and presented as relative expression compared with the lowest observed value according to the $\Delta\Delta$Ct method. FV3 loads in PLs (**D**) and kidneys (**E**) were measured using absolute qPCR with primers against FV3 polymerase II. For PLs, each dot represents a pool of 3 tadpoles, while for kidneys, each dot represents a single tadpole; * $p < 0.05$ denotes statistically significant differences between the indicated groups; NS indicates no significant differences (One way ANOVA followed by Tukey's multiple comparison test).

Consistent with XNC10 tetramers inducing iVα6 T cell death, iVα6-Jα1.43 transcripts were markedly reduced, albeit not significantly, in the peritoneal cavity (site of infection) at 2 dpi, and remained lower than controls at 3 dpi. In the kidney, iVα6-Jα1.43 transcripts were significantly lower in the XNC10 tetramer treated group at 2 dpi, reaching levels equivalent to controls by 3 dpi. It is noteworthy that consistent with previous reports, iVα6-Jα1.43 transcripts were either undetected or near threshold levels in uninfected tadpoles. Thus, XNC10 tetramer treatment provides an efficient transient depletion of iVα6 T cells in vivo, providing us with a system to directly investigate the functional roles of iVα6 T cells. No significant difference was observed in XNC10 expression between the two groups (Figure S3).

In regard to viral loads, while there was no significant difference in viral genome copy numbers in the peritoneal cavity, XNC10 tetramer treated tadpoles displayed a significantly ($p = 0.005$) higher viral load in the kidneys by 3 dpi (Figure 5B). By 6 dpi, using the same injection scheme as described in Figure 4A, there were significantly higher viral loads in the kidneys of the XNC10-tetramer treated group compared to vehicle control (Figure 5A). Notably, viral loads were not increased in tadpoles injected with XNC10 monomers, which do not bind to iVα6 T cells [10]. In addition, XNC10 tetramer treatment had only a transitory effect on iVα6 T antiviral activity, as shown by comparable viral loads detected when FV3 infection occurred 3 days rather than 1 day after XNC10 tetramer injection (Figure 5A). To assess whether this increase in viral loads had an impact on host resistance, we determined the cumulative mortality and mean survival of XNC10 tetramer injected and FV3 challenged tadpoles (Figure 6). Notably, transitory depletion of antiviral iVα6 T cells within 24 h. of FV3 infection resulted in increased mortality rate compared with FV3 infected controls ($p = 0.05$). The seven-fold increase in post-mortem viral loads found in infected XNC10 tetramer-treated tadpoles that died within 6 to 20 dpi compared to vehicle treated infected controls (Table S2) further suggests that the increased mortality was due to an inefficient control of viral infection. These data support the conclusion that iVα6 T cells are intricately involved in *X. laevis* anti-FV3 immune responses and highlight the importance of an efficient onset of appropriate anti-viral immune response in order to control and efficiently combat FV3 infection.

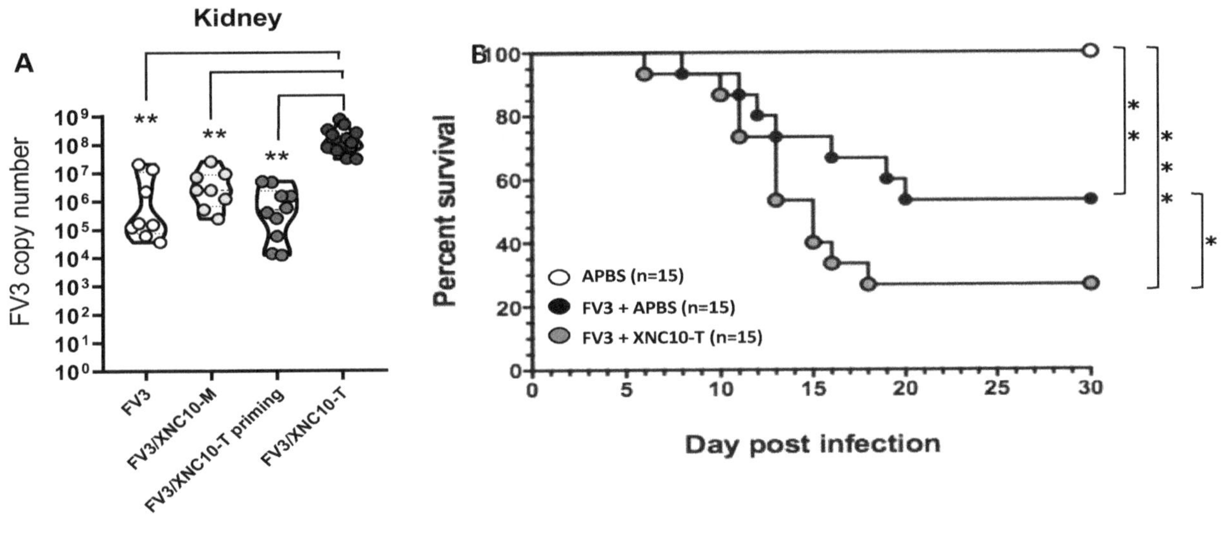

Figure 5. Specificity and long term impact of transitory iVα6T cell depletion. (**A**) Three week-old (stage 55) tadpoles were injected with either 1 µg XNC10-tetramers (FV3/XNC10-T), 1 µg XNC10-monomers (FV3/XNC10-M), or vehicle control (FV3/APBS) 1 day pre and 1 day post i.p. infection with 10,000 PFUs of FV3 ($n = 8$), and kidneys were collected at 6 dpi. The last group (FV3/XNC10-T priming) was first injected with 1 µg XNC10-tetramers, then 3 days later infected with FV3, and kidneys were collected at 6 dpi. Viral loads were assessed by absolute qPCR with primers against FV3 polymerase II. The results are combined from two separate experiments, and each dot represents an individual tadpole; ** $p < 0.01$ above the line denotes statistically significant differences between the indicated groups (One way ANOVA followed by Tukey's multiple comparison test). (**B**) Three week-old (stage 55) tadpoles were injected with either 1 µg XNC10 tetramers (FV3/XNC10-T) or vehicle control (FV3/APBS) 1 day pre, and 1 day post were i.p injected with 10,000 PFUs of FV3 ($n = 15$) and survival was monitored daily over a 30-day period. Survival was determined using Kaplan-Meier, * $p < 0.05$, ** $p < 0.005$, and *** $p < 0.0005$. Uninfected controls (white circle), FV3 infected tadpoles (black circle), and XNC10 tetramer treated FV3 infected tadpoles (grey circle).

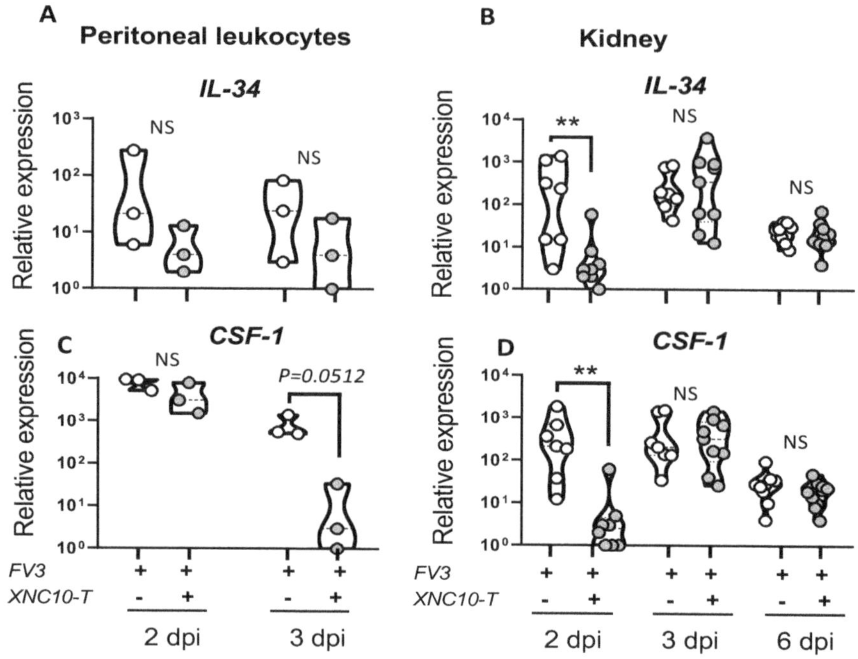

Figure 6. Effect of XNC10 tetramer treatment on the expression of the macrophage stimulating factor genes CSF-1 and IL-34 in the peritoneal cavity and kidneys of FV3 infected tadpoles. PLs and kidneys were collected from three week-old (stage 55) tadpoles that had been injected with 1 µg XNC10 tetramers or vehicle control 1 day pre- and 1 day post-i.p. injection with 10,000 PFUs of FV3, at the indicated time points (n = 8–9). Quantitative gene expression analysis of IL-34 (**A**,**B**) and CSF-1(**C**,**D**) were determined relative to a endogenous control (GAPDH), and relative expression was calculated against the lowest observed expression according to the $\Delta\Delta$Ct method (n = 9); ** $p < 0.005$ above the line denotes statistically significant differences between the different treatment groups; NS indicates no significant differences (One way ANOVA followed by Tukey's multiple comparison test).

3.4. *Effects of iVα6T cell depletion on PLs and kidney antiviral responses in tadpoles*

We have previously shown that the compromised anti-FV3 response observed in XNC10-deficient transgenic frogs correlates with the induction of macrophages, exhibiting a less potent antiviral state [10]. In particular, XNC10 deficiency hampered the expression of the macrophage growth factor IL-34. Notably, IL-34 has been shown to elicit the differentiation of mononuclear phagocytes into robust type I interferon (type I IFN), producing macrophages exhibiting strong FV3 antiviral activity [19]. These findings prompted us to hypothesize that iVα6 T cells promote timely and efficient type I IFN-mediated anti-FV3 responses by influencing the polarization of macrophages into a more potent antiviral state. Thus, to further delineate the putative functional roles of iVα6 T cells during FV3 infection, we next determined if direct and transient iVα6 T cell depletion has an impact on macrophage effector functions. To address this possibility, we monitored the expression profiles of the two macrophage growth factors CSF-1 [20] and IL-34 [19] in PLs and kidneys (Figure 6) from FV3 infected tadpoles, with or without XNC10 tetramer treatment. Both IL-34 and CSF-1 transcription levels were significantly lower in the kidneys of XNC10 tetramer treated tadpoles at 2 dpi, correlating with the reduced iVα6-Jα1.43 expression, and levels comparable to controls were reached by 3 dpi.

Given that the type I IFN response is critical in controlling viral infections and is compromised in XNC10-deficient animals [13], we postulated that a direct impairment of iVα6 T cells should affect type I IFN gene expression to a similar extent. Indeed, unlike control animals, both type I IFN and type III IFN gene expression were magnitudes lower at 3 dpi in PLs of XNC10 tetramer treated animals compared to controls, whereas the levels of type II IFN were highly variable between animals, with no apparent difference in expression between the two groups (Figure 7A–C). Similarly, in kidneys the expression of both type I and type III IFN genes was significantly impaired in XNC10 tetramer treated animals at 2 dpi, whereas by 3 dpi both treated and control groups exhibited high transcript levels of type I and type III IFN (Figure 7D,E).

In response to a viral infection, protective host response relies on the onset of appropriate cell-mediated immunity to mediate viral clearance. Thus, we next examined the expression of interleukin18 (IL-18) and interleukin 12 (IL-12) upon FV3 infection in PLs and kidneys of control and iVα6 T cell deficient tadpoles. Both IL-18 and IL-12 are produced by macrophages and act in synergy to promote the development of innate and specific Th1 type immune responses [21–23]. Given that iVα6 T cell deficiency hampers the induction of macrophage differentiating factors, resulting in a delayed interferon response, we postulated that IL-18 and IL-12 cytokine production would be similarly impaired. Indeed, unlike control animals, IL-18 expression was significantly lower at 2 dpi in kidneys of XNC10 tetramer treated animals, whereas at 3 dpi XNC10 tetramer treated tadpoles exhibited elevated IL-18 transcript levels on par with that of controls that were significantly reduced by 6 dpi in both groups (Figure 8). While there was a slight decrease in IL-12 gene expression at 2 dpi in XNC10 tetramer treated animals, it was not statistically significant.

Overall, these findings are in accordance with what had previously been reported for XNC10-deficient transgenic animals (that lack both XNC10 and iVα6 T cells, [13]), suggesting that iVα6 T cells are directly and critically involved in mounting an anti-FV3 response, and that acute loss or impairment of these cells result in both a delayed type I and type III IFN antiviral response, as well as a delayed IL-18 response.

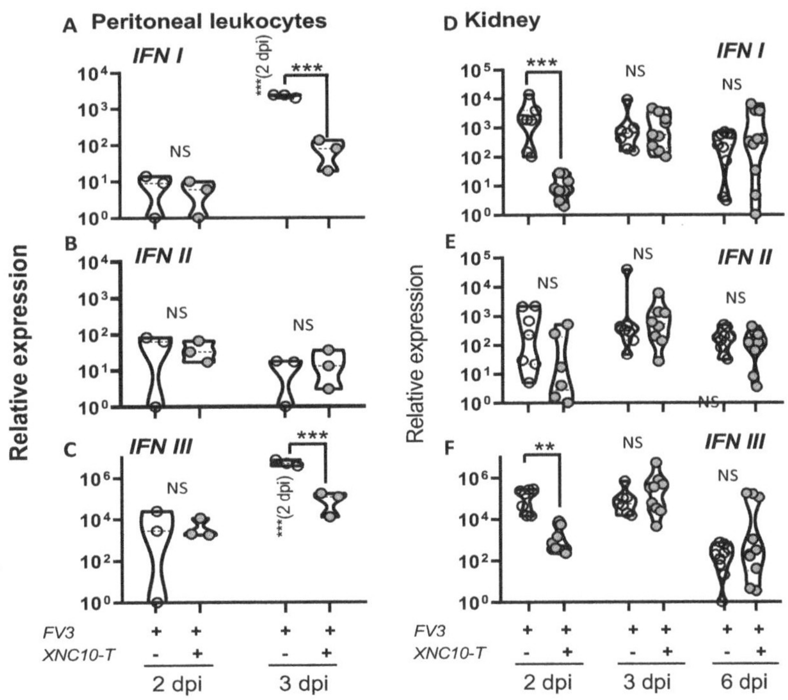

Figure 7. XNC10 tetramer treatment results in a delayed antiviral response in the peritoneal cavity and kidneys of FV3 infected tadpoles. PLs and kidneys were collected from three week-old (stage 55) tadpoles that had been injected with 1 µg XNC10 tetramers (XNC10-T) or vehicle control 1 day pre- and 1 day post-i.p. injection with 10,000 PFUs of FV3, at the indicated time points (n = 8–9). Quantitative gene expression analysis of type I (**A**,**D**), type II (**B**,**E**), and type III IFN (**C**,**F**) were determined relative to a endogenous control (GAPDH), and relative expression was calculated against the lowest observed expression according to the $\Delta\Delta$Ct method (n = 9); * $p < 0.05$, ** $p < 0.005$, and *** $p < 0.001$ above the line denote statistically significant differences between the different treatment groups; significant differences between time points within each treatment group is indicated within parentheses; NS indicates no significant differences (One way ANOVA followed by Tukey's multiple comparison test).

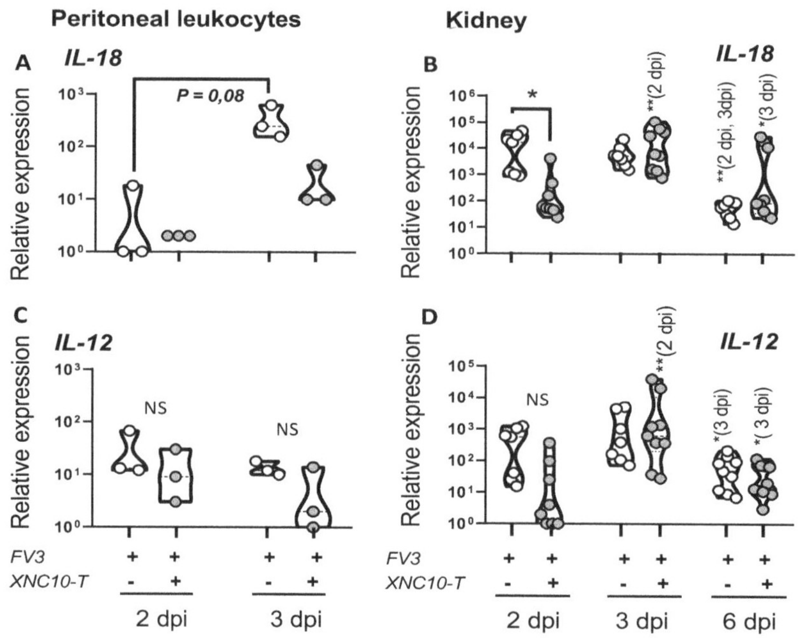

Figure 8. Effects of iVα6T cell depletion on IL-18 and IL-12 responses. PLs and kidneys were collected from three week-old (stage 55) tadpoles that had been injected with 1 µg XNC10 tetramers (XNC10-T) or vehicle control 1 day pre- and 1 day post-i.p. injection with 10,000 PFUs of FV3, at the indicated time points (n = 8–9). Quantitative gene expression of IL-18 (**A**,**B**) and type IL-12 (**C**,**D**) was determined relative to an endogenous control (GAPDH) and relative expression was calculated against the lowest observed expression according to the $\Delta\Delta$Ct method (n = 9); * $p < 0.05$; ** $p < 0.005$, above the line denotes statistically significant differences between the different treatment groups; significant differences between time points within each treatment group is indicated within parentheses; NS indicates no significant differences (One way ANOVA followed by Tukey's multiple comparisons test).

4. Discussion

The present study provides new insights into the requirement for an effective amphibian host response to the ranavirus FV3 trough, an early detection by the MHC class I-like XNC10, and the rapid recruitment of the activated XNC10-restricted innate T cell subset iVα6. A remarkable feature of this novel antiviral immune surveillance system is that it is prominent in naturally classical MHC class Ia-deficient tadpoles, whereas it is a component in class Ia and conventional T cell competent adults. Nevertheless, while this antiviral system is central for tadpole survival at early stage of FV3 infection, it is also critical for a timely control of viral burden in adult frogs. To better functionally define this novel antiviral immune surveillance system, we examined the initial conditions of FV3 infection that triggers the iVα6 T cell response by using deficient FV3 recombinants, and we defined the antiviral immune response initiated by iVα6 T cells.

The critical role of iNKT cells in antiviral immunity is well documented in mammals, where their pre-activated or memory phenotypes allow them to serve as early responders (within hours following infection) without requiring expansion as conventional T cells [7,24]. Similarly, in *X. laevis*, we have shown a rapid increase of iVα6-Jα1.43 transcript levels (within 1 dpi) at the site of FV3 infection in the peritoneum and in the main site of viral replication in the kidney, concomitant with a decrease in the spleen. Notably, the recruitment of these iVα6 T cells occurs days before significant proliferation of conventional T cells, which become significant only from 3 dpi onward in the spleen [25]. This strongly suggests that iVα6 T cells are rapidly activated and recruited from the spleen by FV3 infection [13].

Here, we provide further evidence that iVα6 T cell recruitment occurs in direct response to FV3 infection and is not indirectly triggered by inflammation induced by heat-killed bacteria or by a replication-defective Δ64-FV3 recombinant. Furthermore, while both attenuated Δ64-FV3 and heat-killed bacterial stimulation induced an early transcriptional induction of the MHC class I-like

XNC10 that restricts iVα6 T cells, these elevated expression levels were only maintained in WT-FV3 infected frogs. This suggests a coordinated role of XNC10 and iVα6 T cells in detecting and eliciting a rapid response against ranavirus pathogens. This coordinate response of XNC10 and iVα6 T cells only induced by a productive viral infection is also consistent with the detection of virally-derived products or ligands. It is unknown whether XNC10 binds ligands, and if so the identity of such putative XNC10 ligands is currently unknown, but nucleotide sequence comparisons and 3D-modeling of the putative ligand binding regions indicate that XNC10 could accommodate lipid ligands in a way reminiscent of CD1d [3,26,27]. Thus, it is tempting to speculate that XNC10 can be upregulated and bind products and ligands from early FV3 replication or release of assembled viruses to activate iVα6 T cell response. In this scenario, infection by Δ64-FV3 would not lead to sufficient production and release of FV3-derived products and ligands to initiate iVα6 T cell recruitment and activation.

Interestingly, the correlation between defective viral replication, blunted XNC10 gene expression response, and impaired iVα6 T cell recruitment was more robust in tadpoles. This may be due to the heavier reliance on iT cells in tadpoles. In absence of an efficient classical MHC-I restricted conventional CD8 T cell response, iVα6 T cells are likely to be more directly and critically involved in tadpole host response against FV3. This is evidenced by marked vulnerability of XNC10-deficient transgenic tadpoles lacking iVα6 T cell to FV3 infection, especially at early stages of infection [10]. Our findings using XNC10 tetramer treatment substantiate this initial requirement of iVα6 T cell response for controlling viral infection and host resistance to FV3. Indeed, a partial and transient depletion of iVα6 T cells prompted by injection of XNC10 tetramers shortly before and after infection impairs tadpole immune defenses, which not only leads to more severe viral burden but also decreased survival rate.

While the effector functions of iVα6 T cells are likely to be multifaceted, our previous findings support a role of these cells as immune modulators that influence the polarization of peritoneal macrophages into a more robust anti-viral state [13]. Although XNC10 gene expression is detectable on peritoneal macrophages, in absence of XNC10-specific antibodies, it is currently unknown whether all macrophages express XNC10 molecules at the cell surface and whether macrophages can modulate XNC10 surface expression upon FV3 infection. It is noteworthy that the expression of type I and III IFN, as well as cytokines, such as CSF-1 and IL-18, are transiently altered in parallel to iVα6 T cell depletion. Little is known about the iVα6 T cell mechanism of action, but previous findings have hinted at a modulation of macrophages [10,13]. While type I and III IFN is likely produced and released by many cell types, the rapid increase of their transcript levels in the peritoneum and kidneys upon FV3 infection coincides with the recruitment of monocytic phagocytes, both in tadpoles and adult frogs [18,28,29]. Infiltration of MHC class II+ leukocytes in kidneys at early stages of FV3 infection is also consistent with infiltration of macrophages [25]. CSF-1 and IL-34 are two key macrophage growth factors acting through the same CSF-1 receptor across jawed vertebrates [30]. In *X. laevis*, CSF-1 is required for the commitment and maturation of mature macrophage lineage cells [20], whereas IL-34 stimulates a stronger antiviral effector function by macrophages both in tadpoles and adult frogs [19,31]. The acute decrease in type I and III IFN, as well as CSF-1 and IL-34 at 2 dpi concomitantly with the XNC10 tetramer-mediated iVα6 T cell defect, reinforces the postulated functional link between iVα6 T cells and macrophages. This, of course, does not preclude a possible interaction of iVα6 T cells with other immune cell effectors, as well as a reciprocal effect of macrophages on iVα6 T cells. Indeed, macrophages in mammals are known to be important producers of IL-12 and IL-18, two cytokines that stimulate iNKT cells and promote Th1 adaptive immune response [21,22,32]. The transitory blunted expression of IL-18, and to a lesser extent IL-12, that accompanied iVα6 T cell transient defect, besides being consistent with decreased macrophage function, may have amplified XNC10 tetramer-mediated iVα6 T cell impairment. In any case, it is remarkable that the brief interference of iVα6 T cell function has a long-lasting negative impact on the tadpole ability to control FV3 infection, and ultimately to survive infection.

Author Contributions: Conceptualization, E-S.I.E., F.D.J.A. and J.R.; methodology, E-S.I.E., F.D.J.A., J.Y., K.W. and J.R.; validation, E-S.I.E., F.D.J.A., J.Y. and K.W.; formal analysis, E-S.I.E., F.D.J.A., J.Y., K.W. and J.R.; investigation, E-S.I.E., F.D.J.A., J.Y., and K.W.; resources, J.R.; data curation, E-S.I.E., F.D.J.A.; writing—original draft preparation, E-S.I.E., F.D.J.A., and J.R.; writing—review and editing, E-S.I.E., F.D.J.A., J.Y., K.W. and J.R.; project administration, F.D.J.A and J.R.; funding acquisition, J.R.

Acknowledgments: We thank Tina Martin for animal husbandry.

References

1. Adams, E.J.; Luoma, A.M. The adaptable major histocompatibility complex (MHC) fold: Structure and function of nonclassical and MHC class I-like molecules. *Annu. Rev. Immunol.* **2013**, *31*, 529–561. [CrossRef]
2. Borg, N.A.; Wun, K.S.; Kjer-Nielsen, L.; Wilce, M.C.; Pellicci, D.G.; Koh, R.; Besra, G.S.; Bharadwaj, M.; Godfrey, D.I.; McCluskey, J.; et al. CD1d-lipid-antigen recognition by the semi-invariant NKT T-cell receptor. *Nature* **2007**, *448*, 44–49. [CrossRef] [PubMed]
3. Rossjohn, J.; Pellicci, D.G.; Patel, O.; Gapin, L.; Godfrey, D.I. Recognition of CD1d-restricted antigens by natural killer T cells. *Nat. Rev. Immunol.* **2012**, *12*, 845–857. [CrossRef] [PubMed]
4. Wesley, J.D.; Tessmer, M.S.; Chaukos, D.; Brossay, L. NK Cell–Like Behavior of Vα14i NK T Cells during MCMV Infection. *PLoS Pathog.* **2008**, *4*, e1000106. [CrossRef] [PubMed]
5. Diana, J.; Griseri, T.; Lagaye, S.; Beaudoin, L.; Autrusseau, E.; Gautron, A.-S.; Tomkiewicz, C.; Herbelin, A.; Barouki, R.; von Herrath, M.; et al. NKT Cell-Plasmacytoid Dendritic Cell Cooperation via OX40 Controls Viral Infection in a Tissue-Specific Manner. *Immunity* **2009**, *30*, 289–299. [CrossRef] [PubMed]
6. Raftery, M.J.; Winau, F.; Giese, T.; Kaufmann, S.H.; Schaible, U.E.; Schonrich, G. Viral danger signals control CD1d de novo synthesis and NKT cell activation. *Eur. J. Immunol.* **2008**, *38*, 668–679. [CrossRef] [PubMed]
7. Juno, J.A.; Keynan, Y.; Fowke, K.R. Invariant NKT cells: Regulation and function during viral infection. *PLoS Pathog.* **2012**, *8*, e1002838. [CrossRef]
8. Abboud, G.; Tahiliani, V.; Desai, P.; Varkoly, K.; Driver, J.; Hutchinson, T.E.; Salek-Ardakani, S. Natural Killer Cells and Innate Interferon Gamma Participate in the Host Defense against Respiratory Vaccinia Virus Infection. *J. Virol.* **2016**, *90*, 129–141. [CrossRef] [PubMed]
9. Edholm, E.S.; Banach, M.; Robert, J. Evolution of innate-like T cells and their selection by MHC class I-like molecules. *Immunogenetics* **2016**, *68*, 525–536. [CrossRef]
10. Edholm, E.S.; Albertorio Saez, L.M.; Gill, A.L.; Gill, S.R.; Grayfer, L.; Haynes, N.; Myers, J.R.; Robert, J. Nonclassical MHC class I-dependent invariant T cells are evolutionarily conserved and prominent from early development in amphibians. *Proc. Natl. Acad. Sci. USA* **2013**, *110*, 14342–14347. [CrossRef]
11. Price, S.J.; Garner, T.W.; Nichols, R.A.; Balloux, F.; Ayres, C.; Mora-Cabello de Alba, A.; Bosch, J. Collapse of amphibian communities due to an introduced Ranavirus. *Curr. Biol. CB* **2014**, *24*, 2586–2591. [CrossRef] [PubMed]
12. Gray, M.J.; Chinchar, V.G. *Ranaviruses*; Springer Open: Heidelberg, Germany; New York, NY, USA; Dordrecht, The Netherland; London, UK, 2015.
13. Edholm, E.S.; Grayfer, L.; De Jesus Andino, F.; Robert, J. Nonclassical MHC-Restricted Invariant Valpha6 T Cells Are Critical for Efficient Early Innate Antiviral Immunity in the Amphibian Xenopus laevis. *J. Immunol.* **2015**, *195*, 576–586. [CrossRef] [PubMed]
14. Nieuwkoop, P.; Faber, J. *Normal Table of Xenopus Laevis (Daudin): A Systematical and Chronological Survey of the Development from the Fertilized Egg Till the End of Metamorphosis*, 2nd ed.; North Holland: Amsterdam, The Netherland, 1967.
15. Andino Fde, J.; Grayfer, L.; Chen, G.; Gregory Chinchar, V.; Edholm, E.S.; Robert, J. Characterization of Frog Virus 3 knockout mutants lacking putative virulence genes. *Virology* **2015**, *485*, 162–170. [CrossRef] [PubMed]
16. Chen, G.; Ward, B.M.; Yu, K.H.; Chinchar, V.G.; Robert, J. Improved knockout methodology reveals that frog virus 3 mutants lacking either the 18K immediate-early gene or the truncated vIF-2alpha gene are defective for replication and growth in vivo. *J. Virol.* **2011**, *85*, 11131–11138. [CrossRef]
17. Jacques, R.; Edholm, E.S.; Jazz, S.; Odalys, T.L.; Francisco, J.A. Xenopus-FV3 host-pathogen interactions and immune evasion. *Virology* **2017**, *511*, 309–319. [CrossRef]
18. Morales, H.D.; Abramowitz, L.; Gertz, J.; Sowa, J.; Vogel, A.; Robert, J. Innate immune responses and permissiveness to ranavirus infection of peritoneal leukocytes in the frog Xenopus laevis. *J. Virol.* **2010**, *84*, 4912–4922. [CrossRef]

19. Grayfer, L.; Robert, J. Divergent antiviral roles of amphibian (Xenopus laevis) macrophages elicited by colony-stimulating factor-1 and interleukin-34. *J. Leukoc. Biol.* **2014**, *96*, 1143–1153. [CrossRef]
20. Grayfer, L.; Robert, J. Colony-Stimulating Factor-1-Responsive Macrophage Precursors Reside in the Amphibian (Xenopus laevis) Bone Marrow Rather than the Hematopoietic Sub-Capsular Liver. *J. Innate Immun.* **2013**, *5*, 531–542. [CrossRef]
21. Barbarin, A.; Cayssials, E.; Jacomet, F.; Nunez, N.G.; Basbous, S.; Lefevre, L.; Abdallah, M.; Piccirilli, N.; Morin, B.; Lavoue, V.; et al. Phenotype of NK-Like CD8(+) T Cells with Innate Features in Humans and Their Relevance in Cancer Diseases. *Front. Immunol.* **2017**, *8*, 316. [CrossRef]
22. Huber, C.M.; Doisne, J.M.; Colucci, F. IL-12/15/18-preactivated NK cells suppress GvHD in a mouse model of mismatched hematopoietic cell transplantation. *Eur. J. Immunol.* **2015**, *45*, 1727–1735. [CrossRef]
23. Zajonc, D.M.; Girardi, E. Recognition of Microbial Glycolipids by Natural Killer T Cells. *Front. Immunol.* **2015**, *6*, 400. [CrossRef]
24. Kohlgruber, A.C.; Donado, C.A.; LaMarche, N.M.; Brenner, M.B.; Brennan, P.J. Activation strategies for invariant natural killer T cells. *Immunogenetics* **2016**, *68*, 649–663. [CrossRef] [PubMed]
25. Morales, H.D.; Robert, J. Characterization of primary and memory CD8 T-cell responses against ranavirus (FV3) in Xenopus laevis. *J. Virol.* **2007**, *81*, 2240–2248. [CrossRef]
26. Edholm, E.S.; Goyos, A.; Taran, J.; De Jesus Andino, F.; Ohta, Y.; Robert, J. Unusual evolutionary conservation and further species-specific adaptations of a large family of nonclassical MHC class Ib genes across different degrees of genome ploidy in the amphibian subfamily Xenopodinae. *Immunogenetics* **2014**, *66*, 411–426. [CrossRef] [PubMed]
27. Goyos, A.; Sowa, J.; Ohta, Y.; Robert, J. Remarkable conservation of distinct nonclassical MHC class I lineages in divergent amphibian species. *J. Immunol.* **2011**, *186*, 372–381. [CrossRef]
28. De Jesús Andino, F.; Chen, G.; Li, Z.; Grayfer, L.; Robert, J. Susceptibility of Xenopus laevis tadpoles to infection by the ranavirus Frog-Virus 3 correlates with a reduced and delayed innate immune response in comparison with adult frogs. *Virology* **2012**, *432*, 435–443.
29. Grayfer, L.; De Jesus Andino, F.; Robert, J. Prominent amphibian (Xenopus laevis) tadpole type III interferon response to the frog virus 3 ranavirus. *J. Virol.* **2015**, *89*, 5072–5082. [CrossRef]
30. Droin, N.; Solary, E. Editorial: CSF1R, CSF-1, and IL-34, a "menage a trois" conserved across vertebrates. *J. Leukoc. Biol.* **2010**, *87*, 745–747. [CrossRef]
31. Grayfer, L.; Robert, J. Distinct functional roles of amphibian (Xenopus laevis) colony-stimulating factor-1- and interleukin-34-derived macrophages. *J. Leukoc. Biol.* **2015**, *98*, 641–649. [CrossRef]
32. Yasuda, K.; Nakanishi, K.; Tsutsui, H. Interleukin-18 in Health and Disease. *Int. J. Mol. Sci.* **2019**, *20*, E649. [CrossRef]

17

eDNA Increases the Detectability of Ranavirus Infection in an Alpine Amphibian Population

Claude Miaud [1,*], Véronique Arnal [1], Marie Poulain [1], Alice Valentini [2] and Tony Dejean [2]

[1] CEFE, EPHE-PSL, CNRS, Univ. Montpellier, Univ Paul Valéry Montpellier 3, IRD, Biogeography and Vertebrate Ecology, 1919 route de Mende, 34293 Montpellier, France; veronique.arnal@cefe.cnrs.fr (V.A.); marie.poulain@cefe.cnrs.fr (M.P.)

[2] SPYGEN, 17 Rue du Lac Saint-André, 73370 Le Bourget-du-Lac, France; alice.valentini@spygen.com (A.V.); tony.dejean@spygen.com (T.D.)

* Correspondence: claude.miaud@cefe.cnrs.fr

Abstract: The early detection and identification of pathogenic microorganisms is essential in order to deploy appropriate mitigation measures. Viruses in the Iridoviridae family, such as those in the *Ranavirus* genus, can infect amphibian species without resulting in mortality or clinical signs, and they can also infect other hosts than amphibian species. Diagnostic techniques allowing the detection of the pathogen outside the period of host die-off would thus be of particular use. In this study, we tested a method using environmental DNA (eDNA) on a population of common frogs (*Rana temporaria*) known to be affected by a *Ranavirus* in the southern Alps in France. In six sampling sessions between June and September (the species' activity period), we collected tissue samples from dead and live frogs (adults and tadpoles), as well as insects (aquatic and terrestrial), sediment, and water. At the beginning of the breeding season in June, one adult was found dead; at the end of July, a mass mortality of tadpoles was observed. The viral DNA was detected in both adults and tadpoles (dead or alive) and in water samples, but it was not detected in insects or sediment. In live frog specimens, the virus was detected from June to September and in water samples from August to September. Dead tadpoles that tested positive for *Ranavirus* were observed only on one date (at the end of July). Our results indicate that eDNA can be an effective alternative to tissue/specimen sampling and can detect *Ranavirus* presence outside die-offs. Another advantage is that the collection of water samples can be performed by most field technicians. This study confirms that the use of eDNA can increase the performance and accuracy of wildlife health status monitoring and thus contribute to more effective surveillance programs.

Keywords: eDNA; *Ranavirus*; Common frog; *Rana temporaria*; early detection; virus surveillance

1. Introduction

Amphibians are often considered ecosystem health indicators ("the canaries in the coal mine", [1]) due to their permeable skin, their sensitivity to environmental disturbance, and their often biphasic (requiring water and land) life cycle (e.g., [2]). Over 40% of the more than 7000 species that have been evaluated are classified as threatened today, and the main cause of the disappearance or decline of amphibian populations is the degradation and destruction of their habitats [3]. Emerging infectious diseases (EIDs) are also causing drastic amphibian declines worldwide [4–6], and the chytridiomycosis panzootic has been involved in the decline of at least 501 amphibian species over the past half-century, including 90 presumed extinctions [7]. *Ranaviruses* (family Iridoviridae)—double-stranded DNA viruses that infect fish, reptiles, and amphibians [8]—are considered the second most common infectious cause of mortality in amphibians [9,10]. These viruses have caused amphibian die-offs on five continents [11]. The common midwife toad virus (CMTV) [12] is a *Ranavirus* that has been causing amphibian die-offs since 2005 in Spain [10]. Since then, CMTV-related mortality has been recorded in other European countries, affecting several amphibian species in both Anura and Caudata [13,14].

The rapid and accurate identification of pathogenic microorganisms is essential for the early detection of infection and the deployment of appropriate mitigation measures [15]. Several field studies have reported *Ranavirus*-infected amphibians that do not present clinical signs or histological changes (e.g., [16]). The use of molecular techniques to identify microbial pathogens (e.g., see reviews in [17] for fungus and [18] for viruses) has considerably improved detection possibilities. Molecular tools applied to environmental samples are now widely used to identify infectious agents [19] and are particularly promising for the early detection of aquatic pathogens that can be introduced by non-native species (e.g., [20]). In the last decade, the DNA detection of pathogens in environmental samples (e.g., in water) has been performed successfully to identify several metazoan parasites, fungi (Chytrid and Oomycota), and *Ranaviruses* [21–28]. By comparing water samples and the *Ranavirus* infection status of wood frog tadpoles (*Lithobates sylvaticus*) in several ponds in north-eastern Connecticut in the United States, Hall et al. (2016) [24] demonstrated a strong relationship between the viral load in environmental DNA (eDNA) and larval tissues, indicating the effectiveness of eDNA-based *Ranavirus* detection in the field.

In Europe, *Ranavirus* infections and die-offs have been described in four amphibian families and ten species [8,13]. One of these species is the common frog (*Rana temporaria*), which is experiencing population die-offs in alpine lakes [13]. This study had two key aims: (1) To describe the infection status of different potential hosts of *Ranavirus* (common frog adults, tadpoles, and insects) and ecosystem compartments (sediment and water) in an alpine lake during the activity period (summer), both prior to and after an observed die-off; and (2) to use the eDNA method to detect *Ranavirus* during this infection event. We also compared previous studies using eDNA to detect Chytrids and *Ranavirus* to provide some recommendations for a more effective implementation of water sampling in monitoring programs.

2. Materials and Methods

2.1. Study Area

The study was conducted in Mercantour National Park in the south-eastern Alps in France. The sampled area consisted of several small lakes and a pond. Balaour pond (44.1082 N, 7.3742 E) is 25 × 20 m, and 2355 m a.s.l. It lies at a distance approximately 100 m from the closest lake. The maximum depth is 1.5 m, and the bottom is granitic rock, partly covered by a shallow (max 0.10 m) sedimentary layer of mud. There is no macrophytic vegetation.

The common frog *R. temporaria* uses these kinds of water bodies for breeding and hibernation in this alpine region [29]. In the study area, the common frog is the only amphibian species present, and it breeds only in Balaour pond (no breeding in the neighboring lakes). Many lakes are stocked with brown trout (*Salmo trutta*), and the common minnow (*Phoxinus phoxinus*), which is used as bait, has also been introduced in the lakes. However, there are no trout or minnows in the studied pond.

2.2. Water Temperature

The water temperature during the study was recorded using two data loggers (model iBCod 21G, Maxim/Dallas Semiconductor Inc., San Jose, CA, USA, accuracy +/− 1 °C). These were attached to a PVC tube (50 mm diameter; 40 cm in length) with one end glued to a 20 × 20 cm polystyrene float. One logger was positioned just below the float, and the other was positioned 25 cm from the first logger. They were programmed to record the temperature every two hours. The device was set in the field at 14:00 on the first day of the survey (11 June 2016) and remained there until 10:00 on the last day (19 September 2016). To do this, a 2 m high metallic pole was staked into the ground in the middle of Balaour pond (depth about 1 m). The PVC tube was then fitted to the pole with the float at the top. The data loggers were at a depth of about 5 and 30 cm, respectively. The PVC tube and float were designed to follow changes in water level so the data loggers recorded temperatures at a constant depth.

2.3. *Amphibian Sampling*

Sampling was carried out at Balaour pond on six dates between June and September in 2016 (Table 1), covering the developmental period of embryos and tadpoles in this region and at this altitude.

Table 1. Sampling at Balaour pond in 2016.

Samples	Date					
	10–11 June	23–24 June	08 July	27 July	15–16 August	19 September
Tadpole alive	5 (9–11) [1]	5 (24–25)	5 (30)	5 (35–39)	5 (39–41)	5 (43–45)
Tadpole dead	-	-	-	10 (35–39)	-	
Adult alive [2]	5	5	5	0	1	5
Adult dead	1	-	-	-	-	-
Terrestrial insects [3]	1	1	1	1	1	1
Aquatic insects [4]	1	1	1	1	1	1
Sediment [5]	1	1	1	1	1	1
Water [6]	1	1	1	1	1	1

[1] Stage of development (based on Gosner stages [30]); [2] the distal phalange of the second forelimb toe was collected from adults caught in water; [3] one sample per date, consisting of all flying insects caught with three yellow plates placed around the pond and then pooled together; [4] one sample per date, consisting of all aquatic insects caught by dip-netting in the pond and then pooled together; [5] one sample of sediment per date, consisting of 20 subsamples collected around the pond and then pooled together; [6] one sample of water per date, consisting of 20 subsamples collected around the pond, filtered and pooled together (see M&M for more details).

The pond was visually scanned to detect tadpoles and adults resting on the bottom. Live tadpoles were caught by a dip-net (mesh 1 mm). They were stored in individual plastic bottles (1.5 mL) filled with 95% ethanol. Dead tadpoles were collected and stored similarly. Live adults were caught by a dip-net (5 mm mesh), and the distal phalange of the second forelimb toe was collected and stored individually in a snap-cap tube (1.5 mL) filled with 95% ethanol. Disposable gloves were used and changed between each sampling, and the material (scissors and pliers) was soaked in ethanol and passed through a flame between each sample collection. The dead adults were collected with the dip-net and stored in a plastic bottle (300 mL) filled with 95% ethanol. Table 1 shows the sample size and developmental stages [30] of the tadpoles and adults collected. The toe clipping allowed us to check that unique adult specimens were sampled during the study period.

2.4. *Insect Sampling*

Both terrestrial and aquatic insects were collected in the six sampling sessions (Table 1). Terrestrial insects were collected using three yellow plastic plates placed on the ground about 10 m from the edge of the pond. These plates were filled with water and one drop of domestic detergent. After one hour of sampling, the flying insects found dead in the liquid were collected and stored in a 15 mL plastic bottle filled with 95% ethanol. Aquatic insects were caught by dip-netting (1 mm mesh), mostly in the shallow area of the pond (less than 0.60 m deep). To limit contamination with the surrounding water, each specimen was removed with needle-nose pliers and washed with distilled water. The aquatic insects collected on the same date were stored in the same 15 mL plastic bottle filled with 95% ethanol.

2.5. *Water Sampling*

The field survey method was modified from that used in [31]. Using a sterile water-sample dipper, a 100 mL water sample was collected at 20 locations equally spaced around the edge of Balaour pond, resulting in a pooled sample of approximately 2 L contained in a sterile self-supporting plastic bag. Samples were collected from the top 0.10 m of the water column preceded by a gentle circular movement with the sampling ladle. Surveyors stood on the pond bank without entering the water to avoid possible contamination from their boots or from stirring up sediment. The 2 L water sample was homogenized by gently shaking the bag to ensure the eDNA was evenly mixed throughout the sample, and the whole 2 L water sample was then filtered directly in the field through a VigiDNA 0.45 µm filter (SPYGEN, Le Bourget du Lac, France) using a sterile 100 mL syringe. The filter was filled with 80 mL of a CL1 conservative buffer (SPYGEN) and stored at room temperature before DNA extraction.

2.6. Sediment Sampling

At each of the six sampling sessions, sediment (approximately 10 mL) was collected from 20 locations evenly distributed around Balaour pond. Each sediment sample was collected approximately 0.5 m from the shoreline using a sterile syringe (20 mL) pushed into the top 5 cm of sediment. The 20 samples were placed into a shared sterile bag, mixed together, and then stored in a sterile wide-neck barrel (1 L).

2.7. Ranavirus Detection in Frogs, Tadpoles and Insects

From the adult frogs, a small piece (2 mm^3) of tissue was collected: Liver tissue from dead specimens (which were dissected) and toe tissue from live specimens. For small tadpoles, the total tadpole was used (e.g., individuals <10 mm, Gosner stage 25). For larger tadpoles (Gosner stages 30–45), individuals were dissected, and a total tissue volume of about 2 mm^3 was collected, mostly composed of the liver and heart. For both aquatic and terrestrial insects, specimens were grouped per date of sampling and crushed together in 5 mL of 95% ethanol. A subsample of 1.5 mL was collected for DNA analysis.

The DNA was extracted with a REDExtract-N-Amp Plant Kit (Sigma-Aldrich-Merck, Darmstadt, Germany) by incubating a piece of tissue or the bulk of insects in 50 µL of extraction solution at 95 °C for 20 min after ethanol evaporation (56 °C for 30 min). An equal volume of dilution solution was added to the extract to neutralize inhibitory substances before a polymerase chain reaction (PCR).

The Taqman real-time quantitative PCRs were performed following the protocol described in detail in [13], using primers and probes designed by Leung et al. (2017) [32]. All assays were performed in triplicate. *Ranavirus* DNA, provided in several densities by Stephen Price (Zoological Society, London, UK), was used to calibrate the standard curve.

A sample was considered positive if the amplification curves were similar to those in positive controls (i.e., shape, cycle threshold, values, and > 0.1 genomic equivalent), and at least two replicates gave positive amplification. If only one well resulted in a positive signal, the sample was rerun and was considered positive if at least two wells gave a positive amplification signal. The results were expressed in terms of genomic equivalent (GE) and prevalence (number of positive samples/total number of samples).

2.8. Ranavirus Detection in Sediment

The sediment samples were weighed, and a similar weight of a saturated phosphate buffer (Na_2HPO_4; 0.12 m; pH $\approx$ 8) was added, as described in [33]. The DNA extraction was performed using a commercial kit for soil DNA (NucleoSpin® Soil; Macherey-Nagel, Düren, Germany), following the manufacturer's instructions. The PCR amplifications were performed following the protocol previously described for amphibian tissue and insects.

2.9. Ranavirus Detection in Water

For the water samples, DNA extraction was performed following the protocol described in [34] in a dedicated room for water DNA extraction equipped with positive air pressure, UV treatment, and frequent air renewal. Before entering this extraction room, personnel used a connecting zone to change into full protective clothing, comprising a disposable body suit with hood, mask, laboratory shoes, overshoes, and gloves. All workbenches were decontaminated with commercial bleach, diluted to achieve a 0.5% sodium hypochlorite solution, before and after each manipulation. For DNA extraction, each filtration capsule, containing the CL1 buffer, was agitated for 15 min on an S50 shaker (cat Ingenieurbüro™) at 800 rpm, and then the buffer was emptied into a 50 mL tube before being centrifuged for 15 min at 15,000× *g*. The supernatant was removed with a sterile pipette, leaving 15 mL of liquid at the bottom of the tube. Subsequently, 33 mL of ethanol and 1.5 mL of 3 M sodium acetate were added to each 50 mL tube and stored for at least one night at −20 °C. The tubes were centrifuged at 15,000× *g* for 15 min at 6 °C, and the supernatants were discarded. After this step, 720 µL of an ATL buffer from the DNeasy Blood & Tissue Extraction Kit (Qiagen, Hilden, Germany) was added. The tubes were then vortexed, and the supernatants were transferred to 2 mL tubes containing 20 µL of Proteinase K. The tubes were then incubated at 56 °C for two hours. Subsequently, DNA extraction was performed using NucleoSpin®Soil (MACHEREY-NAGEL GmbH & Co., Düren, Germany) starting from step 6 and following the manufacturer's instructions. The elution was performed by adding 100 µL of a SE buffer twice. After DNA extraction, the samples were tested for inhibition using real-time amplification following the protocol described in Biggs et al. (2015) [31], which involved adding a synthetic DNA sequence to each sample and then trying to amplify it. None of the samples were found to be inhibited. The samples were amplified using primers and probes designed by Leung et al. (2017) [32]. The qPCR was carried out in 12 replicates on a final volume of 25 µL, using 3 µL of template DNA, 12.5 µL of TaqMan Environmental Master Mix 2.0 (Life Technologies, Carlsbad, CA, USA), 6.5 µL of ddH2O, 1 µL of forward primer and reverse primer, and 1 µL of probe (MCP_probe) using thermal cycling at 50 °C for 5 min and 95 °C for 10 min, followed by 50 cycles at 95 °C for 30 s and 60 °C for 1 min. To detect potential contamination, qPCR negative controls and DNA extraction controls (with 12 replicates) were amplified in parallel. Standard curve calculations were based on three standards (4.5×10^7, 4.5×10^4, and 4.5×10^1 target copies per 3 µL) made from a plasmid containing the viral MCP target. Samples were run on a BIO-RAD CFX96 Touch real-time PCR detection system in a room dedicated to amplified DNA analysis with negative air pressure and physically separated from the DNA extraction room.

The results were expressed in terms of the number of positive replicates/total number of replicates per sample and the mean number of DNA copies (using the number of copies per Rv+ replicates) per sample.

2.10. *Ranavirus Identification*

Viral DNA was obtained from 1 dead adult, 3 live adults, 5 dead tadpoles, and 5 live tadpoles. These 14 samples were sequenced following the Mao et al. (1999) PCR method [35], using the BigDye Terminator Cycle Sequencing kit (PE Biosystems, Thermo Fisher Scientific, Bleiswijk, Netherlands) on an ABI Prism 3100 (Applied Biosystems, Foster City, CA, USA). The electropherogram was exported and converted to Kodon (Applied Maths, Sint-Martens-Latem, Belgium), and, using the BLAST program (default settings), the sequences were compared to *Ranavirus* sequences previously identified in this region [15].

3. Results

3.1. *Field Observations*

At altitudes above 2000 m, the common frog breeds as soon as a small open body of water is available. The breeding population in Balaour pond has long been observed, but the population size has not been precisely estimated. The number of egg-masses is regularly around several hundred (M.-F. Lecchia, pers.comm.).

At Sampling Session 1 (10–11 June 2016), the breeding season had started, and about 200 egg-masses were counted. The embryos were at stages 9–11. Sixteen adults were observed in the water. One dead frog was observed at the bottom of the pond. At Session 2 (23–24 June), the tadpoles (stages 24 and 25) formed large aggregates on the remains of the spawn (jelly) along the pond shore. Fewer than 10 adults were observed in the water. At Session 3 (8 July), the tadpoles (stage 30) were grouped along the pond shore where the water temperature was highest. Five adults were observed in the water. At Session 4

(27 July), the tadpoles (stages 35–39) were grouped along the pond shore. Dead tadpoles (about 250 counted) were observed on the bottom of the pond. Several individuals were observed with atypical behaviour, such as slow movements when stimulated, lateral swimming, or lying on their back on the pond bottom (about 50 counted). Most of the dead tadpoles were eaten by their congeners (Figure 1). Only two adults were observed in the water. At Session 5 (15–16 August), the tadpoles (stages 39–41) were more widespread in the pond, i.e., observed swimming beyond the 2 m strip near the shoreline. While we did not estimate the (relative) density at each session, the number of live tadpoles observed at Session 5 was clearly lower than at Session 4. No dead tadpoles were observed, nor tadpoles with the previously described atypical behaviour. Only two adults were observed in the pond. At Session 6 (19 September), most of the tadpoles had metamorphosed or were very close to metamorphosis (stages 43–45) and were widespread along the pond shore, both in the water and on land. No dead tadpoles or froglets were observed. Eight adults were observed in the pond.

Figure 1. Common frog *Rana temporaria* tadpoles feeding on their dead congeners (photo L. Miaud).

Terrestrial insects (Diptera, Chironomids, Hymenoptera, Rhopalocera, and ants) and aquatic insects (Diptera, Heteroptera, and Odonata) were present around and in Balaour pond from June to September.

The two data loggers allowed the water temperature to be recorded at the surface and at a depth of 30 cm (Figure 2). The temperature increased from about 7 °C in mid-June to 14 °C at the beginning of July. The mean temperature during the July–August sampling period was 14 °C ± 4.3, with a maximum of 20.5 °C (at the surface) and 20 °C (at a depth of 30 cm) in July. In September, the temperature decreased to about 7 °C.

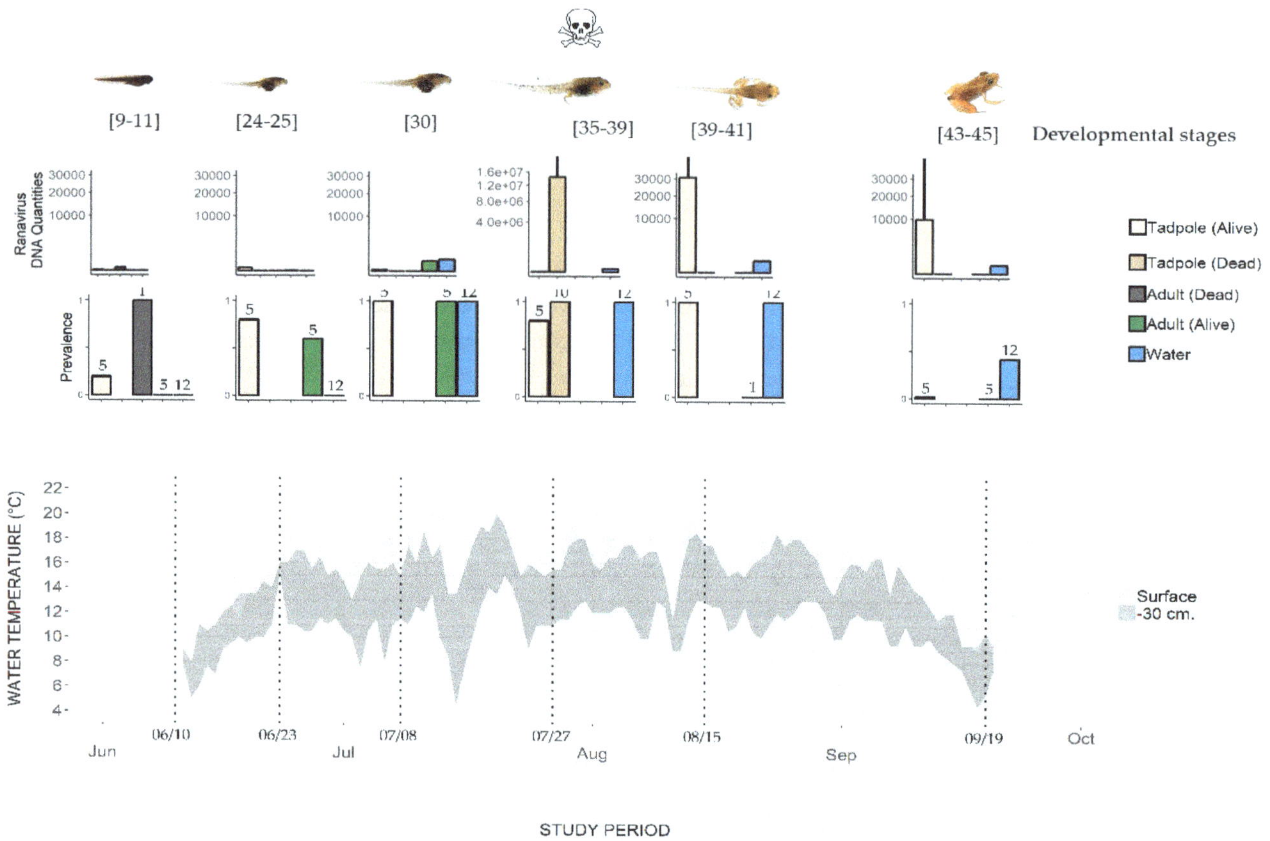

Figure 2. Change in water temperature and *Ranavirus* prevalence and load in a pond where the mass mortality of common frog tadpoles was observed. Water temperature was recorded at the surface and 30 cm below the surface. Top row of histograms: *Ranavirus* DNA quantities in common frogs (adults and tadpoles) and water samples (log scales, mean value, and SD). Bottom row of histograms: Prevalence (number of *Ranavirus* positive specimens/total number of sampled specimens, with sample size indicated above the bars). Developmental stages are based on Gosner stages; mortality was observed only on July 27.

3.2. *Ranavirus* Detection in Organisms

Ranavirus was detected in several compartments of the studied ecosystem (Supplementary Materials S1 and Figure 2) during the common frog activity period (June–September).

At Sampling Session 1 (10–11 June), the adult frogs present in the pond tested negative for *Ranavirus* (hereafter Rv-). The dead frog found on the pond bottom tested positive for *Ranavirus* (hereafter Rv+) with a genomic equivalent (GE) load of 2.83×10^1 (Supplementary Materials S1 and Figure 2). At Session 2, 12 days later (23–24 June), three out of five adults were Rv+, with a mean GE of 0.43 ±. 1.45 ($n = 3$). Adults caught in the water continued to be Rv+ (four out of five testing positive) at Session 3 (8 July), with the highest mean GE (345 ± 14.5, $n = 4$, Figure 2). No adults were sampled at Session 4 (27 July). The individual caught at Session 5 (15–16 August) and the five caught at Session 6 (19 September) were all Rv- (Figure 2). Most of the frogs that reproduce in Balaour pond leave the water after spawning to reach their surrounding terrestrial summer habitats. We did not catch adult frogs on land, so we do not know the infection status of these adults.

One recently hatched tadpole, i.e., hatched for less than one week in the prevailing environmental conditions at this altitude (water temperature about 6 °C, Figure 2 and [29]), of the five tested was Rv+ (Table 2). Thereafter, live tadpoles were Rv+ until metamorphosis (Session 6; 19 September). The prevalence of *Ranavirus* increased with time, reaching 100% ($n = 5/5$) at Session 5 (15–16 August), and then dropped to 20% ($n = 1/5$) at Session 6 (19 September) (Figure 2). The variation in mean GE across the tadpole developmental period shows a rather low viral load until Session 4 (27 July) in live tadpoles, while the load in dead tadpoles was higher by an order of magnitude of 7 (10^7). After this date, the GE values in live tadpoles stayed high (10^4).

None of the terrestrial and aquatic insects collected from June to September tested positive for *Ranavirus*.

3.3. *Ranavirus* Detection in Water

Water samples from the first two sampling sessions were Rv- (Figure 2). From Session 3 (8 July) to Session 5 (15–16 August), the 12 qPCR replicates performed for each water sample were Rv+. *Ranavirus* was still detected at Session 6 (19 September), though it had the lowest detectability (5/12 replicates). The number of DNA copies (Figure 2) reached its maximum on 27 July (by an order of magnitude of 2), when the dead tadpoles were also observed with a high GE load. However, the GE values in infected live tadpoles and the number of DNA *Ranavirus* copies in water were not significantly correlated (Spearman rank correlation $R = 0.25, p > 0.05$).

3.4. *Ranavirus* in Sediment

All the qPCR replicates were negative for the samples collected at Sessions 1, 2, 4, and 6. At Sessions 3 and 5, 1 qPCR of the three was positive for both samples; these two samples were rerun. Only one positive replicate was observed again, so the two sediment samples were also considered negative.

3.5. *Ranavirus* Identification

The sequences obtained from the DNA extracted from dead tadpoles ($n = 5$), live tadpoles ($n = 5$), dead adults ($n = 1$), and live adults ($n = 3$) were 100% identical to the CMTV (GenBank accession number JQ231222) isolated from a common midwife toad (*Alytes obstetricans*; [12]) and an alpine newt (*Ichthyosaura alpestris*) in Spain [36]. This *Ranavirus* has been identified as the etiologic agent of the mass mortality of the common frogs observed in the region [13].

4. Discussion

4.1. Seasonal Dynamics of *Ranavirus* Epidemics in a Common Frog Population

The common frogs (tadpoles and adults, dead and alive) testing positive for *Ranavirus* (CMTV, [12]) in Balaour pond in 2016 confirm the widespread distribution of this pathogen in the south-eastern Alps [13].

At the first sampling session (10 June), egg-masses had been deposited in the pond, but only 16 adults were observed in the water, and they did not exhibit breeding behavior. Spawning in the common frog is synchronous and short in duration [37]. On this date, one dead adult was observed in the water. This specimen was Rv+. Of the live adults, 5 specimens were Rv-, while 1 out of 5 live tadpoles was Rv+. Of the few adults that remained in the pond until July, some were Rv+, but no further adult mortality was observed in the pond. In contrast, tadpoles suffered mass mortality at the end of July, when they reached developmental stages 35–39. *Ranavirus* was not detected in the sampled insects (aquatic or terrestrial), nor in the sediment collected during the summer activity period.

Sampling different life stages of organisms, as well as biotic and abiotic components, is necessary to provide a comprehensive view of *Ranavirus* dynamics (e.g., [38]). The observed dynamics in Balaour pond raise several questions: *Ranavirus* (including CMTV) is often highly pathogenic, and adult mass mortalities in common frog populations have been described in this region [13]. Yet adult mass mortality was not observed in Balaour pond in 2016. The breeding population is estimated to consist of at least 400 breeding adults (based on the observation of approximately 200 egg-masses in the pond and assuming an unbiased sex ratio [39]). Frog mass mortality is easily detectable in these small alpine ponds, so it is unlikely that an adult die-off went undetected. While frog aggregation during breeding can potentially foster pathogen transmission [40,41], the small fraction of adults that remained in the water after breeding was not infected at the beginning of the activity season. We have no data on the infection status of adults coming to breed or leaving the water after breeding, and the source and timing of the *Ranavirus* introduction remain unknown. It is unlikely that *Ranavirus* persists in the pond water from one year to the next [42], so it may be that sub-lethally infected adults (hibernating in water or on land) expose hatchlings to *Ranavirus* each year. Other reservoir species may also contribute to seasonal epidemics [43]; *Ranavirus* has been detected in fish in neighboring lakes [13]. However, the studied pond is free of fish, and the other communities we tested (aquatic and terrestrial insects) were Rv-. The characteristics of *Ranavirus* persistence from year to year in this pond remain to be studied.

In contrast to adults, tadpoles did suffer mass mortality. Rapid and synchronous mass mortality of tadpoles in frog populations is well known [38]. The sudden introduction of a virus can be the cause of such a die-off, but this does not seem to be the case in Balaour pond, as *Ranavirus* was detected from the very beginning of the activity season. The existence of a window of host susceptibility mediated by environmental conditions (e.g., temperature) may contribute to this pattern (e.g., [38]). *Ranavirus* epidemics often occur during late spring or summer and can begin and end within weeks [38,44–47]. Infected tadpoles in Balaour pond were observed throughout the activity season (from June to September), i.e., from early developmental to pre-metamorphosis stages, while tadpoles died in mass only at the end of July, at stages 35–39 (Figure 2). Hall et al. (2018) [38] observed that *Ranavirus* prevalence reached high levels (>50%) in wood frog tadpoles up to six weeks before mass mortality. High prevalence was also observed in the common frog population one month before the mass mortality in our study. In the wood frog [38], mortality occurred when tadpoles reached developmental stages (hind limb formation) that coincide with higher water temperatures (>15 °C). Several studies indicate that the pathogenicity of *Ranavirus* depends on the individual's developmental stage, with the most susceptible stage varying between species [10,11,43,44]. In our study, the tadpole die-off was found to coincide with the highest virus load in tadpole tissue (Figure 2, [48,49]). The accumulation of infectious dead tadpoles may also facilitate transmission (Figure 1, [50–52]). Water temperature, independent of developmental stage, has been shown to increase virus pathogenicity [53]. Common frog tadpoles suffer greater mortality at 20 °C than at 15 °C, whether exposed to *Ranavirus* (FV3) or not [54]. The mass mortality of tadpoles observed in Balaour pond corresponded with the highest temperature recorded during the sampling period (Figure 2, 20 °C at the surface). However, no mortality was observed at the beginning of July or the end of August, when similar high temperatures were recorded. The respective and interacting roles of water temperature and tadpole developmental stage on *Ranavirus* pathogenicity remain to be studied in this common frog population.

4.2. Designing Pathogen Surveys Using eDNA and Occupancy Models

Diagnosing *Ranavirus* infection in amphibians requires collecting samples from either live individuals (e.g., a piece clipped from the tail or toe or skin swabs [55–57]) or dead individuals (e.g., a sample from the liver) [55]). However, *Ranavirus* can also resist adverse conditions (e.g., the drying out or freezing of a host carcass) and can then be shed into the water from infected individuals [58,59]. While viruses are rapidly degraded by microbes and zooplankton predation in water [42], they can remain detectable for at least seven days [60]. Water sampling and eDNA testing can thus indicate the presence of the pathogen; Hall et al. (2016) demonstrated that in ponds where wood frog tadpoles were Rv+, collected water samples also revealed the presence of *Ranavirus* DNA [24]. Other pathogens have also been detected using water samples—several metazoan parasites [26,61,62], chytrid fungus [21–25,27,63,64], and the Oomycota fungus *Aphanomyces astaci*, the causative agent of crayfish plague [28]—demonstrating the usefulness of eDNA for detecting pathogen infections in wildlife.

Table 2. Water sampling for Chytris and *Ranavirus* detection.

Pathogen	Filter Mesh Water Collection	Volume of Water Sample	Field Sampling	Date	Occupancy Design	Reference
Chytrid	0.45 µm Peristaltic pump	0.05 to 2.3 L per site (filter clogged)	In shallow water (0.1 to 0.75 m deep) in known or likely amphibian habitats. 3 locations per site, 4 sites	1 date	no	[21]
Chytrid	0.45 µm 50 mL syringe	<1 L per site (filter clogged)	Within 10 cm of the edge 1 location per site, 42 sites	1 date	no	[22]
Chytrid	0.22 µm 60 mL syringe	600 mL per site	Samples spaced evenly along the entire site circumference, but taken only from areas where frogs or tadpoles were present 30 samples of 20 mL combined, 20 sites	4 dates	No, yes *	[23,63]
Chytrid	1.2 µm Hand pump	500 to 1500 mL per site	Every 40 m along the shoreline. 5 samples of 50 mL combined, 13 sites	1 date	no	[26]
Chytrid	0.22 µm 60 mL syringe	20 mL to 2.4 L per site	In shallow water (5 and 20 cm below the water surface) 3 spatial replicates per site, 41 sites	1 date	yes	[64]
Ranavirus	0.2 µm Disposable paper cup	750 mL 250 mL × 3 per site	At 3 distinct locations (north, east and west) along the shore and surface (ca. 10 cm deep) 3 spatial replicates per site, 20 sites	1 date	yes	[24]
Ranavirus	0.2 µm Disposable paper cup	750 mL 250 mL × 3 per site	As above 8 sites	16 dates	no	[27]
Chytrids *Ranavirus*	0.2 µm Hand pump	0.6–1 L 150–250 mL × 4 per site	4 locations approximately equidistant around the site (within 2 m of the edge and 20–40 cm below the surface), 4 spatial replicates per site; 21 sites	3 dates	yes	[65]
Ranavirus	0.45 µm Sterile water-sample dipper	2 L 100 mL × 20 combined	20 locations equidistant around the site (within 0.5 m of the edge and 10 cm below the surface), 1 site	6 dates	no	This study

* the same data set was used in the two studies.

Table 2 summarizes different methods used to collect water for pathogen detection (Chytrids and *Ranavirus*). The most usual method is to collect water samples that are then filtered in the field. The filters are then stored until DNA extraction and amplification in the laboratory. The sampling design varies, e.g., discrete samples may be filtered separately or combined before filtering (Table 2). The location of the water sampling also differs, from rather exhaustive sampling (e.g., at approximately equidistant locations around the site) to solely where the presence of the pathogen would be expected (e.g., locations where frogs or tadpoles have been present). Water collection and filtering procedures are simple and have been used by non-experts in a citizen science program [65]. Clearly, the detection of a pathogen in water samples depends on its density (DNA quantities) and distribution (e.g., continuous or patchy) in the studied site. Several estimates concerning sampling effort are available; Julian et al. (2019) evaluated that as few as five water samples (Table 2) taken in June or July can detect both Bd and *Ranavirus* with 95% confidence [65]. In another study, every pond with *Ranavirus*-infected tadpoles tested positive with just three 250 mL water samples (Table 2, [27]). In our study, *Ranavirus* detection from water samples was effective in four out of six sessions, i.e., 66.6% detectability. In contrast, mass mortality was observed on only one date, i.e., 16.6% detectability. When live tadpoles were collected and molecular diagnostics performed, the presence of the pathogen was detected throughout the six sampling dates (100% detectability). Increasing the number of water samples will increase the cost of the molecular diagnostics, but pooling the samples before filtration allows a good sampling coverage of the study area at a reasonable cost. When the distribution of the pathogen (or the potential hosts) is known, we recommend to perform a pilot study to optimize the number (or volume) of water samples. Without preliminary knowledge, the screening of the site can be based on samples spaced evenly along the entire site circumference, then combined in one water samples which is then filtered to collect DNA.

Nonetheless, most pathogen detection tests are imperfect, resulting in false negatives when a pathogen is present but not detected. False negatives can occur when specimens or water samples are collected in the field, from organs in infected hosts (histological diagnostics) or in PCR replicates (molecular diagnostics). To address this, multi-scale occupancy models [66] can be used to develop a species distribution model to estimate the proportion of sites where a species or a pathogen occurs [67], even when detection is imperfect. This statistical framework is particularly suitable for eDNA analyses in which replicates (and thus detection probability) are present at the different steps of the method [68]; for example, it has been recommended in disease ecology [69]. Table 2 provides some studies that used eDNA to detect Bd or *Ranavirus* as well as occupancy models to estimate pathogen prevalence. Naïve prevalence (i.e., estimated prevalence without taking into account imperfect detection) can lead to the underestimation of Bd prevalence in ponds [23]; estimating correct prevalence requires taking multiple samples per site (Table 2). Replicates to evaluate detection probabilities can be temporal (i.e., sampling the same site at least three occasions) or spatial (i.e., sampling in at least three locations in the site) [66].

5. Conclusions

Inferring the health status of common frogs or ecologically similar amphibian species based only on the observation of mass mortality events in breeding sites runs the risk of missing potential infections. A better strategy is to sample amphibian tissue (adults and/or tadpoles) during, for example, the tadpole developmental phase and then perform molecular diagnostics. As this requires invasive techniques, administrative authorizations for protected species and/or protected areas and specific materials to collect and store the samples are needed. In some cases, catching frogs and tadpoles can be challenging, and tissue sampling requires a certain level of technical competence (and, sometimes, certified training). In light of these challenges, our results confirm previous findings that eDNA can be an effective alternative, since the temporal window for detecting the pathogen is rather large, collecting water samples is simple and can be performed by most field technicians, and, moreover, no specimen has to be caught, manipulated, or sacrificed. Moreover, applying a sampling design based on occupancy modelling (e.g., replicates allowing the evaluation of detection probability) provides prevalence estimates that are comparable between sites or years. As a biodiversity inventory method in marine and freshwater environments, eDNA has been extraordinarily successful; likewise, it holds clear potential for improving disease surveillance programs and increasing the performance of wildlife health status monitoring.

Author Contributions: C.M. conceived and designed the experiment. C.M. conducted the experiment in the field. V.A., A.V., M.P., and T.D. performed the pathogen detection analysis. C.M., A.V., and T.D. wrote the paper.

Acknowledgments: We would like to thank Coralie Barbier, Léo Miaud, Benjamin Viel, and the technicians of the Mercantour National Park for their help in the field, Marie-Laure Leccia for her support in studying the health status of the park's amphibian population, and Stephen Price for the MCP standards and help with the qPCR protocol. Discussions with A. Duffus, T. Garner, W. Leung, S. Price, and F. Pasmans on the design of the longitudinal study were greatly appreciated. Jules Chiffard and Antoine Miaud kindly created Figure 2. Authorisation to collect the tadpoles and to sample tissue from the adults was provided by Mercantour National Park (authorisation n° 2016-408, 12 May 2016) and the Alpes-de-Haute-Provence Regional Department of the Environment, Development and Housing (prefectural decree, 20 May 2016). Many thanks to Elise Bradbury (elisebrad@gmail.com) for the text review and editing. A first draft of this manuscript was highly improved by the reviewer's constructive comments.

References

1. Miaud, C.; Guillaume, O. Are amphibian good biological indicators? In Proceedings of the Abstract of the Fifth World Congress of Herpetology 2005, Stellenbosch, South Africa, 19–24 June 2005; p. 72.
2. Welsh, H.H.; Ollivier, L.M. Stream Amphibians as Indicators of Ecosystem Stress: A Case Study from California's Redwoods. *Ecol. Appl.* **1998**, *8*, 1118–1132. [CrossRef]
3. Temple, H.J.; Cox, N.A. *European Red List of Amphibians*; Office for Official Publications of the European Communities: Luxembourg, 2009.
4. Daszak, P.; Berger, L.; Cunningham, A.A.; Hyatt, A.D.; Green, D.E.; Speare, R. Emerging infectious diseases and amphibian population declines. *Emerg. Infect. Dis* **1999**, *5*, 735–748. [CrossRef] [PubMed]
5. Lips, K.R.; Brem, F.; Brenes, R.; Reeve, J.D.; Alford, R.A.; Voyles, J.; Carey, C.; Livo, L.; Pessier, A.P.; Collins, J.P. Emerging infectious disease and the loss of biodiversity in a Neotropical amphibian community. *Proc. Natl. Acad. Sci. USA* **2006**, *103*, 3165–3170. [CrossRef] [PubMed]
6. Rachowicz, L.J.; Knapp, R.A.; Morgan, J.A.T.; Stice, M.J.; Vredenburg, V.T.; Parker, J.M.; Briggs, C.J. Emerging infectious disease as a proximate cause of amphibian mass mortality. *Ecology* **2006**, *87*, 1671–1683. [CrossRef]
7. Scheele, B.C.; Pasmans, F.; Skerratt, L.F.; Berger, L.; Martel, A.; Beukema, W.; Acevedo, A.A.; Burrowes, P.A.; Carvalho, T.; Catenazzi, A.; et al. Amphibian fungal panzootic causes catastrophic and ongoing loss of biodiversity. *Science* **2019**, *363*, 1459–1463. [CrossRef] [PubMed]
8. Gray, M.J.; Chinchar, V.G. *Ranaviruses: Lethal Pathogens of Ectothermic Vertebrates*; Springer Science + Business Media: Berlin, Germany, 2015; ISBN 978-3-319-13755-1.
9. Chinchar, V.G. Ranaviruses (family Iridoviridae): Emerging cold-blooded killers—Brief review. *Arch. Virol.* **2002**, *147*, 447–470. [CrossRef] [PubMed]
10. Price, S.J.; Garner, T.W.J.; Nichols, R.A.; Balloux, F.; Ayres, C.; Mora-Cabello de Alba, A.; Bosch, J. Collapse of Amphibian Communities Due to an Introduced Ranavirus. *Curr. Biol.* **2014**, *24*, 2586–2591. [CrossRef]
11. Miller, D.; Gray, M.; Storfer, A. Ecopathology of Ranaviruses Infecting Amphibians. *Viruses* **2011**, *3*, 2351–2373. [CrossRef]
12. Balseiro, A.; Dalton, K.P.; del Cerro, A.; Marquez, I.; Cunningham, A.A.; Parra, F.; Prieto, J.M.; Casais, R. Pathology, isolation and molecular characterisation of a ranavirus from the common midwife toad Alytes obstetricans on the Iberian Peninsula. *Dis. Aquat. Organ.* **2009**, *84*, 95–104. [CrossRef]
13. Miaud, C.; Pozet, F.; Gaudin, N.C.G.; Martel, A.; Pasmans, F.; Labrut, S. Ranavirus Causes Mass Die-Offs of Alpine Amphibians in the Southwestern Alps, France. *J. Wildl. Dis.* **2016**, *52*, 242–252. [CrossRef]
14. Saucedo, B.; Hughes, J.; Spitzen-van der Sluijs, A.; Kruithof, N.; Schills, M.; Rijks, J.M.; Jacinto-Maldonado, M.; Suarez, N.; Haenen, O.L.M.; Voorbergen-Laarman, M.; et al. Ranavirus genotypes in the Netherlands and their potential association with virulence in water frogs (Pelophylax spp.). *Emerg. Microbes Infect.* **2018**, *7*, 56. [CrossRef] [PubMed]
15. Anonymous. Preventing emerging infectious diseases: A strategy for the 21st century. Overview of the updated CDC plan. *MMWR Recomm. Rep.* **1998**, *47*, 1–14.

16. Brunner, J.L.; Schock, D.M.; Davidson, E.W.; Collins, J.P. Intraspecific reservoirs: Complex life history and the persistence of a lethal ranavirus. *Ecology* **2004**, *85*, 560–566. [CrossRef]
17. Tibpromma, S.; Hyde, K.D.; Jeewon, R.; Maharachchikumbura, S.S.N.; Liu, J.-K.; Bhat, D.J.; Jones, E.B.G.; McKenzie, E.H.C.; Camporesi, E.; Bulgakov, T.S.; et al. Fungal diversity notes 491–602: Taxonomic and phylogenetic contributions to fungal taxa. *Fungal Divers.* **2017**, *83*, 1–261. [CrossRef]
18. Mokili, J.L.; Rohwer, F.; Dutilh, B.E. Metagenomics and future perspectives in virus discovery. *Curr. Opin. Virol.* **2012**, *2*, 63–77. [CrossRef] [PubMed]
19. Guy, R.A.; Payment, P.; Krull, U.J.; Horgen, P.A. Real-time PCR for quantification of Giardia and Cryptosporidium in environmental water samples and sewage. *Appl. Environ. Microbiol.* **2003**, *69*, 5178–5185. [CrossRef] [PubMed]
20. Ganoza, C.A.; Matthias, M.A.; Collins-Richards, D.; Brouwer, K.C.; Cunningham, C.B.; Segura, E.R.; Gilman, R.H.; Gotuzzo, E.; Vinetz, J.M. Determining risk for severe leptospirosis by molecular analysis of environmental surface waters for pathogenic Leptospira. *PLoS Med.* **2006**, *3*, 1329–1340. [CrossRef]
21. Walker, S.F.; Salas, M.B.; Jenkins, D.; Garner, T.W.J.; Cunningham, A.A.; Hyatt, A.D.; Bosch, J.; Fisher, M.C. Environmental detection of Batrachochytrium dendrobatidis in a temperate climate. *Dis. Aquat. Organ.* **2007**, *77*, 105–112. [CrossRef] [PubMed]
22. Kirshtein, J.D.; Anderson, C.W.; Wood, J.S.; Longcore, J.E.; Voytek, M.A. Quantitative PCR detection of batrachochytrium dendrobatidis DNA from sediments and water. *Dis. Aquat. Organ.* **2007**, *77*, 11–15. [CrossRef]
23. Schmidt, B.R.; Kery, M.; Ursenbacher, S.; Hyman, O.J.; Collins, J.P. Site occupancy models in the analysis of environmental DNA presence/absence surveys: A case study of an emerging amphibian pathogen. *Methods Ecol. Evol.* **2013**, *4*, 646–653. [CrossRef]
24. Hall, E.M.; Crespi, E.J.; Goldberg, C.S.; Brunner, J.L. Evaluating environmental DNA-based quantification of ranavirus infection in wood frog populations. *Mol. Ecol. Resour.* **2016**, *16*, 423–433. [CrossRef] [PubMed]
25. Kamoroff, C.; Goldberg, C.S. Using environmental DNA for early detection of amphibian chytrid fungus Batrachochytrium dendrobatidis prior to a ranid die-off. *Dis. Aquat. Organ.* **2017**, *127*, 75–79. [CrossRef] [PubMed]
26. Sato, M.O.; Rafalimanantsoa, A.; Ramarokoto, C.; Rahetilahy, A.M.; Ravoniarimbinina, P.; Kawai, S.; Minamoto, T.; Sato, M.; Kirinoki, M.; Rasolofo, V.; et al. Usefulness of environmental DNA for detecting Schistosoma mansoni occurrence sites in Madagascar. *Int. J. Infect. Dis.* **2018**, *76*, 130–136. [CrossRef] [PubMed]
27. Mosher, B.A.; Huyvaert, K.P.; Bailey, L.L. Beyond the swab: Ecosystem sampling to understand the persistence of an amphibian pathogen. *Oecologia* **2018**, *188*, 319–330. [CrossRef] [PubMed]
28. Robinson, C.V.; Uren Webster, T.M.; Cable, J.; James, J.; Consuegra, S. Simultaneous detection of invasive signal crayfish, endangered white-clawed crayfish and the crayfish plague pathogen using environmental DNA. *Biol. Conserv.* **2018**, *222*, 241–252. [CrossRef]
29. Miaud, C.; Guyetant, R.; Elmberg, J. Variations in life-history traits in the common frog Rana temporaria (Amphibia: Anura): A literature review and new data from the French Alps. *J. Zool.* **1999**, *249*, 61–73. [CrossRef]
30. Gosner, K.L. A Simplified Table for Staging Anuran Embryos and Larvae with Notes on Identification. *Herpetologica* **1960**, *16*, 183–190.
31. Biggs, J.; Ewald, N.; Valentini, A.; Gaboriaud, C.; Dejean, T.; Griffiths, R.A.; Foster, J.; Wilkinson, J.W.; Arnell, A.; Brotherton, P.; et al. Using eDNA to develop a national citizen science-based monitoring programme for the great crested newt (Triturus cristatus). *Biol. Conserv.* **2015**, *183*, 19–28. [CrossRef]
32. Leung, W.T.M.; Thomas-Walters, L.; Garner, T.W.J.; Balloux, F.; Durrant, C.; Price, S.J. A quantitative-PCR based method to estimate ranavirus viral load following normalisation by reference to an ultraconserved vertebrate target. *J. Virol. Methods* **2017**, *249*, 147–155. [CrossRef]
33. Taberlet, P.; Prud'homme, S.M.; Campione, E.; Roy, J.; Miquel, C.; Shehzad, W.; Gielly, L.; Rioux, D.; Choler, P.; Clement, J.-C.; et al. Soil sampling and isolation of extracellular DNA from large amount of starting material suitable for metabarcoding studies. *Mol. Ecol.* **2012**, *21*, 1816–1820. [CrossRef]
34. Pont, D.; Rocle, M.; Valentini, A.; Civade, R.; Jean, P.; Maire, A.; Roset, N.; Schabuss, M.; Zornig, H.; Dejean, T. Environmental DNA reveals quantitative patterns of fish biodiversity in large rivers despite its downstream transportation. *Sci. Rep.* **2018**, *8*, 10361. [CrossRef] [PubMed]

35. Mao, J.H.; Green, D.E.; Fellers, G.; Chinchar, V.G. Molecular characterization of iridoviruses isolated from sympatric amphibians and fish. *Virus Res.* **1999**, *63*, 45–52. [CrossRef]
36. Mavian, C.; López-Bueno, A.; Balseiro, A.; Casais, R.; Alcamí, A.; Alejo, A. The Genome Sequence of the Emerging Common Midwife Toad Virus Identifies an Evolutionary Intermediate within Ranaviruses. *J. Virol.* **2012**, *86*, 3617. [CrossRef] [PubMed]
37. Haapanen, A. Breeding of the Common Frog (rana-Temporaria L). *Ann. Zool. Fenn.* **1982**, *19*, 75–79.
38. Hall, E.M.; Goldberg, C.S.; Brunner, J.L.; Crespi, E.J. Seasonal dynamics and potential drivers of ranavirus epidemics in wood frog populations. *Oecologia* **2018**, *188*, 1253–1262. [CrossRef] [PubMed]
39. Elmberg, J. Long-Term Survival, Length of Breeding-Season, and Operational Sex-Ratio. *Can. J. Zool.-Rev. Can. Zool.* **1990**, *68*, 121–127. [CrossRef]
40. Cunningham, A.A.; Langton, T.E.S.; Bennett, P.M.; Lewin, J.F.; Drury, S.E.N.; Gough, R.E.; MacGregor, S.K. Pathological and microbiological findings from incidents of unusual mortality of the common frog (Rana temporaria). *Philos. Trans. R. Soc. Lond. Ser. B Biol. Sci.* **1996**, *351*, 1539–1557.
41. Price, S.J.; Garner, T.W.J.; Cunningham, A.A.; Langton, T.E.S.; Nichols, R.A. Reconstructing the emergence of a lethal infectious disease of wildlife supports a key role for spread through translocations by humans. *Proc. R. Soc. B Biol. Sci.* **2016**, *283*, 20160952. [CrossRef] [PubMed]
42. Johnson, A.F.; Brunner, J.L. Persistence of an amphibian ranavirus in aquatic communities. *Dis. Aquat. Org.* **2014**, *111*, 129–138. [CrossRef]
43. Duffus, A.L.J.; Pauli, B.D.; Wozney, K.; Brunetti, C.R.; Berrill, M. FROG VIRUS 3-LIKE INFECTIONS IN AQUATIC AMPHIBIAN COMMUNITIES. *J. Wildl. Dis.* **2008**, *44*, 109–120. [CrossRef]
44. Green, D.E.; Converse, K.A.; Schrader, A.K. Epizootiology of sixty-four amphibian morbidity and mortality events in the USA, 1996-2001. In *Domestic Animal/Wildlife Interface: Issue for Disease Control, Conservation, Sustainable Food Production, and Emerging Diseases*; Gibbs, E.P.J., Bokma, B.H., Eds.; New York Acad Sciences: New York, NY, USA, 2002; Volume 969, pp. 323–339. ISBN 978-1-57331-439-8.
45. Gray, M.J.; Miller, D.L.; Hoverman, J.T. Ecology and pathology of amphibian ranaviruses. *Dis. Aquat. Org.* **2009**, *87*, 243–266. [CrossRef] [PubMed]
46. Rosa, G.M.; Sabino-Pinto, J.; Laurentino, T.G.; Martel, A.; Pasmans, F.; Rebelo, R.; Griffiths, R.A.; Stoehr, A.C.; Marschang, R.E.; Price, S.J.; et al. Impact of asynchronous emergence of two lethal pathogens on amphibian assemblages. *Sci. Rep.* **2017**, *7*, 43260. [CrossRef] [PubMed]
47. Rijks, J.M.; Saucedo, B.; Spitzen-van der Sluijs, A.; Wilkie, G.S.; van Asten, A.J.A.M.; van den Broek, J.; Boonyarittichaikij, R.; Stege, M.; van der Sterren, F.; Martel, A.; et al. Investigation of Amphibian Mortality Events in Wildlife Reveals an On-Going Ranavirus Epidemic in the North of the Netherlands. *PLoS ONE* **2016**, *11*, e0157473. [CrossRef] [PubMed]
48. Kik, M.; Martel, A.; Spitzen-van der Sluijs, A.; Pasmans, F.; Wohlsein, P.; Grone, A.; Rijks, J.M. Ranavirus-associated mass mortality in wild amphibians, The Netherlands, 2010: A first report. *Vet. J.* **2011**, *190*, 284–286. [CrossRef] [PubMed]
49. Hoverman, J.T.; Gray, M.J.; Miller, D.L.; Haislip, N.A. Widespread Occurrence of Ranavirus in Pond-Breeding Amphibian Populations. *EcoHealth* **2012**, *9*, 36–48. [CrossRef] [PubMed]
50. Pearman, P.B.; Garner, T.W.J.; Straub, M.; Greber, U.F. Response of the Italian agile frog (Rana latastei) to a Ranavirus, frog virus 3: A model for viral emergence in naive populations. *J. Wildl. Dis.* **2004**, *40*, 660–669. [CrossRef] [PubMed]
51. Harp, E.M.; Petranka, J.W. Ranavirus in wood frogs (Rana sylvatica): Potential sources of transmission within and between ponds. *J. Wildl. Dis.* **2006**, *42*, 307–318. [CrossRef] [PubMed]
52. Torrence, S.M.; Green, D.E.; Benson, C.J.; Ip, H.S.; Smith, L.M.; McMurry, S.T. A New Ranavirus Isolated from Pseudacris clarkii Tadpoles in Playa Wetlands in the Southern High Plains, Texas. *J. Aquat. Anim. Health* **2010**, *22*, 65–72. [CrossRef] [PubMed]
53. Echaubard, P.; Leduc, J.; Pauli, B.; Chinchar, V.G.; Robert, J.; Lesbarreres, D. Environmental dependency of amphibian-ranavirus genotypic interactions: Evolutionary perspectives on infectious diseases. *Evol. Appl.* **2014**, *7*, 723–733. [CrossRef]
54. Bayley, A.E.; Hill, B.J.; Feist, S.W. Susceptibility of the European common frog Rana temporaria to a panel of ranavirus isolates from fish and amphibian hosts. *Dis. Aquat. Org.* **2013**, *103*, 171–183. [CrossRef] [PubMed]
55. Greer, A.L.; Collins, J.P. Sensitivity of a diagnostic test for amphibian ranavinus varies with sampling protocol. *J. Wildl. Dis.* **2007** *43*, 525–532. [CrossRef] [PubMed]

56. St-Ainour, V.; Lesbarreres, D. Genetic evidence of Ranavirus in toe clips: An alternative to lethal sampling methods. *Conserv. Genet.* **2007**, *8*, 1247–1250. [CrossRef]
57. Gray, M.J.; Miller, D.L.; Hoverman, J.T. Reliability of non-lethal surveillance methods for detecting ranavirus infection. *Dis. Aquat. Org.* **2012**, *99*, 1–6. [CrossRef] [PubMed]
58. Brunner, J.L.; Collins, J.P. Testing assumptions of the trade-off theory of the evolution of parasite virulence. *Evol. Ecol. Res.* **2009**, *11*, 1169–1188.
59. Robert, J.; George, E.; Andino, F.D.J.; Chen, G. Waterborne infectivity of the Ranavirus frog virus 3 in Xenopus laevis. *Virology* **2011**, *417*, 410–417. [CrossRef] [PubMed]
60. Grizzle, J.M.; Brunner, C.J. Review of largemouth bass virus. *Fisheries* **2003**, *28*, 10–14. [CrossRef]
61. Huver, J.R.; Koprivnikar, J.; Johnson, P.T.J.; Whyard, S. Development and application of an eDNA method to detect and quantify a pathogenic parasite in aquatic ecosystems. *Ecol. Appl.* **2015**, *25*, 991–1002. [CrossRef] [PubMed]
62. Rusch, J.C.; Hansen, H.; Strand, D.A.; Markussen, T.; Hytterod, S.; Vralstad, T. Catching the fish with the worm: A case study on eDNA detection of the monogenean parasite Gyrodactylus salaris and two of its hosts, Atlantic salmon (Salmo salar) and rainbow trout (Oncorhynchus mykiss). *Parasit. Vect.* **2018**, *11*, 333. [CrossRef]
63. Hyman, O.J.; Collins, J.P. Evaluation of a filtration-based method for detecting Batrachochytrium dendrobatidis in natural bodies of water. *Dis. Aquat. Org.* **2012**, *97*, 185–195. [CrossRef]
64. Chestnut, T.; Anderson, C.; Popa, R.; Blaustein, A.R.; Voytek, M.; Olson, D.H.; Kirshtein, J. Heterogeneous Occupancy and Density Estimates of the Pathogenic Fungus Batrachochytrium dendrobatidis in Waters of North America. *PLoS ONE* **2014**, *9*, e106790. [CrossRef]
65. Julian, J.; Glenney, G.; Rees, C. Evaluating observer bias and seasonal detection rates in amphibian pathogen eDNA collections by citizen scientists. *Dis. Aquat. Organ.* **2019**, *134*, 15–24. [CrossRef]
66. MacKenzie, D.I.; Nichols, J.D.; Lachman, G.B.; Droege, S.; Andrew Royle, J.; Langtimm, C.A. Estimating Site Occupancy Rates When Detection Probabilities Are Less Than One. *Ecology* **2002**, *83*, 2248–2255. [CrossRef]
67. Colvin, M.E.; Peterson, J.T.; Kent, M.L.; Schreck, C.B. Occupancy Modeling for Improved Accuracy and Understanding of Pathogen Prevalence and Dynamics. *PloS ONE* **2015**, *10*, e0116605. [CrossRef] [PubMed]
68. Yoccoz, N.G. The future of environmental DNA in ecology: NEWS AND VIEWS: OPINION. *Mol. Ecol.* **2012**, *21*, 2031–2038. [CrossRef] [PubMed]
69. McClintock, B.T.; Nichols, J.D.; Bailey, L.L.; MacKenzie, D.I.; Kendall, W.L.; Franklin, A.B. Seeking a second opinion: Uncertainty in disease ecology: Uncertainty in disease ecology. *Ecol. Lett.* **2010**, *13*, 659–674. [CrossRef] [PubMed]

Permissions

All chapters in this book were first published by MDPI; hereby published with permission under the Creative Commons Attribution License or equivalent. Every chapter published in this book has been scrutinized by our experts. Their significance has been extensively debated. The topics covered herein carry significant findings which will fuel the growth of the discipline. They may even be implemented as practical applications or may be referred to as a beginning point for another development.

The contributors of this book come from diverse backgrounds, making this book a truly international effort. This book will bring forth new frontiers with its revolutionizing research information and detailed analysis of the nascent developments around the world.

We would like to thank all the contributing authors for lending their expertise to make the book truly unique. They have played a crucial role in the development of this book. Without their invaluable contributions this book wouldn't have been possible. They have made vital efforts to compile up to date information on the varied aspects of this subject to make this book a valuable addition to the collection of many professionals and students.

This book was conceptualized with the vision of imparting up-to-date information and advanced data in this field. To ensure the same, a matchless editorial board was set up. Every individual on the board went through rigorous rounds of assessment to prove their worth. After which they invested a large part of their time researching and compiling the most relevant data for our readers.

The editorial board has been involved in producing this book since its inception. They have spent rigorous hours researching and exploring the diverse topics which have resulted in the successful publishing of this book. They have passed on their knowledge of decades through this book. To expedite this challenging task, the publisher supported the team at every step. A small team of assistant editors was also appointed to further simplify the editing procedure and attain best results for the readers.

Apart from the editorial board, the designing team has also invested a significant amount of their time in understanding the subject and creating the most relevant covers. They scrutinized every image to scout for the most suitable representation of the subject and create an appropriate cover for the book.

The publishing team has been an ardent support to the editorial, designing and production team. Their endless efforts to recruit the best for this project, has resulted in the accomplishment of this book. They are a veteran in the field of academics and their pool of knowledge is as vast as their experience in printing. Their expertise and guidance has proved useful at every step. Their uncompromising quality standards have made this book an exceptional effort. Their encouragement from time to time has been an inspiration for everyone.

The publisher and the editorial board hope that this book will prove to be a valuable piece of knowledge for researchers, students, practitioners and scholars across the globe.

List of Contributors

Ayca Zeynep Ilter-Akulke
Department of Medical Microbiology, School of Medicine, Acıbadem Mehmet Ali Aydınlar University, Atasehir, Istanbul 34752, Turkey

İkbal Agah İnce
Department of Medical Microbiology, School of Medicine, Acıbadem Mehmet Ali Aydınlar University, Atasehir, Istanbul 34752, Turkey
Department of Biostatistics and Medical Informatics, Acıbadem Mehmet Ali Aydınlar University, Atasehir, Istanbul 34752, Turkey

Orhan Özcan
Department of Medical Microbiology, School of Medicine, Acıbadem Mehmet Ali Aydınlar University, Atasehir, Istanbul 34752, Turkey
Department of Biostatistics and Medical Informatics, Acıbadem Mehmet Ali Aydınlar University, Atasehir, Istanbul 34752, Turkey
Evolutionary Genetics, Groningen Institute for Evolutionary Life Sciences, University of Groningen, 9700 CC Groningen, The Netherlands

Erin D. Scully
Stored Product Insect and Engineering Research Unit, USDA-Agricultural Research Service, Center for Grain and Animal Health Research, Manhattan, KS 66502, USA

Arzu Özgen
Department of Medical Microbiology, School of Medicine, Acıbadem Mehmet Ali Aydınlar University, Atasehir, Istanbul 34752, Turkey
Vocational School of Health, Istanbul Gelisim University, Avcılar, Istanbul 34310, Turkey

Hsin-I Liu
Department of Aquaculture, National Taiwan Ocean University, Keelung 20224, Taiwan

Pinwen Peter Chiou, Hong-Yi Gong and Hsin-Yiu Chou
Department of Aquaculture, National Taiwan Ocean University, Keelung 20224, Taiwan
Center of Excellence for the Oceans, National Taiwan Ocean University, No. 2, Pei-Ning Rd., Keelung 20224, Taiwan

Joy A. Becker and Alison Tweedie
Sydney School of Veterinary Science, The University of Sydney, Camden 2570, Australia

Dean Gilligan
NSW Industry and Investment, Batemans Bay Fisheries Office, Batemans Bay 2536, Australia

Martin Asmus
NSW Industry and Investment, Narrandera Fisheries Centre, Narrandera 2700, Australia

Richard J. Whittington
Sydney School of Veterinary Science, The University of Sydney, Camden 2570, Australia
OIE Reference Laboratory for Epizootic Haematopoietic Necrosis Virus and Ranavirus Infection of Amphibians, Sydney School of Veterinary Science, The University Sydney, Camden 2570, Australia

Veronica A. Pagowski and Gideon J. Mordecai
Department of Earth, Ocean and Atmospheric Sciences, University of British Columbia, 2207 Main Mall, Vancouver, BC V6T 1Z4, Canada

Kristina M. Miller, Angela D. Schulze, Karia H. Kaukinen, Tobi J. Ming, Shaorong Li and Amy Tabata
Pacific Biological Station, Fisheries and Oceans Canada, 3190 Hammond Bay Rd, Nanaimo, BC V9T 6N7, Canada

Amy K. Teffer
Biology Department, University of Victoria, 3800 Finnerty Rd, Victoria, BC V8P 5C2, Canada

Curtis A. Suttle
Department of Earth, Ocean and Atmospheric Sciences, University of British Columbia, 2207 Main Mall, Vancouver, BC V6T 1Z4, Canada
Department of Microbiology and Immunology, University of British Columbia, 2207 Main Mall, Vancouver, BC V6T 1Z4, Canada
Department of Botany, University of British Columbia, 2207 Main Mall, Vancouver, BC V6T 1Z4, Canada
Institute for the Oceans and Fisheries, University of British Columbia, 2207 Main Mall, Vancouver, BC V6T 1Z4, Canada

Xiao-Tao Zeng
State Key Laboratory of Freshwater Ecology and Biotechnology, Institute of Hydrobiology, Wuhan 430072, China

Qi-Ya Zhang
State Key Laboratory of Freshwater Ecology and Biotechnology, The Innovation Academy of Seed Design, Institute of Hydrobiology, Chinese Academy of Sciences, Wuhan 430072, China
College of Modern Agriculture Sciences, University of Chinese Academy of Sciences, Beijing 100049, China
The Innovation Academy of Seed Design, Chinese Academy of Sciences, Beijing 100101, China

Estefania J. Valverde, Alejandro M. Labella, Juan J. Borrego and Dolores Castro
Departamento de Microbiología, Facultad de Ciencias, Universidad de Málaga, 29017 Málaga, Spain

Hailiang Wang, Peizhuo Zou and Xuan Dong
Laboratory for Marine Fisheries Science and Food Production Processes, National Laboratory for Marine Science and Technology (Qingdao); Key Laboratory of Maricultural Organism Disease Control, Ministry of Agriculture and Rural Affairs; Qingdao Key Laboratory of Mariculture Epidemiology and Biosecurity; Yellow Sea Fisheries Research Institute, Chinese Academy of Fishery Sciences, Qingdao 266071, China

Xing Chen
Laboratory for Marine Fisheries Science and Food Production Processes, National Laboratory for Marine Science and Technology (Qingdao); Key Laboratory of Maricultural Organism Disease Control, Ministry of Agriculture and Rural Affairs; Qingdao Key Laboratory of Mariculture Epidemiology and Biosecurity; Yellow Sea Fisheries Research Institute, Chinese Academy of Fishery Sciences, Qingdao 266071, China
College of Fisheries and Life Science, Shanghai Ocean University, Shanghai 201306, China

Fuhua Li
Key Laboratory of Experimental Marine Biology, Institute of Oceanology, Chinese Academy of Sciences, Qingdao 266071, China

Emily Penny and Craig R. Brunetti
Biology Department, Trent University, 1600West Bank Dr, Peterborough, ON K9J 7B8, Canada

Zi-Hao Wang and Cheng-Yue Ming
State Key Laboratory of Freshwater Ecology and Biotechnology, The Innovation Academy of Seed Design, Institute of Hydrobiology, Chinese Academy of Sciences, Wuhan 430072, China

Fei Ke
State Key Laboratory of Freshwater Ecology and Biotechnology, The Innovation Academy of Seed Design, Institute of Hydrobiology, Chinese Academy of Sciences, Wuhan 430072, China
College of Modern Agriculture Sciences, University of Chinese Academy of Sciences, Beijing 100049, China

Cara West, Florentina Rus, Anni Kleino and Neal Silverman
Division of Infectious Diseases and Immunology, Department of Medicine, University of Massachusetts Medical School, Worcester, MA 01605, USA

Ying Chen
RNA Therapeutics Institute, University of Massachusetts Medical School, Worcester, MA 01605, USA

Monique Gangloff
Department of Biochemistry, University of Cambridge, Cambridge CB2 1GA, UK

Don B. Gammon
Department of Microbiology, University of Texas Southwestern Medical Center, Dallas, TX T5390, USA

Liang Qiu and Chen Li
Yellow Sea Fisheries Research Institute, Chinese Academy of Fishery Sciences; Laboratory for Marine Fisheries Science and Food Production Processes, National Laboratory for Marine Science and Technology (Qingdao); Key Laboratory of Maricultural Organism Disease Control, Ministry of Agriculture and Rural Affairs; Qingdao Key Laboratory of Mariculture Epidemiology and Biosecurity, Qingdao 266071, China

Wen Gao and Qing-Li Zhang
Yellow Sea Fisheries Research Institute, Chinese Academy of Fishery Sciences; Laboratory for Marine Fisheries Science and Food Production Processes, National Laboratory for Marine Science and Technology (Qingdao); Key Laboratory of Maricultural Organism Disease Control, Ministry of Agriculture and Rural Affairs; Qingdao Key Laboratory of Mariculture Epidemiology and Biosecurity, Qingdao 266071, China
Shanghai Ocean University, Shanghai 201306, China

Ruo-Heng Zhao
Yellow Sea Fisheries Research Institute, Chinese Academy of Fishery Sciences; Laboratory for Marine Fisheries Science and Food Production Processes, National Laboratory for Marine Science and Technology (Qingdao); Key Laboratory of Maricultural Organism Disease Control, Ministry of Agriculture and Rural Affairs; Qingdao Key Laboratory of Mariculture Epidemiology and Biosecurity, Qingdao 266071, China
Dalian Ocean University, Dalian 116023, China

Jie Huang
Yellow Sea Fisheries Research Institute, Chinese Academy of Fishery Sciences; Laboratory for Marine Fisheries Science and Food Production Processes, National Laboratory for Marine Science and Technology (Qingdao); Key Laboratory of Maricultural Organism Disease Control, Ministry of Agriculture and Rural Affairs; Qingdao Key Laboratory of Mariculture Epidemiology and Biosecurity, Qingdao 266071, China
Shanghai Ocean University, Shanghai 201306, China
Dalian Ocean University, Dalian 116023, China

Bernardo Saucedo, Andrea Gröne and Steven J. van Beurden
Department of Pathobiology, Faculty of Veterinary Medicine, Utrecht University, 3584 CL Utrecht, The Netherlands

José M. Serrano
Laboratorio de Genética y Evolución, Departamento de Ciencias Ecológicas, Facultad de Ciencias, Universidad de Chile, Las Palmeras, Santiago 3425, Chile
Programa de Fisiología y Biofísica, Facultad de Medicina, Universidad de Chile, Santiago 8380453, Chile

Mónica Jacinto-Maldonado
Wildlife and Laboratory Animals, Department of Ethology, Faculty of Veterinary Medicine, National Autonomous University of Mexico, Mexico City 045010, Mexico
C3-Complexity sciences Center, Autonomous University of Mexico, Mexico City 045010, Mexico

Rob S. E. W. Leuven
Department of Animal Ecology and Physiology, Institute for Water and Wetland Research, Radboud University, 6500 GL Nijmegen, The Netherlands
Netherlands Centre of Expertise on Exotic Species, 6500 GL Nijmegen, The Netherlands

Abraham A. Rocha García and Adriana Méndez Bernal
Department of Pathology 04510, Faculty of Veterinary Medicine, National Autonomous University of Mexico, Mexico City 045010, Mexico

César M. Escobedo-Bonilla
Department of Aquaculture, Instituto Politécnico, Nacional-CIIDIR Unidad Sinaloa, Guasave Sinaloa 81101, Mexico

Maya A. Halaly
Department of Animal Sciences, College of Agricultural and Life Sciences, University of Florida, Gainesville, FL 32611, USA

Kuttichantran Subramaniam and Samantha A. Koda and Thomas B. Waltzek
Department of Infectious Diseases and Immunology, College of Veterinary Medicine, University of Florida, Gainesville, FL 32611, USA

Vsevolod L. Popov
Department of Pathology, University of Texas Medical Branch, Galveston, TX 77555, USA

David Stone and Keith Way
Centre for Environment, Fisheries and Aquaculture Science (CEFAS), Weymouth DT4 8UB, UK

Amanda L.J. Duffus and Joshua P. Standridge
Department of Mathematics and Natural Sciences, Gordon State College, Barnesville, GA 30204, USA

Trenton W.J. Garner
Institute of Zoology, Zoological Society of London, London NW1 4RY, UK

Richard A. Nichols
School of Biological and Chemical Sciences, Queen Mary, University of London, London E1 4NS, UK

Julia E. Earl
School of Biological Sciences, Louisiana Tech University, Ruston, LA 71272, USA

Joseph R. Mihaljevic
School of Informatics, Computing and Cyber Systems, Northern Arizona University, Flagstaff, AZ 86011, USA

Amy L. Greer
Department of Population Medicine, University of Guelph, Guelph, ON N1G 2W1, Canada

Jesse L. Brunner
School of Biological Sciences, Washington State University, Pullman, WA 99163, USA

Francisco De Jesús Andino, Jinyeong Yim, Katherine Woo and Jacques Robert
Department of Microbiology and Immunology, University of Rochester Medical Center, Rochester, NY 14642, USA

Eva-Stina Isabella Edholm
Department of Microbiology and Immunology, University of Rochester Medical Center, Rochester, NY 14642, USA
The Norwegian College of Fishery Science, University of Tromsø, the Arctic university of Norway, 9037, Tromsø, Norway

Claude Miaud, Véronique Arnal and Marie Poulain
CEFE, EPHE-PSL, CNRS, Univ. Montpellier, Univ Paul Valéry Montpellier 3, IRD, Biogeography and Vertebrate Ecology, 1919 route de Mende, 34293 Montpellier, France

Alice Valentini and Tony Dejean
SPYGEN, 17 Rue du Lac Saint-André, 73370 Le Bourget-du-Lac, France

Index

A

Alphairidovirinae, 3, 37, 80, 91, 105, 169

Amphibians, 6, 8-9, 19, 22, 26, 36-37, 67, 80, 91, 105, 113, 153-154, 156, 158, 164-167, 169, 178, 188, 190-191, 193-194, 201-203, 214, 216-217, 224, 227, 229

Antisera, 26, 28-29, 31

Antiviral Immunity, 9, 23, 202, 212, 214

Apoptosis, 9, 13, 15-16, 24-25, 124, 136, 174, 194

Aqueous Phase, 106, 156

B

Betairidovirinae, 1, 3, 80, 91, 103, 105, 150, 169

Biosecurity, 36-37, 90, 102, 104, 139, 147, 149, 151

C

Chloriridovirus, 1, 3, 6, 56, 80, 91, 105, 170

Clinical Symptoms, 27-29, 60, 90, 101, 142

Co-immunoprecipitation, 67-68

Cross-species Transmission, 67-68, 77

Cytoplasm, 13, 51, 67, 73, 75, 77, 102, 105, 108, 110, 130-131, 144, 148, 159, 168, 172, 174

D

Decapodiridovirus, 80, 90-91, 140, 147

Disease Dynamics, 178-179, 182

E

Early Detection, 212, 216-217, 228

Ecosystem, 50, 66, 161, 166, 216-217, 222, 227-228

Electron Microscopy, 27, 51-52, 90-91, 94, 100, 103, 139, 141, 169-170, 172

Envelope Protein, 79, 113-115, 119-120, 122-123

Epidemiology, 34, 36-37, 49-50, 52-54, 58, 61, 63, 90-91, 139, 148, 189, 200

Erythrocytic Necrosis Virus, 51-52, 56, 63

Exopalaemon Carinicauda, 90-91, 94-97, 99-100, 103-104, 148, 151

G

Genomic Sequence, 18-19, 52, 93, 121, 175

Genomics, 1, 16-17, 19, 111, 165

Grouper Iridovirus, 19, 21, 24, 26-27, 34, 56, 63, 65, 79, 111, 123, 149, 152, 171, 173

H

Hemocytes, 8-9, 22-23, 90, 96, 102, 139-140, 143-144, 148-149

Heparan Sulfate, 113-117, 120-123

Hepatopancreas, 90, 93-94, 97, 99, 101, 139-144, 146-150

Histopathology, 96, 102, 143, 153, 155, 158

Host Cells, 6, 14, 16, 24, 68, 110, 113, 120

I

Immune Evasion, 78, 124, 132, 135, 203, 214

Immune Response, 9, 23, 124, 132, 134, 136-138, 192-193, 197-200, 202, 209, 212-213, 215

Immunofluorescence, 75, 105, 107-109

Immunohistochemistry, 153, 155, 159

Infected Tissue, 42, 92

Infectious Spleen, 20, 26, 34-35, 56, 63, 111, 169, 171, 174, 176-177

Interferon, 79, 192-194, 198, 200, 202-203, 210-211, 214-215

Invertebrate Iridoviruses, 1, 20, 22, 65

Iridoviridae, 1, 6, 12-13, 17-22, 31, 34, 37, 49, 52, 54-55, 60, 62, 67-68, 78-80, 87, 91, 102-103, 105-106, 108, 110-113, 121, 140, 150-151, 154, 156, 165-166, 168-169, 173-176, 178, 216, 227

K

Kidney Necrosis Virus, 26, 34-35, 63, 169, 171, 174-177

L

Lymphocystivirus, 1, 6, 10, 26, 56, 80, 87-88, 91, 105, 169, 171

M

Macrobrachium Rosenbergii, 86, 89, 91, 104, 139-140, 146, 150-151

Major Capsid Protein, 1, 12, 15, 20-21, 24, 26, 32, 35, 37, 49, 52, 57, 63, 65, 79-80, 82, 84, 115, 129, 137, 156, 158, 168-170, 188

Mammalian Cell, 110, 112

Megalocytivirus, 1, 20, 26, 31, 33, 56, 80, 91, 168-169, 171, 173-176

Microscopic Lesions, 153-154, 158, 168, 175

Mortality, 1, 8, 17, 20, 26-27, 33, 36-38, 51, 62-63, 84, 91-95, 101-104, 139-140, 142, 147, 150-151, 153, 164-167, 169, 178-179, 181-183, 185-191, 193, 198-199, 201, 203, 209, 216-217, 222-224, 226-227, 229

N

Nucleic Acids, 23, 52, 65, 136, 156, 166, 176-177, 204

O

Oligonucleotide, 106-107

Open Reading Frames, 9, 105, 124, 168, 172

P

Pacific Herring, 51-52, 55, 60, 62-65

Pacific Salmon, 51-52, 60-64

Paracrystalline Array, 144-145, 149

Pereiopods, 90, 94, 100, 102, 140-141, 146-147, 149

Persistent Infection, 61, 200

Polyclonal Antibodies, 26, 28-29, 31

Polymerase, 9, 12-13, 15, 22, 24, 32-33, 37, 49, 52, 54-57, 64-65, 67-68, 77-78, 88, 93, 98, 106, 125, 132, 134, 137, 141, 144, 156, 204-205, 208-209, 219

Population Dynamics, 52, 67, 175, 178, 188

Procambarus Clarkii, 101, 139-140

Proteomics, 1, 8, 17, 123

R

Rana Grylio Virus, 19, 67-68, 79, 113-114, 116, 122, 160, 165

Rana Temporaria, 163, 166-167, 178-180, 182, 189-190, 216-217, 221, 228-229

Ranavirus, 1, 3, 10, 12-13, 18-20, 26, 31, 36-37, 49-50, 56, 67-68, 77, 79-80, 87, 91, 105, 111-114, 116, 120-121, 123, 148, 153-154, 156-161, 163-167, 169, 171, 175, 177-183, 185-192, 194, 198-203, 212-217, 219-220, 222-224, 226-230

Risk Assessment, 153, 156-157, 166

S

Sparus Aurata, 80, 87-88

Survival Rate, 30, 33, 213

Susceptible Host, 80, 86, 90, 100-101, 113, 163, 178

Susceptible Species, 61, 91, 101-102, 139, 148-149, 178

T

Taxonomy, 1, 17-18, 37, 49, 63, 78, 87, 91, 102-103, 111, 121, 140, 151, 167, 174-176

Tissue Culture, 42, 180

U

Uninfected Controls, 132, 205-206, 209

V

Viral Infection, 12, 15, 17, 31, 62, 68, 80, 109, 113-114, 117, 134, 193, 203, 209, 211, 213-214

Viral Isolates, 1, 87, 184

Viral Load, 52-53, 58, 60, 62, 80, 83-85, 139, 148, 198, 204-205, 209, 217, 222, 228

Viral Pathogens, 26, 86, 101, 140, 202

Viral Replication, 8-9, 22, 81, 87, 100, 105, 110, 124-125, 128-129, 132, 191-194, 197-199, 202, 205-208, 212-213

Viral Transmission, 80-81, 87

Virulence, 1, 16-17, 19, 61, 65, 112, 135, 137, 147, 151, 164, 166-167, 191, 201, 205, 214, 227, 230

Virus Binding, 113, 115, 118

Virus Isolation, 36-37, 41-43, 48-49

Virus Population, 192-193, 197, 199

Y

Yeast Two-hybrid, 67-68, 71, 78

www.ingramcontent.com/pod-product-compliance
Lightning Source LLC
LaVergne TN
LVHW060521210926
848485LV00005B/1123
* 9 7 8 1 6 3 9 2 7 7 3 2 2 *